Deductive Systems
in Traditional and Modern Logic

Deductive Systems in Traditional and Modern Logic

Editors

Alex Citkin
Urszula Wybraniec-Skardowska

MDPI • Basel • Beijing • Wuhan • Barcelona • Belgrade • Manchester • Tokyo • Cluj • Tianjin

Editors
Alex Citkin
Metropolitan Telecommunications
USA

Urszula Wybraniec-Skardowska
Cardinal Stefan Wyszyński
University in Warsaw,
Department of Philosophy
Poland

Editorial Office
MDPI
St. Alban-Anlage 66
4052 Basel, Switzerland

This is a reprint of articles from the Special Issue published online in the open access journal *Axioms* (ISSN 2075-1680) (available at: http://www.mdpi.com/journal/axioms/special_issues/deductive_systems).

For citation purposes, cite each article independently as indicated on the article page online and as indicated below:

LastName, A.A.; LastName, B.B.; LastName, C.C. Article Title. *Journal Name* **Year**, *Article Number*, Page Range.

ISBN 978-3-03943-358-2 (Pbk)
ISBN 978-3-03943-359-9 (PDF)

© 2020 by the authors. Articles in this book are Open Access and distributed under the Creative Commons Attribution (CC BY) license, which allows users to download, copy and build upon published articles, as long as the author and publisher are properly credited, which ensures maximum dissemination and a wider impact of our publications.

The book as a whole is distributed by MDPI under the terms and conditions of the Creative Commons license CC BY-NC-ND.

Contents

About the Editors vii

Alex Citkin and Urszula Wybraniec-Skardowska
Deductive Systems in Traditional and Modern Logic
Reprinted from: *Axioms* 2020, 9, 108, doi:10.3390/axioms9030108 1

Piotr Kulicki
Aristotle's Syllogistic as a Deductive System
Reprinted from: *Axioms* 2020, 9, 56, doi:10.3390/axioms9020056 5

Peter Simons
Term Logic
Reprinted from: *Axioms* 2020, 9, 18, doi:10.3390/axioms9010018 21

J.-Martín Castro-Manzano
Distribution Tableaux, Distribution Models
Reprinted from: *Axioms* 2020, 9, 41, doi:10.3390/axioms9020041 31

Eugeniusz Wojciechowski
The *Zahl-Anzahl* Distinction in Gottlob Frege: Arithmetic of Natural Numbers with *Anzahl* as a Primitive Term
Reprinted from: *Axioms* 2020, 9, 6, doi:10.3390/axioms9010006 41

Valentin Goranko
Hybrid Deduction–Refutation Systems
Reprinted from: *Axioms* 2019, 8, 118, doi:10.3390/axioms8040118 49

Krystyna Mruczek-Nasieniewska and Marek Nasieniewski
A Kotas-Style Characterisation of Minimal Discussive Logic
Reprinted from: *Axioms* 2019, 8, 108, doi:10.3390/axioms8040108 69

Janusz Ciuciura
A Note on Fernández–Coniglio's Hierarchy of Paraconsistent Systems
Reprinted from: *Axioms* 2020, 9, 35, doi:10.3390/axioms9020035 87

Alex Citkin
Deductive Systems with Multiple-Conclusion Rules and the Disjunction Property
Reprinted from: *Axioms* 2019, 8, 100, doi:10.3390/axioms8030100 99

Dariusz Surowik
Minimal Systems of Temporal Logic
Reprinted from: *Axioms* 2020, 9, 67, doi:10.3390/axioms9020067 123

Joanna Golińska-Pilarek and Magdalena Welle
Deduction in Non-Fregean Propositional Logic SCI
Reprinted from: *Axioms* 2019, 8, 115, doi:10.3390/axioms8040115 151

Sopo Pkhakadze and Hans Tompits
Sequent-Type Calculi for Three-Valued and Disjunctive Default Logic
Reprinted from: *Axioms* 2020, 9, 84, doi:10.3390/axioms9030084 171

Henrique Antunes, Walter Carnielli, Andreas Kapsner and Abilio Rodrigues
Kripke-Style Models for Logics of Evidence and Truth
Reprinted from: *Axioms* **2020**, *9*, 100, doi:10.3390/axioms9030100 . . . . **201**

Andrzej Malec
Deontic Logics as Axiomatic Extensions of First-Order Predicate Logic: An Approach Inspired by Wolniewicz's Formal Ontology of Situations
Reprinted from: *Axioms* **2019**, *8*, 109, doi:10.3390/axioms8040109 . . . . **217**

Dorota Leszczyńska-Jasion and Szymon Chlebowski
Synthetic Tableaux with Unrestricted Cut for First-Order Theories
Reprinted from: *Axioms* **2019**, *8*, 133, doi:10.3390/axioms8040133 . . . . **231**

Urszula Wybraniec-Skardowska
On Certain Axiomatizations of Arithmetic of Natural and Integer Numbers
Reprinted from: *Axioms* **2019**, *8*, 103, doi:10.3390/axioms8030103 . . . . **257**

Jean-Pierre Desclés and Anca Christine Pascu
Logic of Typical and Atypical Instances of a Concept—A Mathematical Model
Reprinted from: *Axioms* **2019**, *8*, 104, doi:10.3390/axioms8030104 . . . . **271**

Alfredo Roque Freire
Review of "The Significance of the New Logic" Willard Van Orman Quine. Edited and Translated by Walter Carnielli, Frederique Janssen-Lauret, and William Pickering. Cambridge University Press, Cambridge, UK, 2018, pp. 1–200. ISBN-10: 1107179025 ISBN-13: 978-1107179028
Reprinted from: *Axioms* **2019**, *8*, 64, doi:10.3390/axioms8020064 . . . . **285**

About the Editors

Alex Citkin spent 25 years of his career as a research fellow of Uzhgorod State University (Ukraine) studying non-classical and algebraic logics, as well as logical methods of pattern recognition. He was among the pioneers investigating the admissibility of inference rules and structural completeness of propositional logics. After relocating in 1994 to the United States, he was made CIO of Metropolitan Telecommunications (New York, USA). He has authored over 70 research papers in logic, universal algebra, group theory and informatics. Citkin is still actively involved in the research in propositional, modal and algebraic logics.

Urszula Wybraniec-Skardowska is a retired professor of logic but is still actively working both scientifically and organizationally. She was named a Prof. of Humanities Science by the President of Poland in 1992. For many years she was a full professor of logic at Opole University and a co-founder and co-chairperson of the Group of Logic, Language and Information established there. Wybraniec-Skardowska is affiliated as a professor at Cardinal Stefan Wyszyński University in Warsaw. Her interdisciplinary research interests include: logic, philosophy, logic and philosophy of language, logical theory of communication, formal linguistics, information sciences and mathematics. She was a visiting professor at many outstanding universities. She is also a member of many Polish and international scientific associations, including: The Association for Symbolic Logic, The European Association for Logic, Language and Information, The Association of Logic and Philosophy of Science, Polish Society of Philosophy, Polish Society of Mathematics, Polish Association for Semiotic Studies. She is the author of about 130 publications. Her book "Theory of Language Syntax. Categorial Approach" was awarded for scientific research by the Polish Ministry of Education (1992). She is a recipient of many national awards.

Editorial

Deductive Systems in Traditional and Modern Logic

Alex Citkin [1,*] and Urszula Wybraniec-Skardowska [2,*]

1. Metropolitan Telecommunications, New York, NY 10041, USA
2. Institute of Philosophy, Cardinal Stefan Wyszynski University in Warsaw, Dewajtis 5, 01-815 Warsaw, Poland
* Correspondence: acitkin@gmail.com (A.C.); skardowska@gmail.com (U.W.-S.)

Received: 3 September 2020; Accepted: 9 September 2020; Published: 13 September 2020

Since its inception, logic has studied the acceptable rules of reasoning, the rules that allow us to pass from certain statements, serving as premises or assumptions, to a statement taken as a conclusion. The first kinds of such rules were distilled by Aristotle and are known as moduses. Stoics ramified Aristotle's system, and for centuries, the syllogistic remained the main tool for logical deduction. With the birth of formal logic, new types of deduction emerged, and to support this new kind of inference, the deductive systems were used. Since then, the deductive systems have been at the heart of logical investigations. In one form or the other, they are used in all branches of logic.

Contemporary understanding of science, as a theory of a high degree of exactness, requires treating it as a deductive theory (a deductive system). Generally speaking, such a theory (system) is a set of its sentential expressions which are derivable (deducible) from some expressions of the set selected as axioms, by means of deduction (inference) rules. The expressions obtained as a result of derivation from a given set of expressions are consequences; that is, they have a proof. The principal feature of deductive systems (theories) is the deducibility or provability of their theorems. From a very general point of view, there are two methods of deduction: (a) the axiomatic method (Hilbert style method) and (b) the natural deduction method (Jaśkowski–Słupecki–Borkowski, Gentzen or semantic tableaux). Method (b) leads to natural deduction systems, while the most often used method (a) leads to presentation (or characterization) of logical and mathematical theories as axiomatic deductive systems. Methods (a) and (b) are used in different scientific disciplines, such as physics, chemistry, sociology, philosophical and psychological sciences, information sciences, discursive sciences, computer science, and some technical sciences.

Deductive sciences have not always been built explicitly as axiomatic systems. Depending on the degree of methodological precision, three of their forms have been distinguished: pre-axiomatic, non-formalized axiomatic, and formalized axiomatic. As we know, a pre-axiomatic form was commonly used in arithmetic and geometry, and later in set theory and probability theory. Their axiomatization was carried out only at the end of the 19th century and the beginning of the 20th century. In contrast, such mathematical theories as the Boolean system and theories of groups, rings, and fields were built as formalized axiomatic systems since inception. The deductive method (calculi) is most often used for formalizations of theories, but these theories also admit formalization as natural deductive systems.

Formalized axiomatic systems are rooted in a tradition originated by G. Frege (1891, 1903), but the first axiomatic system (non-formalized) in the history of science—as it was disclosed by Jan Łukasiewicz in his seminal monograph on Aristotle's syllogistic (1951)—was Aristotle's syllogistic system. J. Łukasiewicz initiated the construction of the first systems of syllogistic satisfying the contemporary requirements, and thus, the requirements of formalized axiomatic systems. He constructed a formalization of syllogistic logic on two levels using (in addition to the commonly used axiomatic method by means of proof) a new axiomatic method by means of rejection—the so-called axiomatic rejection, or refutation, method. He and his disciples (mainly J. Słupecki and his collaborators) applied this method to the bi-level formalization of some classical and non-classical logical deductive systems of sentences or names. This approach allows one to define two disjoint sets of language expressions of a given system: the set of all its theses (theorems), which are asserted, accepted, intuitively true expressions (called the assertion system), and the set of all the other expressions—non-accepted, or intuitively false, refuted, rejected expressions of the system (called the rejection or the refutation system). In such a way, the bi-level formalization of deductive systems provides some new inspiration to build different sciences.

This book is a collection of articles included in the special issue "Deductive Systems" of Axioms regarding mainly the logical deductive system. They are ordered in accordance with the well-known division of logic into term logic (logic of names) and propositional logic (propositional calculus), which correspond to two historical stages of the development of logic, namely, Aristotelian logic and the logic of stoics, with the latter being a contemporary counterpart of propositional logic. Deductive systems for classical propositional logic are broadly known, and one of them is most often assumed for the term logics. Systems for non-classical propositional logics, which are inspired by philosophy, are introduced in the book later than systems related to term logics. Term logic can be interpreted in predicate logic, that is, the second part of contemporary logic. Predicate logic is the basis of mathematical deductive systems (theories).

The volume is opened with paper [1] by P. Kulicki in which he looks back to the roots of Western logic and compares what we have achieved today with the legacy of Aristotle. Somehow surprisingly, we can find many features of today's mature deductive systems in Aristotle's system of syllogism. The paper discusses some of these features, focusing on Aristotle's approach to the issue of completeness reconstructed by J. Łukasiewicz.

In [2], P. Simons considers term logic (logic of names) which is a successor of Aristotle's syllogistic along with 19th century algebraic logic. This is a very natural medium for representing many inferences of ordinary discourse. The axiomatic term logic proposed by P. Simons is intuitive and easy to understand without deeper knowledge of predicate logic.

The paper [3] by J.-M.Castro-Manzano introduces an idea of a distribution model for Sommers' and Englebretsen's term logic. It provides some alternative formal semantics to aforementioned logic.

In his paper [4], E. Wojciechowski makes a reference to the differentiation between Zahl and Anzahl, which is present in the works of Frege and formulates Peano's axiomatic for arithmetic of natural numbers, following Leśniewski on the grounds of the names calculus. This differentiation corresponds syntactically to the name (of natural number)-functor (category n/n). This functor (equivalent of Anzahl) is a primitive term of the proposed axiomatic system.

In [5], V. Goranko introduces hybrid deduction–refutation systems, which are deductive systems intended to derive both valid and non-valid, i.e., semantically refutable, formulae of a given logical system, by employing together separate derivability operators for each of these and combining "hybrid derivation rules" that involve both deduction and refutation. The concept is illustrated with a hybrid deduction–refutation system of natural deduction for classical propositional logic, for which soundness and completeness for both deductions and refutations are proved.

In [6], K. Mruczek-Nasieniewska and M. Nasieniewski analyze the so called discussive logic introduced by Stanisław Jaśkowski, and this is probably the first fully formally formulated system of paraconsistent logic. In 1974 Jerzy Kotas gave an axiomatization of discussive logic. In the paper, Kotas' style axiomatization of the minimal discussive logic is presented.

In [7], J. Ciuciura presents an alternative axiomatization for the hierarchy of paraconsistent systems. The main idea behind it is to focus explicitly on the (in)validity of the principle of ex contradictione sequitur quodlibet. This makes the hierarchy less complex and more transparent, especially from the paraconsistency standpoint.

In [8], A. Citkin studies the deductive systems with multiple conclusion rules which admit the introduction of meta-disjunction. Using the defined notion of the inference with multiple-conclusion rules, it is shown that in the logics enjoying the disjunction property, any derivable rule can be inferred from the single-conclusion rules and a single multiple-conclusion rule, which represents the disjunction property. Additionally, the conversion algorithm of single- and multiple-conclusion deductive systems into each other is studied.

In his paper [9], D. Surowik constructs and studies properties of the minimal temporal logic systems built on the basis of classical logic and intuitionistic logics.

In [10], J. Golińska-Pilarek and M. Welle study deductive systems defining the weakest, extensional two-valued, non-Fregean propositional logic, the language of which is obtained by endowing the language of classical propositional logic with a new binary connective that expresses the identity of two statements.

In [11], S. Pkhakadze and H. Tompits present axiomatizations in terms of the well-known sequent method for two variants of default logic, which is a nonmonotonic formalism relevant for artificial intelligence. The distinguishing feature of the calculi is the usage of rejection systems which axiomatize non-theorems.

In [12], H. Antunes, W. Carnielli, A. Kapsner, and A.Rodrigues construct Kripke-style semantics for the natural deduction systems of the logics of evidence and truth LET_J and LET_F introduced earlier by W. Carnielli and A. Rodrigues. Such logics were conceived to express the deductive behavior of positive and negative evidence, which can be conclusive or non-conclusive. Here, the logics are interpreted in terms of positive and negative information, which can be either reliable or unreliable.

The paper [13] by A. Malec studies the classical first-order predicate logic. This logic is a sufficient and desirable basis for deontic theories which are free-from paradoxes inherent in propositional deontic logics that are adequate to the domain of law. The specific axioms of these theories proposed in the paper refer to Bogusław Wolniewicz's "Ontology of Situations" and reflect: (i) relations between sets of legal events, (ii) properties of simple acts, and (iii) properties of compound acts.

In [14], D. Leszczyńska-Jasion and S. Chlebowski develop a proof method (synthetic tableaux method) for a class of the first-order theories axiomatized by universal axioms. Completeness of the system is demonstrated, and some similarities between the method of synthetic tableaux and the axiomatic method are discussed.

The paper [15] by U. Wybraniec-Skardowska presents two equivalent axiomatic systems of arithmetic of natural numbers: Peano's (P) and Wilkosz's (W), and two intuitive axiomatic extensions of integer arithmetic modeled on them. All these systems of arithmetic are based on second-order predicate calculus, and the systems P and W differ mainly in that while in both categorical systems P and W, the primitive concept is a set of natural numbers, in the former, the primitive concepts are also zero and a successor of the natural number; in the latter, the primitive concept is the inequality relation.

In [16], J-P. Desclés and A. Pascu study mathematical models of the logic of the determination of objects (LDO) and the logic of typical and atypical instances of concept (LTA). The novelty of the model presented in this book is that it describes the structural level of LDO in the framework of preordered sets and lattices. A mathematical model of LTA is constructed as an extension of LDO model. In the case of LTA, a set of objects related to a concept gets equipped with a quasi-topological structure.

A review [17] of the book "The Significance of the New Logic" by Willard Van Orman Quine, contributed by R. Freire, completes the volume.

Conflicts of Interest: The authors declare no conflict of interest.

References

1. Kulicki, P. Aristotle's Syllogistic as a Deductive System. *Axioms* **2020**, *9*, 56. [CrossRef]
2. Simons, P. Term Logic. *Axioms* **2020**, *9*, 18. [CrossRef]
3. Castro-Manzano, J.-M. Distribution Tableaux, Distribution Models. *Axioms* **2020**, *9*, 41. [CrossRef]
4. Wojciechowski, E. The Zahl-Anzahl Distinction in Gottlob Frege: Arithmetic of Natural Numbers with Anzahl as a Primitive Term. *Axioms* **2020**, *9*, 6. [CrossRef]
5. Goranko, V. Hybrid deduction-refutation systems. *Axioms* **2019**, *8*, 118. [CrossRef]
6. Mruczek-Nasieniewska, K.; Nasieniewski, M. A Kotas-Style Characterisation of Minimal Discussive Logic. *Axioms* **2019**, *8*, 108. [CrossRef]

7. Ciuciura, J. A Note on Fernández–Coniglio's Hierarchy of Paraconsistent Systems. *Axioms* **2020**, *9*, 35. [CrossRef]
8. Citkin, A. Deductive Systems with Multiple-Conclusion Rules and the Disjunction Property. *Axioms* **2019**, *8*, 100. [CrossRef]
9. Surowik, D. Minimal Systems of Temporal Logic. *Axioms* **2020**, *9*, 67. [CrossRef]
10. Golińska-Pilarek, J.; Welle, M. Deduction in Non-Fregean Propositional Logic SCI. *Axioms* **2019**, *8*, 115. [CrossRef]
11. Pkhakadze, S.; Tompits, H. Sequent-Type Calculi for Three-Valued and Disjunctive Default Logic. *Axioms* **2020**, *9*, 84. [CrossRef]
12. Antunes, H.; Carnielli, W.; Kapsner, A.; Rodrigues A. Kripke-Style Models for Logics of Evidence and Truth. *Axioms* **2020**, *9*, 100. [CrossRef]
13. Malec, A. Deontic Logics as Axiomatic Extensions of First-Order Predicate Logic: An Approach Inspired by Wolniewicz's Formal Ontology of Situations. *Axioms* **2019**, *8*, 109. [CrossRef]
14. Leszczyńska-Jasion, D.; Chlebowski, S. Synthetic Tableaux with Unrestricted Cut for First-Order Theories. *Axioms* **2019**, *8*, 133. [CrossRef]
15. Wybraniec-Skardowska, U. On Certain Axiomatizations of Arithmetic of Natural and Integer Numbers. *Axioms* **2019**, *8*, 103. [CrossRef]
16. Desclés, J.-P.; Pascu, A.C. Logic of Typical and Atypical Instances of a Concept—A Mathematical Model. *Axioms* **2019**, *8*, 104. [CrossRef]
17. Freire, A.R. Review of "The Significance of the New Logic" Willard Van Orman Quine. Edited and Translated by Walter Carnielli, Frederique Janssen-Lauret, and William Pickering. Cambridge University Press, Cambridge, UK, 2018, pp. 1–200. ISBN-10: 1107179025 ISBN-13: 978-1107179028. *Axioms* **2019**, *8*, 64.

© 2020 by the authors. Licensee MDPI, Basel, Switzerland. This article is an open access article distributed under the terms and conditions of the Creative Commons Attribution (CC BY) license (http://creativecommons.org/licenses/by/4.0/).

Article

Aristotle's Syllogistic as a Deductive System

Piotr Kulicki

Institute of Philosophy, The John Paul II Catholic University of Lublin, 20-950 Lublin, Poland; kulicki@kul.pl

Received: 9 April 2020; Accepted: 15 May 2020; Published: 19 May 2020

Abstract: Aristotle's syllogistic is the first ever deductive system. After centuries, Aristotle's ideas are still interesting for logicians who develop Aristotle's work and draw inspiration from his results and even more from his methods. In the paper we discuss the essential elements of the Aristotelian system of syllogistic and Łukasiewicz's reconstruction of it based on the tools of modern formal logic. We pay special attention to the notion of completeness of a deductive system as discussed by both authors. We describe in detail how completeness can be defined and proved with the use of an axiomatic refutation system. Finally, we apply this methodology to different axiomatizations of syllogistic presented by Łukasiewicz, Lemmon and Shepherdson.

Keywords: Aristotle's logic; syllogistic; Jan Łukasiewicz; axiomatic system; axiomatic refutation; completeness

1. Introduction

Deductive systems of different kinds are the heart of contemporary logic. One could even state that logic itself, as it is understood nowadays, is just a collection of deductive systems appropriate for different kinds of reasoning. Even when ways of reasoning that are usually distinguished from deduction, such as induction or abduction, are considered, they are finally presented in a deduction-like form of a strict system. The theory and methodology of deductive systems is established and well developed, and so is the folklore spread through the community of logicians.

While discussing deductive systems in contemporary logic, it is however still interesting to look back to the roots of Western logic and compare what we have achieved today with the legacy of Aristotle. Somehow surprisingly, we can find many features of today's mature deductive systems in his system of syllogistic. Robin Smith in his entry in the Stanford Encyclopedia of Philosophy [1] notices that "scholars trained in modern formal techniques have come to view Aristotle with new respect, not so much for the correctness of his results as for the remarkable similarity in spirit between much of his work and modern logic. As Jonathan Lear has put it, 'Aristotle shares with modern logicians a fundamental interest in metatheory': his primary goal is not to offer a practical guide to argumentation but to study the properties of inferential systems themselves." Thus, analysing Aristotle's syllogistic allows us to reflect on the most essential features of a deductive system and abstract them from their exact content, context and the terminology used.

No wonder that in recent decades we can observe a significant interest in the logical works of Aristotle. Klaus Glashoff in 2005 [2] (p. 949) stated that "[u]nlike several decades ago, Aristotelian logic meets with growing interest today. Not only philosophers, but also specialists in information and communication theory employ ideas which can be explicitly traced back to Aristotle's work on categories and syllogisms. [...] Independently of these rather recent developments, there has been a renewed interest in matters of formalization of Aristotelian logic by a small group of logicians, philosophers and philologists." Since then, many new works have been published either directly on the writings of Aristotle [3–5] or on extensions or technical aspects of his syllogistic [6–19], to mention only a few.

After Aristotle, syllogistic was for many centuries the dominant form of logic attracting interest of many generations of scholars. There were at least a few important contributions to the theory before the rise of modern mathematical logic in the twentieth century, including the medieval systematisation of traditional syllogistic, several mathematical interpretations of syllogistic presented by Gottfried Wilhelm Leibniz and the diagrammatic approach to the theory introduced by Leonard Euler and John Venn. In this paper we are, however, interested mostly in modern reconstructions of syllogistic starting from the works of Jan Łukasiewicz and some of the ideas inspired by Aristotle presented in this context.

We will start our considerations with some remarks on the original presentation of syllogistic given by Aristotle mainly to trace his methodology of deductive systems. Then, we will look at the system presented by Łukasiewicz. From the perspective of almost a century we trace and assess the choices he made while formalizing syllogistic. We will be especially interested in the way Łukasiewicz developed the Aristotelian discussion of the completeness of the system of syllogistic. Moreover, we will compare this approach with theory and practice of completeness investigations in contemporary logic. Finally we will present how Łukasiewicz's methodology works on the several variants of the system of syllogistic.

The technical results presented in the paper are not novel. The most interesting from the technical point of view is perhaps the refutation counterpart of Shepherdson's axiomatization of syllogistic. The main contribution of the paper lies in its methodological discussion of the issue of correctness of a deductive system. The paper is also rich in references covering sources that present different attempts at the formalization of syllogistic, as well as selected recent works on the subject.

2. Original Presentation

To obtain the right perspective in order to discuss some details of the modern formalizations of syllogistic let us start from a few remarks on its original, Aristotelian presentation. Innocenty M. Bocheński expressed a very strong, but in principle right, opinion on its role in the history of thought that "[t]he assertoric syllogism is probably the most important discovery in all the history of formal logic, for it is not only the first formal theory with variables, but it is also the first axiomatic system ever constructed" [20] (p. 46). This claim takes into account the significance of the Aristotelian system not only for logic. Formal theories and formal modelling are ubiquitous in modern science. Mathematics and mathematically founded physics have been using these tools for the longest time but many other disciplines of natural and social science build their own formalized theories which share the same crucial features. It was Euclidean geometry that in the modern era gained the position of the icon of a deductive system (c.f. famous Spinoza's *more geometrico* but it was syllogistic that earlier had set the standard and prepared the basic conceptual framework for formal techniques in science.

Bocheński justifying his claim on the importance of syllogistic mentioned two issues: the use of variables and the form of an axiomatic system. While the former is simple, understanding the latter requires a reflection on what an axiomatic system is. To acknowledge that a theory forms an axiomatic system two things are required. One is a division of the elements of the system into two groups: axioms and theorems. Some propositions (let us at this point skip the issue, to which we will come back in the following section, of whether syllogisms are indeed propositions, since the same construction can be designed for objects other than propositions, like valid rules or designated modes of reasoning) are treated in a special way and are accepted as axioms and other propositions are derived from them. The other requirement concerns the relation between axioms and theorems. Theorems are derived and the derivation must be deductive. This is the point where maturity of deduction methodology can be observed. In mature systems rules of deduction are explicit and formal.

In the Aristotelian presentation of syllogistic the syllogisms of the first figure are perfect (they are axioms) and the syllogisms of the two other figures are imperfect (i.e., derived from axioms). Bocheński [20] (pp. 46–47) points out three rules of deduction used by Aristotle in his axiomatic system of syllogistic: the direct reduction, the *reductio ad impossibile* and the ecthesis. These rules are deductively valid and recognized in contemporary logic. In modern terminology we can call the direct reduction strengthening of a premise, the *reductio ad impossibile*—transposition, and the ecthesis—reasoning by example. Since we are interested only in the fact of axiomatization and the level of formalization we are not going to present the precise formulation of rules and details of derivations here (for the reconstructions of proofs of all syllogism see [20] (pp. 49–54)).

In a series of loose notes placed throughout *Posterior Analytics*, we can also find general rules of construction of an axiomatic theory. Bocheński reconstructs them in the following way [20] (p. 46):

1. there must be some undemonstrated claims: axioms, and other claims: theorems are deduced ([21], 72b),
2. axioms must be intuitively evident ([21], 99b),
3. the number of steps of deduction in proofs of theorems must be finite ([21], 81b).

Aristotle's approach to axiomatization is similar but not identical to the contemporary one. The main difference lies in the above point 2 regarding axioms. Conditions such as self-evidence, certainty and ontological priority are no longer imposed on them. An axiom differs from other statements of a system only in the fact that it is not derived (c.f. [22] (pp. 70–71)). There are different axiomatizations of the same theories and they are equally correct provided they define the same set of accepted objects. Still some choices of axioms may be evaluated higher than others. What are the criteria applied by contemporary logicians here? The answer is not straightforward. Surely, most of them are not strict and formal. Some of them are similar to what Aristotle required. Sometimes we value higher axioms that are intuitively clear or self-evident. Similar to these criteria is the simplicity of axioms, which is sometimes stressed as an advantage. On the other hand, sometimes axiomatizations with a smaller number of axioms are evaluated higher.

There are some more metalogical notions whose presence (or at least traces) in the Aristotelian system of syllogistic is pointed out by some authors. We will discuss the notion of completeness in detail in the following sections, now let us just briefly mention the notion of compactness.

The issue of compactness of Aristotle's syllogistic was raised by Lear [23]. He claimed that in *Posterior Analytics* I.19-22 Aristotle discusses a proof-theoretic analogue of compactness. Compactness itself is a model theoretical property of a system stating that if a proposition α is a semantic consequence of an infinite set of propositions φ, then there exists a finite set $\varphi_1 \subset \varphi$ such that α is a semantic consequence of φ_1. What is then the proof-theoretic analogue of compactness? It is a property stating that every demonstrable conclusion can be demonstrated from finitely many premises. In other words, there are no valid ways of deductive reasoning that effectively use infinitely many premises.

The question arises whether what Aristotle discusses is really related to compactness in the sense used in contemporary metalogic or it is just a misinterpretation of Aristotle. The second opinion is presented by Michael Scanlan [24], who states that introducing compactness in the context of syllogistic is anachronistic since Aristotle did not use model theory at all. An interesting and balanced discussion of the issue is presented by Adam Crager in [4]. In the context of the present paper it is enough to ascertain that some contemporary logicians want to find traces of modern logical ideas in Aristotelian works even if they are not quite clear there, and that these logicians might be right.

3. Łukasiewicz's Reconstruction of Syllogistic: Formalization Choices

3.1. Preliminaries

It is hard to tell whether Łukasiewicz was aware of the different possibilities he could use when he was formalizing syllogistic using the tools of modern formal logic. "the" in the title of his book: *Aristotle's Syllogistic from the Standpoint of Modern Formal Logic* may suggest that in his opinion his point of view concerning the theory was the only one.

Now, taking into account later works on syllogistic we can see that it is not that simple. There are many possible formal tools that can be applied to construct a system of syllogistic and many variants of the content of the theory. Looking from today's perspective the most fundamental decision is the choice of a kind of object a syllogism should be. In the later literature (see e.g., [25,26]) at least three interpretations of a (correct) syllogism are discussed: (1) valid premise-conclusion argument, (2) true proposition or (3) cogent argumentation or deduction. In terms of the formal structure that leads to two clear possibilities: inference rules for (1) and implication propositions for (2). Interpretation (3) requires a less direct formal account of syllogism.

Łukasiewicz constructed a theory where syllogisms are represented as propositions. This approach seems to be in accordance with the spirit of the 30s in logic. The hype was for axiomatisation in, what we would call now, the Hilbert style. Natural deduction, being an alternative to it, had just been invented by Gentzen and Jaśkowski and only budding. It is less known that Gödel in his Notre Dame lectures in 1939 [27] also presented a formalization of syllogistic with the use of mathematical logic and his system was constructed in a way similar to Łukasiewicz's system. The main difference was in the choice of axioms and in the fact that while Łukasiewicz presented a full-fledged theory, Gödel presented only a sketch.

Another important issue where approaches to syllogistic may vary is connected with the sort of names that can be used within categorical sentences that are the components of syllogisms. Two distinctions are relevant here for individual/common names and empty/nonempty ones. This issue was also discussed extensively after Łukasiewicz and different proposals are now available here. In the following sections we will discuss Łukasiewicz's approach in detail.

3.2. Axiomatic Theory Based on the Classical Propositional Logic

What is shared by all the aforementioned interpretations of Aristotelian logic is that the purpose of syllogistic is to study reasoning in which categorical propositions are both premises and conclusions. As we have mentioned, such reasoning can be formalized with the tools available to modern formal logic, in several ways, for example as sentences of language with the implication structure or as inference rules or schemata.

In the former case, the premises for reasoning can be treated as factors of the conjunction constituting the antecedent of implication, and the conclusion as the consequent of implication. What results are formulas that can be converted into rules in a natural way. To view syllogisms as such implications requires an interpretation of implication and conjunction. Łukasiewicz adopted the simplest solution available for him, where these operators are taken from the classical propositional calculus. The classical interpretation of operators is, however, by no means obvious. The definition of syllogism itself, as derived from *Prior Analytics*: "[a] syllogism is an argument in which, certain things being posited, something other then what was laid down results by necessity because these things are so", [21] (24b, 20) suggests two features of syllogisms that the classical calculus ignores—non-tautologicality "something other then what was laid down" and relevance: "because these things are so" (see e.g., [3] for a discussion of the issue of relevance).

Łukasiewicz went further to assume that syllogistic is built over the whole classical propositional calculus and thus allows structures other than those in the form of syllogism. In this way, the direct relationship with rules is lost. This element of his approach to Aristotle's syllogistic seems to be particularly controversial.

Therefore, to provide a better understanding of the essence of Łukasiewicz's approach to syllogistic, three elements can be separated: (1) the formalization of reasoning by sentences of language with an implication structure, (2) the use of the classical understanding of propositional operators, (3) the use of propositional calculus operators in any configuration to build complex formulas.

Łukasiewicz's approach was strongly criticized by John Corcoran [28,29]. His criticism concerned mainly point (1) above. Instead, Corcoran proposed to formalize syllogisms as rules within a system of natural deduction. From the further perspective, however, the difference between the two approaches is not that essential. When propositions in the form of implication are considered, there is a close relation between the truth of sentences and the soundness of inference rules. True implications are the basis of correct rules, and correct rules can be transformed into corresponding true sentences. Such a proposition-rule duality of implications reveals itself especially in the context of logic programming. Logic programs are sets of Horn clauses. In a declarative interpretation clauses are implications in which ancetedents are conjunctions of atoms and consequents are atoms. In a procedural interpretation they are rules with multiple premises that are triggered in certain situations. It is easy to see that syllogisms have the same structure and therefore they also can be interpreted dually.

Indeed, the other aspects of Łukasiewicz's approach seems to be more controversial. Classical propositional calculus is probably not the logic which can adequately describe the Aristotelian way of thinking. Moreover, Aristotle did not use any structures other than standard syllogisms and sorites (syllogism with more than two premises). Again these discrepancies between the Aristotelian theory and its reconstruction by Łukasiewicz does not seriously undermine Łukasiewicz's practice. That is because he does not really make use of propositional logic more than it is necessary to reconstruct arguments confluent with the ones acceptable by Aristotle, mainly proving one syllogisms on the basis of others.

3.3. Admissible Types of Names

In logical semiotics, there are two divisions of names that are interesting from the point of view of the formalization of the syllogistic. The first is made by the type of reference and distinguishes between common and proper (individual) names. Common names designate objects because they meet some conditions and can always be linked with appropriate predicates that represent these conditions. Individual names designate specific objects under a language convention. The other division is made by the number of designates and distinguishes between empty names, i.e., having no designates, particular names, i.e., having exactly one designate, and general names, i.e., those having more than one designate. Note that empty and particular names can be common or proper. As an example of a common empty name we can take "unicorn" or "square circle", as an example of a proper empty name we can take "Pegasus" or "Santa Claus".

While building a system of logic of names such as syllogistic one can narrow the range of names that can be used to selected categories based on the above divisions. Such postulates have had various motivations and justifications. In his famous work "On Sense and Reference" [30], Gottlob Frege proposed that empty names be eliminated from the language of science. In his justification, Frege uses reasoning that can be summarized in the claim that the use of names without denotation leads to pointless discussion and manipulation. Aristotle permits names without denotation, assuming that atomic sentences in which such names appear are false. Łukasiewicz's formalization of syllogistic assumes that all names are non-empty. However, many systems built in his style, like [31–34], allow the use of empty names.

Łukasiewicz also eliminates individual names from the language of syllogistic. A similar narrowing can be observed in Peter Geach, who in combining individual and proper names in traditional syllogistic sees an important source of the "corruption of logic" [35]. The position of Aristotle himself on this issue is not quite clear. When he presents valid syllogisms in *Prior Analitics* (26a–46b) he always uses general names like "animal", "man" and "white". However further in *Prior Analitics* 47b while discussing some invalid forms, where "no syllogism is possible" he puts in these forms individual (proper) names: Aristomenes and Miccalus. It is not obvious, and Aristotle does not state clearly, whether the use of proper names is the reason why *syllogisms are not possible* or it is a coincidence.

The admissibility of propositions in which the same argument appears twice raises yet another type of doubt. In modern logic, such formulas are natural and can be created by substituting the same value (constant or variable) in any expression. In his system formalizing the syllogistic of Aristotle, Łukasiewicz even uses the formulas "each S is S" and "certain S is S" as axioms. Such sentences, however, do not appear in the description of syllogistic modes given by Aristotle.

This fact can be associated with the requirement mentioned above, according to which syllogism should lead to new knowledge. On its basis, one can derive "something else than assumed", and, on the other hand, what is derived "must result because it was assumed". In this context, the sentences "every S is S" and "certain S is S" are not useful, because with the normal use of syllogisms nothing new results from them, nor can they constitute new knowledge resulting from certain assumptions. Łukasiewicz does not set out any restrictions on substitution.

4. Completeness

The basic criterion for assessing the quality of a formal system is its adequacy with respect to underlying intuitions. Adequacy consists of two properties: soundness and completeness. In the case of an axiomatic system, soundness means that all theorems of the system follow underlying intuitions, and completeness that all formulas that are intuitively accepted are also accepted in the system. In the literature one can find many embodiments of this fundamental intuition which differ in important details. We will cite some of them below. In the first two, completeness directly refers to sets of formulas, and the truth of these formulas (sentences) is adopted as the intuitive acceptance criterion. Kazimierz Ajdukiewicz uses the concept of completeness understood as follows:

"each true sentence that can be formulated in the language of this theory can be proved (unless it is an axiom of this theory) by the means of evidence at its disposal." [36] (p. 215)

Ludwik Borkowski gives the following definitions of completeness:

"The S system is complete if and only if each true expression of the S system is a thesis of the S system." [37] (p. 378)

In metalogical considerations, the classical correspondence concept of truth is usually used. In Ajdukiewicz's formulation it is as follows:

"Any declarative sentence is true when it is just as it says; it is false when it is not what it proclaims." [36] (p. 29)

The above framings have, however, a disadvantage. Not all formal systems that logic deals with refer to truthfulness. An example would be intuitionistic logic, where the goal is to capture what is constructively provable rather than true. However, the same method can still be applied to consider completeness in relation to formal approaches to this type of logic. In general, the concept of completeness in the above approaches may be retained, only the term "true" should be replaced with the term "accepted" or "admissible".

Other definitions associate completeness with sets of inference or reasoning methods and refer to their soundness or reliability, which corresponds to the truthfulness of sentences. Andrzej Grzegorczyk writes about completeness:

"The natural, historical development of logic has indeed led to the creation of such a logic system, which can be proved to contain all logical methods (schemata) of correct inference on any subject. We call this property completeness." [38] (p. 121)

Witold A. Pogorzelski phrases this concept as follows:

"The problem of completeness can be formulated as a question of whether all reliable ways of reasoning are actually based on the laws of formal logic." [39] (p. 366)

As we noted earlier, there is a close relationship between the truth (acceptance) of sentences and the soundness of reasoning (rules of inference). True sentences can in fact form the basis for the construction of correct reasoning, and correct reasoning can be transformed into corresponding true sentences. This fact allows us to assume that all the above definitions express in their own way the same intuition, which does not give rise to controversy. In all definitions, completeness is semantic in the sense that it refers a formal system to something external, to reality, or at least to a way of thinking about reality.

How to verify which sentences are true remains a problem, and in particular, how to do it precisely enough to be able to use this validation in logical research. Most often formal models are used for this purpose and truthfulness is defined as truthfulness in a model, which is defined by the formal conditions imposed on objects in the model. As a result, in practice the truthfulness of sentences is usually equated with their truthfulness in a formal model or a class of models. Consequently models directly appear in the definition of completeness, as in the definition below from the Small Encyclopedia of Logic edited by Witold Marciszewski:

> "The deductive logic system is complete if and only if all sentences that are true sentences in each model can be derived from its axioms." [40] (p. 236)

However, by adopting this position, we give up the semantic character of completeness. We consider the mutual relations between the two formal systems, i.e., the axiomatic system and the system defining the formal model. Two formal approaches undoubtedly give a more complete picture of a formal theory, but it does not connect the theory with reality. The problem of completeness of the axiomatic system with respect to the underlying intuitions is not solved by demonstrating completeness in relation to a model, but is only put aside. Another problem arises, one of adequacy of the formal model in relation to reality or the way of thinking which the formal system under consideration is to capture. Sometimes a model theoretical structure is intuitive, but in some cases, as relevance logic, linear logic, or even intuitionistic logic, a proof-theoretical approach is much closer to intuitions then models constructed to match the systems.

In the case of syllogistic, set-theoretical models are quite intuitive and seems to be natural, especially for contemporary people, who are accustomed to thinking in terms of sets from kindergarten. However, there are reasons to consider a theory of syllogism that is not dependent on set-theoretical models. One of them is historical: Aristotle himself did not know set theory. Thus, it is good to be able to conduct metalogical considerations concerning syllogistic without sets just to avoid anachronisms. The other reason is that some researchers claim that set-theoretical approach used in modern logic does not fit to the way we use natural language and use alternative approaches like Leśniewski's ontology [41].

Following and referring to Aristotle, Łukasiewicz proposed a different solution. In *Prior Analytics* Aristotle shows that syllogistic schemata other than the syllogisms that he accepted should not be accepted. In this way, he proves completeness of his system of syllogistic. He considers all possible schemata with two premises belonging to each of the three figures. In most cases, he justifies the rejection of a schema by providing a counter-example, as in the following passage:

> "Nor will there ever be a syllogism if both intervals are particular, whether positive or privative, or if one is stated positively, the other privatively, or one indeterminate, the other determinate, or both indeterminate. Common terms for all cases: animal, white, horse; animal, white, stone." [21], (26b)

Showing examples falsifying all unacceptable formulas is labour-intensive, and in many cases impossible due to their unlimited number, e.g., if one were to consider reasoning with any number of premises (factors in the antecedent of implications). Already in Aristotle, however, one can find a hint regarding a different way of rejecting such formulas, which Łukasiewicz extracts and expands. The following text occurs in Aristotle:

> "For since it is true that M does not belong to some X even if it belongs to none, and there was no syllogism when it belonged to none, it is evident that there will not be one in this case either." [21], (27b)

Out of this short note of Aristotle, Łukasiewicz derived an idea of axiomatic refutation that was first further developed by Słupecki and his collaborators and then entered into the wider logic community. A substantial theory concerning the logic of rejected propositions is presented in [42]. Recently the achievements in the field were recapitulated in [43] and the Łukasiewicz-Słupecki approach to the issue was discussed in [44].

The basic ideas of Łukasiewicz are as follows. In addition to usual axioms and rules, rejected axioms and rules of refutation are introduced. Rejected axioms should not be valid and rules of refutation also produce non-valid formulas. A system is said to be refutationaly complete if each formula of its language is either a theorem or a rejected formula. In his writings Łukasiewicz, and also Słupecki, stressed that under certain conditions refutationaly complete systems are decidable.

In the opinion of the Author of the present paper, even more interesting is the argument that refutationaly complete systems are adequate (sound and complete) in the very basic sense of adequacy discussed above. In the following section we will present how refutation works in Łukasiewicz's system and in some other axiomatizations of syllogistic constructed in his style.

5. Axiomatic Systems of Syllogistic with Refutation Counterparts

The language of all the systems discussed in this section is the same. It contains name variables $S, P, M, N, \ldots$, propositional operators and the two primitive operators a and i specific for syllogistic read in a usual way: SaP is read as "every S is P" and SiP—as "some Ss are Ps" or "certain S is P". We will call formulas like SaP and SiP atoms. Formally, a formula of the language can be defined in the following way (using Backus–Naur notation):

$$\alpha = SaS \mid SiS \mid \neg\alpha \mid \alpha \wedge \alpha \mid \alpha \vee \alpha \mid \alpha \rightarrow \alpha \mid \alpha \equiv \alpha$$

The usual negative syllogistic operators e and o (where SeP is read as "no S is P" and SoP—as "some Ss are not Ps" or "certain S in not P") can be defined as negations of the primitive operators:

$$SeP \triangleq \neg SiP,$$

$$SoP \triangleq \neg SaP.$$

In all the systems any substitution of a classical tautology is an axiom. The common derivation rules are *Modus Ponens MP* and substitution *Sub* of the following schemata:

$$\frac{\vdash \alpha \rightarrow \beta; \vdash \alpha}{\vdash \beta}$$

$$\frac{\vdash \alpha}{\vdash e(\alpha)},$$

where e is a substitution for name variables.

5.1. Łukasiewicz

Let us start with the Łukasiewicz's system. Its specific axioms of are as follows:

$$SaS, \tag{1}$$

$$SiS, \tag{2}$$

$$MaP \wedge SaM \rightarrow SaP, \tag{3}$$

$$MaP \wedge MiS \rightarrow SiP. \tag{4}$$

The negative (rejected) part of the system is defined by the following three rules ($\vdash$ marks an accepted formula and $\dashv$ a rejected one):

- rejection by detachment MP^{-1}:
$$\frac{\vdash \alpha \to \beta; \dashv \beta}{\dashv \alpha},$$

- rejection by substitution Sub^{-1}:
$$\frac{\dashv e(\alpha)}{\dashv \alpha},$$

where e is a substitution for name variables,

- decomposition rule $Comp^{-1}$:
$$\frac{\dashv \alpha \to \beta_1; \ldots; \dashv \alpha \to \beta_n}{\dashv \alpha \to \beta_1 \vee \ldots \vee \beta_n}, n \geq 1,$$

where α is a conjunction of atoms and $\beta_i (1 \leq i \leq n)$ are atoms.

The last rule is a variant from [45] of the rule of Słupecki, used by Słupecki, Łukasiewicz and Shepherdson, that reflects in its shape a more general result on Horn theories from [46]

The following formula is the sole rejected axiom:

$$PaM \wedge SaM \to SiP. \tag{5}$$

To see how the axiomatic system works let us give a proof of the following conversion law for general negative sentences (for the application of the laws of propositional calculus we use the abbreviation "PC", to abbreviate substitutions like $e(M) = P$ we will write M/P):

$$SeP \to PeS$$

1. $MaP \wedge MiS \to SiP$ axiom (4)
2. $PaP \wedge PiS \to SiP$ $Sub: 1 (M/P)$
3. $PaP \to (PiS \to SiP)$ PC: 2
4. MaM axiom (1)
5. PaP $Sub: 4 (M/P)$
6. $PiS \to SiP$ MP: 3, 5
7. $\neg SiP \to \neg PiS$ PC: 6
    $SeP \to PeS$ definition of SeP: 7

As an example of a negative derivation let us give a refutation of the analogous conversion of general positive sentences:

$$SaP \to PaS$$

1. $\vdash MaP \wedge SaM \to SaP$ axiom (3)
2. $\vdash MaP \wedge MiS \to SiP$ axiom (4)
3. $\vdash SiS$ axiom (2)
4. $\vdash SaP \wedge SiS \to SiP$ $Sub: 2 (M/S)$
5. $\vdash SiS \to (SaP \to SiP)$ PC: 4
6. $\vdash SaP \to SiP$ MP: 5,3
7. $\vdash MaP \wedge SaM \to SiP$ PC 1, 6
8. $\vdash (MaP \wedge SaM \to SiP) \to ((PaM \to MaP) \to (SaM \wedge PaM \to SiP))$ PC
9. $\vdash (PaM \to MaP) \to (SaM \wedge PaM \to SiP)$ MP: 8, 7
10. $\dashv PaM \wedge SaM \to SiP$ rejected axiom (5)
11. $\dashv PaM \to MaP$ MP^{-1}: 9, 10
    $\dashv SaP \to PaS$ Sub^{-1}: 12 $(S/P, P/M)$

Łukasiewicz shows that all the Aristotelian assertoric syllogisms and all the one-premise valid reasoning schemata mentioned by Aristotle have their counterparts in the form of implications provable in Łukasiewicz's system. Moreover, since the system incorporates the whole classical propositional calculus, many formulas that are not directly connected to the reasoning schemata discussed by Aristotle, or are not implications e.g.,:

$$SaP \rightarrow SaP,$$

$$SiP \vee SoP,$$

$$\neg(SaP \vee SeP)$$

are provable.

Łukasiewicz's system is refutationally complete, i.e., every formula of the language is either a theorem or can be rejected. The proof of that fact is well known (see [47,48]) and we will not repeat it. Let us just mention that the proof relies on the observation that a formula of a form $\alpha \rightarrow SaP$, where α is a conjunction of atoms, is a theorem if and only if α contains a chain connecting S with P, and such a formula is rejected whenever it is not a theorem. A chain is a conjunction of the following form:

$$SaM_1 \wedge M_1aM_2 \wedge \ldots \wedge M_naP \ (n \geq 0).$$

The result is then extended to all formulas of the language on the basis of propositional calculus derivations on the accepted side, and the refutations based on the rules MP^{-1} and $Comp^{-1}$ on the rejected side.

5.2. Lemmon

Let us now look at systems without restriction to non-empty names. Many different axiomatizations adequate for that idea were introduced. They differ, beyond just the choice of axioms, in two main points. One of them is the set of primitive notions: some of them use the same operators *a* and *i* as Łukasiweicz, some other use nominal negation instead of *i*. The other difference is the interpretation of the operator *a*, which can be strong or weak. In both of them, obviously, for SaP to be true S must be contained in P, but in the strong interpretation S must be non-empty, while in the weak one it is not so.

The strong interpretation is adopted, among others, by Wedberg, Menne and Lemmon [32,33,49]. We start from it because the refutation part of this variant of theory is much simpler than the one for the weak interpretation. We use a variant of Lemmon's system from [45] with Equations (3) and (4),

$$SaP \rightarrow SiP, \tag{6}$$

and

$$PiS \rightarrow SaS, \tag{7}$$

as axioms.

The refutational counterpart of the system consists of the same rules as for Łukasiewicz's system, and Equation (5) and

$$PaP \rightarrow SiS \tag{8}$$

as rejected axioms. The proof of refutation completeness of the system similar to the one for Łukasiewicz's system can be found in [50].

5.3. Shepherdson

Now we can pass to Shepherdson's system [34], also called the Brentano style syllogistic [51] (p. 311). The speciffic axioms of the system are in Equations (1), (3) and (4),

$$SiP \rightarrow SiS, \tag{9}$$

and
$$SaP \vee SiS. \tag{10}$$

Assuming the correctness of Lemmon's system we can obtain correctness result for Shepherdson's by embedding it into Lemmon's. To define the relation between Shepherdson's and Lemmon's systems let us for this purpose distinguish two variants of operator a: a_S and a_L occurring in the two systems, respectively. With this convention we can formulate the following equivalences that may be used to mutually define one operator by another:

$$Sa_S P \equiv Sa_L P \vee \neg(SiS),$$

$$Sa_L P \equiv Sa_S P \wedge SiS.$$

Let us, however, introduce the refutational counterpart of Shepherdson's system. The system was first presented in [50] where full proof is given (it is lengthy and laborious but quite predictable). Here we will just sketch the proof and use the final result to comment on the usefulness of the refutation approach for the discussion of completeness.

Let us start with the easier part: the refutation system for the Horn fragment of the Shepherdson's system, i.e., the system with Equations (1), (3), (4) and (9), as axioms.

Here the rules of rejection are the same as in Łukasiewicz's system and the rejected axioms are:

$$SiS \wedge PiP \wedge SaM \wedge PaM \rightarrow SiP \tag{11}$$

$$PiP \wedge SaP \rightarrow SiS. \tag{12}$$

The proof of refutation completeness is analogous to the one for Łukasiewicz's system.

Now, let us come back to the full system of Shepherdson, and its refutation counterpart. To define it we will use the rules MP^{-1}, Sub^{-1}, as in the Łukasiewicz's system and the following modified version of $Comp^{-1}$, $Comp_2^{-1}$:

$$\frac{\dashv \alpha \rightarrow \beta_i \vee \beta_j,\ \text{for each i, j, such that: } 1 \leq i < j \leq n}{\dashv \alpha \rightarrow \beta_1 \vee \ldots \vee \beta_n}, n \geq 2,$$

where α is a conjunction of atoms and β_i ($1 \leq i \leq n$) are atoms.

The sole rejected axiom is as follows:

$$MaS \wedge MaP \wedge MaQ \wedge SaR \wedge PaR \wedge RaN \wedge QaN \wedge SiS \wedge PiP \wedge QiQ \rightarrow SiP \vee RiQ. \tag{13}$$

The choice of rejected axiom is mainly technical: it is chosen as sufficient to prove completeness. In the following section we will show by giving an example that it is not valid and for that reason it should be rejected. The role of the $Comp^{-1}$ rule is similarly technical: to enable the proof of refutational completeness. In the following section we will try to justify its validity.

First let us notice that Equations (11) and (12) are rejected: it can be proved that from each of the formulas separately we can derive the rejected axiom in Equation (13). For Equation (11) to derive Equation (13) from it after substituting R for M we just need to strengthen the antecedent and weaken the consequent. In Equation (12) we first need to substitute S for P and M for S. Then, using the fact, that $MiM \wedge MaS \wedge MaP \rightarrow SiP$ is a theorem of the system we can use classical proposition calculus to derive Equation (13). Thus, each Horn formula of the system is either a theorem or a rejected formula.

Now, we need to prove the same fact for the formulas of the form:

$$\alpha \rightarrow \beta_1 \vee \beta_2, \tag{14}$$

where α is a conjunction of atoms, and β_1 and β_2 are atoms.

For this purpose it will be useful to note that the rejected axiom is equivalent to the following longer formula:

$$\begin{aligned}
& MaM \wedge MaS \wedge MaP \wedge MaR \wedge MaQ \wedge MaN \wedge SaS \wedge SaR \wedge SaN \wedge \\
& PaP \wedge PaR \wedge PaN \wedge RaR \wedge RaN \wedge QaQ \wedge QaN \wedge NaN \wedge \\
& SiS \wedge SiR \wedge RiS \wedge SiN \wedge NiS \wedge PiP \wedge PiR \wedge RiP \wedge PiN \wedge NiP \wedge \\
& RiR \wedge RiN \wedge NiR \wedge QiQ \wedge QiN \wedge NiQ \wedge NiN \rightarrow \\
& SaM \vee SaP \vee SaQ \vee PaM \vee PaS \vee PaQ \vee RaM \vee RaS \vee RaP \vee RaQ \vee \\
& QaM \vee QaS \vee QaP \vee QaR \vee NaM \vee NaS \vee NaP \vee NaR \vee NaQ \vee \\
& MiM \vee MiS \vee SiM \vee MiP \vee PiM \vee MiR \vee RiM \vee MiQ \vee QiM \vee \\
& MiN \vee NiM \vee SiP \vee PiS \vee SiQ \vee QiS \vee PiQ \vee QiP \vee RiQ \vee QiR.
\end{aligned} \tag{15}$$

The intuitive meaning of the formula is not straightforward. It is a maximal combination of atoms built from six variables put on the both sides of the implications that does not allow one to derive consequent from antecedent, needed from the technical point of view to prove the completeness result.

The derivation from Equations (13) to (15) is valid since both the antecedent and consequent of Equation (13) are included in respectively the antecedent and the consequent of Equation (15). The derivation from Equation (15) to Equation (13) is based on the theorems of the system of the form of:

- implications with elements of the antecedent of Equation (15) not present in the antecedent of Equation (13) as the consequent and the antecedent of Equation (13) (or its fragment) as the antecedent, e.g.,

$$MaS \wedge SaR \rightarrow MaR,$$

- and implications with the elements of the consequent of Equation (15) not present in the consequent of Equation (13) and the antecedent of Equation (13) (or its fragments) as antecedents and elements of the antecedent of Equation (15) not present in the antecedent of Equation (13) as consequents, e.g.,

$$SaM \wedge SiS \wedge MaS \wedge MaP \rightarrow SiP.$$

Thus, all the elements of Equation (15) not included in Equation (13) can be eliminated.

Now we can move on to the main point of this part of the proof: the analysis of all possible forms of Equation (14). Its consequent may take one of the three following forms (with possibly different variables): (i) $SiP \vee RiQ$, (ii) $SiP \vee RaQ$, (iii) $SaP \vee RaQ$.

In case (i), any formula of the discussed shape $\alpha \rightarrow SiP \vee RiQ$ is a theorem if $\alpha \rightarrow SiP$ or $\alpha \rightarrow RiQ$ is a theorem. In case (ii), a formula is a theorem when one of the following conditions is fulfilled:

- $\alpha \rightarrow SiP$ is a theorem,
- $\alpha \rightarrow RaQ$ is a theorem,
- the following conditions are satisfied: (I) α contains a chain connecting R with S (or in the place of both S and R the same variable occurs), and (II) α contains a chain connecting R with P (or in the place of both P and R the same variable occurs).

In case (iii), any formula of the discussed shape $\alpha \rightarrow SaP \vee RaQ$ is a theorem if $\alpha \rightarrow SaP$ or $\alpha \rightarrow RaQ$ is a theorem.

In all cases (i)–(iii), if a formula is not the theorem described above, after renaming the variables when needed, it contain only elements of the antecedent of Equation (15) in the antecedent and only elements of the consequent of Equation (15) in the consequent. All formulas fulfilling this condition are rejected. Thus, each formula of the shape seen in Equation (14) is a theorem or is rejected.

It remains to prove that the set of theorems and the set of rejected formulas are disjoint. For that we need to show (a) that the rules of rejection lead from non-theorems to other non-theorems and (b) that the rejected axiom is not a theorem.

As for (a) since for rules MP^{-1} and Sub^{-1} that fact is obvious, the interesting case is the weak version of $Comp^{-1}$. Here we need to show that if a formula $\alpha \to \beta_1 \vee \beta_2 \vee \beta_3$, where α is a conjunction of atoms and β_1, β_2 and β_3 are atoms, is a theorem of the system, then at least one of the $\alpha \to \beta_i \vee \beta_j$ ($i, j \in \{1, 2, 3\}$) is also a theorem. The proof of that fact is based on the constatation that the only non-Horn axiom of the system in Equation (10) cannot be effectively used twice in any derivation in the system, so the only way to obtain a formula of the form $\alpha \to \beta_1 \vee \beta_2 \vee \beta_3$ is by adding a new element of the consequent on the basis of the appropriate law of the classical propositional calculus.

To show that the rejected axiom is not a theorem of the system we can use the following matrices:

a	n_1	n_2	n_3	n_4	n_5	n_6
n_1	1	1	1	1	1	1
n_2	0	1	0	1	0	1
n_3	0	0	1	1	0	1
n_4	0	0	0	1	0	1
n_5	0	0	0	0	1	1
n_6	0	0	0	0	0	1

i	n_1	n_2	n_3	n_4	n_5	n_6
n_1	0	0	0	0	0	0
n_2	0	1	0	1	0	1
n_3	0	0	1	1	0	1
n_4	0	1	1	1	0	1
n_5	0	0	0	0	1	1
n_6	0	1	1	1	1	1

The matrices indicate the truth values of atoms when values n_i ($1 \leq i \leq 6$) are substituted for nominal variables in formulas. Checking that Shepherdson's axioms receive always the value 1 is a usual routine. Rejected axiom in Equation (13) receives the value 0 when we put n_1 for M, n_2 for S, n_3 for P, n_4 for R, n_5 for Q and n_6 for N.

5.4. Refutation and Adequacy

Let us stress that in the presentation of systems in the previous section we did not mention at all set-theoretical models of categorical sentences. That allows us to see syllogistic as a theory on its own, largely independent from the set-theoretical intuitions that are contemporarily usually applied in order to understand syllogistic.

The refutationaly complete axiomatic presentation of syllogistic may be an alternative way to control the correctness of the formalization. To check that a system is correct we need to show that axioms are intuitively correct and rejected axioms are not, and that rules used to deduce theorems from axioms and rejected formulas from rejected axioms (and theorems) work properly. Since correctness is not with respect to another formal system (like a set of models) the argumentation here cannot be strictly formal.

As for axioms of the three systems let us first look at their common part consisting of axioms in Equations (3) and (4). They come from Aristotle and seem to be intuitively very clear and convincing. The remaining parts of the system have much to do with empty names. In Łukasiewicz's system empty names are not allowed. Thus, if we only accept that using the same name twice in positive categorical sentences, like SaS and SiS, makes sense at all, we should also accept that such sentences are true. In the case of Lemmon's system both specific axioms in Equations (6) and (7) express the core of the strong interpretation of SaP in combination with the existential commitment of i. In the case of Shepherdson's system we got SaS accepted in the axiom in Equation (1) as the weak interpretation of a makes it true also for empty names. Equation (9) similarly to Equation (7) is based on the existential commitment of i. Finally, Equation (10) expresses the fact that a name S is either empty (then SaP has to be true under the weak interpretation of a) or non-empty (then SiS is true). Thus, we can say that the axioms of each system correctly reflect their background intuitions.

To show that a rejected axiom should be rejected it is enough to find a counterexample since we just need to confirm that it is not valid. That is the way that Aristotle worked. Let us look at the formulas used as rejected axioms in the axiomatic systems of syllogistic we have discussed.

For Equation (5) we can take a cat for S, a dog for P and an animal for M. Both dogs and cats are animals but no cat is a dog. For Equation (8) we can take a cat for P and a unicorn for S (assuming that unicorns do not exist and the sentence 'some unicorns are unicorns' is therefore false). Even if every cat is a cat it is not the case that some unicorns are unicorns. As Equation (13) is more complicated the counterexample for it is a bit more difficult to follow. Let us put a unicorn for M, a cat for S, a dog for P, a mammal for R, a parrot for Q and an animal for N. Then all the elements of the antecedent of Equation (13) are true: every unicorn is a cat (since there are no unicorns), every unicorn is a dog, every unicorn is a parrot, every cat is a mammal, every dog is a mammal, every mammal is an animal, every parrot is an animal, some cats are cats, some dogs are dogs and some parrots are parrots, but neither any cat is a dog nor any mammal is a parrot.

The rules MP and Sub, and their refutational counterparts MP^{-1} and Sub^{-1} seem to be very natural and common in logic. The justification of decomposition rules is less obvious. In its stronger version $Comp^{-1}$ reflects the simplicity of syllogisms as single-conclusion schemata giving unequivocal result. It's modified version $Comp_2^{-1}$ is weaker, because the Shepherdson's system has Equation (10) as an axiom and, because of this, $Comp^{-1}$ is not valid there. Thus, the system allows for two alternative conclusions of a syllogism but no more than two. This principle can be understood as acknowledging Equation (10) and its consequences as an exception, which is limited by $Comp_2^{-1}$.

To sum up, all conditions required for the discussed systems to be a correct representation of the intuitions they formalize are fulfilled. That gives us an example of how to discuss the correctness (soundness and completeness) of logic without models on the basis of refutation techniques having their sources in Aristotelian logic.

6. Conclusions

Aristotle and contemporary logicians share the same aspiration to give intuitions about correct reasoning a precise formulation. We have presented how Łukasiewicz's ideas of an axiomatic refutation system, inspired by Aristotle, work as a tool of assessment of such a formulation. In particular we have discussed the issue of correctness of three axiomatic systems of syllogistic introduced by Łukasiewicz, Lemmon and Shepherdson.

Let us point out the main conclusions of these considerations. Firstly, we should acknowledge that syllogistic is a fully-fledged deductive system and that Aristotle conducted interesting metalogical studies concerning it. Secondly, when investigating modern reconstructions of syllogistic, it is more interesting, in my opinion, to look at the content of the theories than at the particular tools of formalization. Thirdly, to assess the quality of the content of a theory, including such reconstructions, the most important feature is their adequacy consisting of correctness and completeness. Axiomatic refutation is an interesting method of showing adequacy. It can be used as an alternative to the model-theoretical approach commonly used in contemporary logic.

Finally, as for the three systems of Łukasiewicz's style syllogistic considered within the paper we can see that it is possible to construct refutation counterparts for all of them. All three refutational formalizations allow us to show adequacy of the systems with respect to the intuitions they follow. However, for Shepherdson's system which includes an axiom in a form of disjunction of atomic formulas, the refutation system itself and the proof of adequacy are much more complicated.

Funding: The project is funded by the Minister of Science and Higher Education within the program under the name "Regional Initiative of Excellence" in 2019–2022, project number: 028/RID/2018/19, the amount of funding: 11 742 500 PLN.

Acknowledgments: I would like to thank Urszula Wybraniec-Skardowska for discussions about refutation and encouragement to prepare this paper, anonymous reviewers for their valuable comments and Zdzisław Dywan, who has introduced me to the realm of Aristotelian logic.

Conflicts of Interest: The author declare no conflict of interest.

References

1. Smith, R. Aristotle's Logic. In *The Stanford Encyclopedia of Philosophy*; Zalta, E.N., Ed.; Metaphysics Research Lab, Stanford University: Stanford, CA, USA, 2019.
2. Glashoff, K. Aristotelian Syntax from a Computational-Combinatorial Point of View. *J. Log. Comput.* **2005**, *15*, 949–973. [CrossRef]
3. Steinkrüger, P. Aristotle's assertoric syllogistic and modern relevance logic. *Synthese* **2015**, *192*, 1413–1444. [CrossRef]
4. Crager, A. Meta-Logic in Aristotle's Epistemology. Ph.D. Thesis, Princeton University, Princeton, NJ, USA, 2015.
5. Read, S. Aristotle's Theory of the Assertoric Syllogism. Available online: https://philarchive.org/archive/REAATO-5 (accessed on 1 April 2020).
6. Moss, L. Completeness theorems for syllogistic fragments. In *Logics for Linguistic Structures*; Hamm, S.K., Ed.; Mouton de Gruyter: Berlin, Germany; New York, NY, USA, 2008; pp. 143–174.
7. Pratt-Hartmann, I.; Moss, L. Logics for the relational syllogistic. *Rev. Symb. Log.* **2009**, *2*, 1–37. [CrossRef]
8. Glashoff, K. An Intensional Leibniz Semantics for Aristotelian Logic. *Rev. Symb. Log.* **2010**, *3*, 262–272. [CrossRef]
9. Moss, L.S. Syllogistic Logics with Verbs. *J. Log. Comput.* **2010**, *20*, 947–967. [CrossRef]
10. Kulicki, P. On a Minimal System of Aristotle's Syllogistic. *Bull. Sect. Log.* **2011**, *40*, 129–145.
11. Moss, L.S. Syllogistic Logic with Comparative Adjectives. *J. Log. Lang. Inf.* **2011**, *20*, 397–417. [CrossRef]
12. Rini, A. *Aristotle's Modal Proofs. Prior Analytics A8-22 in Predicate Logic*; Springer: Berlin, Germany, 2011.
13. Kulicki, P. On minimal models for pure calculi of names. *Log. Log. Philos.* **2012**, *1*, 1–16. [CrossRef]
14. Bellucci, F.; Moktefi, A.; Pietarinen, A. Diagrammatic Autarchy: Linear diagrams in the 17th and 18th centuries. In Proceedings of the First International Workshop on Diagrams, Logic and Cognition, Kolkata, India, 28–29 October 2012; Burton, L.C., Ed.; CEUR Workshop Proceedings: Kolkata, India, 2013.
15. Pratt-Hartmann, I. The Syllogistic with Unity. *J. Philos. Log.* **2013**, *42*, 391–407. [CrossRef]
16. Castro-Manzano, J. Re(dis)covering Leibniz's Diagrammatic Logic. *Tópicos Revista de Filosofía* **2017**, *52*, 89–116.
17. Pietruszczak, A.; Jarmużek, T. Pure Modal Logic of Names and Tableau Systems. *Stud. Log.* **2018**, *106*, 1261–1289. [CrossRef]
18. Sautter, F.T.; Secco, G.D. A Simple Decision Method for Syllogistic. In Proceedings of the Diagrammatic Representation and Inference—10th International Conference, Diagrams 2018, Edinburgh, UK, 18–22 June 2018. Lecture Notes in Computer Science; Chapman, P., Stapleton, G., Moktefi, A., Pérez-Kriz, S., Bellucci, F., Eds.; Springer: Berlin, Germany, 2018; Volume 10871, pp. 708–711. [CrossRef]
19. Endrullis, J.; Moss, L.S. Syllogistic logic with "Most". *Math. Struct. Comput. Sci.* **2019**, *29*, 763–782. [CrossRef]
20. Bocheński, I.M. *Ancient Formal Logic*; North-Holland: Oxford, UK, 1951.
21. Aristotle. *Prior Analytics. Book I*; Translated with an Introduction and Commentary by Gisela Striker; Clarendon Press: Oxford, UK, 2014.
22. Bocheński, I.M. *The Methods of Contemporary Thought*; D. Reidel: Dordrecht, The Netherlands, 1965.
23. Lear, J. Aristotle's compactness proof. *J. Philos.* **1979**, *76*, 198–215. [CrossRef]
24. Scanlan, M. On finding compactness in aristotle. *Hist. Philos. Log.* **1983**, *4*, 1–8. [CrossRef]
25. Smiley, T. What is a syllogism? *J. Philos. Log.* **1973**, *2*, 136–154. [CrossRef]
26. Boger, G. Completion, reduction and analysis: Three proof-theoretic processes in Aristotle's Prior Analytics. *Hist. Philos. Log.* **1998**, *19*, 187–226. [CrossRef]
27. Adžic, M.; Došen, K. Gödel's Notre Dame Course. *Bull. Symb. Log.* **2016**, *22*, 469–481. [CrossRef]
28. Corcoran, J. Completeness of an ancient logic. *J. Symb. Log.* **1972**, *37*, 696–702. [CrossRef]
29. Corcoran, J. Aristotle's Natural Deduction System. In *Ancient Logic and Its Modern Interpretations*; Reidel Publishing Co.: Dordrecht, The Netherlands, 1974; pp. 1–100.
30. Frege, G. Sense and Reference. *Philos. Rev.* **1948**, *57*, 209–230. [CrossRef]
31. Słupecki, J. Uwagi o sylogistyce Arystotelesa. *Ann. UMCS* **1946**, *I*, 187–191.
32. Wedberg, A. The Aristotelian theory of classes. *Ajutas* **1948**, *15*, 299–314.
33. Menne, A. *Logik und Existenz*; Westkulturverlag Anton Hain: Berlin, Germany, 1954.
34. Shepherdson, J. On the Interpretation of Aristotelian Syllogistic. *J. Symb. Log.* **1956**, *21*, 137–147. [CrossRef]

35. Geach, P.T. History of the Corruption of Logic (1968). In *Logic Matters*; University of California Press: Berkeley, CA, USA, 1980; pp. 44–61.
36. Ajdukiewicz, K. *Logika Pragmatyczna*; PWN: Warszawa, Poland, 1965.
37. Borkowski, L. *Logika Formalna*; PWN: Warszawa, Poland, 1970.
38. Grzegorczyk, A. *Zarys Logiki Matematycznej*; PWN: Warszawa, Poland, 1969.
39. Pogorzelski, W.A. *Elementarny Słownik Logiki Formalnej*; Dział Wydawnictw Filii Uniwersytetu Warszawskiego: Białystok, Poland, 1992.
40. Marciszewski, W. (Ed.) *Mała Encyklopedia Logiki*; PWN: Warszawa, Poland, 1988.
41. Waragai, T.; Oyamada, K. A System of Ontology Based on Identity and Partial Ordering as an Adequate Logical Apparatus for Describing Taxonomical Structures of Concepts. *Ann. Jpn. Assoc. Philos. Sci.* **2007**, *15*, 123–149. [CrossRef]
42. Słupecki, J.; Bryll, G.; Wybraniec-Skardowska, U. Theory of rejected propositions. I. *Stud. Log.* **1971**, *29*, 75–115. [CrossRef]
43. Goranko, V.; Pulcini, G.; Skura, T. Refutation Systems: An Overview and Some Applications to Philosophical Logics. In *Knowledge, Proof and Dynamics*; Liu, F., Ono, H., Yu, J., Eds.; Springer: Singapore, 2020; pp. 173–197.
44. Wybraniec-Skardowska, U. Rejection in Łukasiewicz's and Słupecki's Sense. In *The Lvov-Warsaw School. Past and Present*; Garrido, Á., Wybraniec-Skardowska, U., Eds.; Springer International Publishing: Cham, Switzerland, 2018; pp. 575–597. [CrossRef]
45. Kulicki, P. The Use of Axiomatic Rejection. In *The Logica Yearbook 1999*; Childers, T., Ed.; Filosofia: Prague, Czech Republic, 2000; pp. 109–117.
46. McKinsey, J. The Decision Problem for some Classes of Sentences without Quantifiers. *J. Symb. Log.* **1943**, *8*, 61–76. [CrossRef]
47. Słupecki, J. *Z badań nad sylogistyką Arystotelesa*; Wrocławskie Towarzystwo Naukowe: Wrocław, Poland, 1948.
48. Lukasiewicz, J. *Aristotle's Syllogistic from the Standpoint of Modern Formal Logic*; Clarendon Press: Oxford, UK, 1952.
49. Lemmon, E.J. Quantifiers and Modal Operators. *Proc. Aristot. Soc.* **1958**, *58*, 245–268. [CrossRef]
50. Kulicki, P. *Aksjomatyczne Systemy Rachunku Nazw*; Wydawnictwo KUL: Warszawa, Poland, 2011.
51. Prior, A.N. *Formal Logic*; Clarendon Press: Oxford, UK, 1962.

© 2020 by the authors. Licensee MDPI, Basel, Switzerland. This article is an open access article distributed under the terms and conditions of the Creative Commons Attribution (CC BY) license (http://creativecommons.org/licenses/by/4.0/).

Article
Term Logic

Peter Simons

Department of Philosophy, Trinity College Dublin, College Green, Dublin 2, Ireland; psimons@tcd.ie

Received: 19 December 2019; Accepted: 6 February 2020; Published: 10 February 2020

Abstract: The predominant form of logic before Frege, the logic of terms has been largely neglected since. Terms may be singular, empty or plural in their denotation. This article, presupposing propositional logic, provides an axiomatization based on an identity predicate, a predicate of non-existence, a constant empty term, and term conjunction and negation. The idea of basing term logic on existence or non-existence, outlined by Brentano, is here carried through in modern guise. It is shown how categorical syllogistic reduces to just two forms of inference. Tree and diagram methods of testing validity are described. An obvious translation into monadic predicate logic shows the system is decidable, and additional expressive power brought by adding quantifiers enables numerical predicates to be defined. The system's advantages for pedagogy are indicated.

Keywords: term logic; Franz Brentano; Lewis Carroll; logic trees; logic diagrams

1. Terminology

A *term logic* is one in which the only categorematic expressions are terms, that is to say, nominal expressions. Examples of terms from ordinary language are: singular terms, such as 'Socrates', 'the North Pole', 'Vulcan'; plural terms, such as 'the Beatles', 'the signatories to the Geneva Convention'; and general terms, such as 'planet', 'black dog', 'negatively charged particle'. From this, it will be seen that the presence or absence of a definite article makes no difference to whether an expression is a term or not. It will further be seen that terms may be simple or complex. In term logic itself, we will employ mainly term variables: there will be only two constant terms, given below. All other expressions in a term logic are formal, or what were once called syncategorematic. They are the logical constants needed to form sentences using terms, and such operators on terms as may form complex terms from simpler ones, and the logical connectives of propositional logic. Quantifiers will be added later.

The syllogistic of Aristotle and his successors was a term logic, as was that of such logical algebraists as Leibniz, Boole, Jevons, Venn and Neville Keynes. Term logic was augmented by relational expressions in De Morgan, Peirce and Schröder, but terms, except for singular terms, disappeared altogether from the predicate logic of Frege, Russell and their successors. An exception was the logical system of Leśniewski, who retained plural and general terms, though Leśniewski's system was also a predicate logic rather than a purely term logic. Term logic is a very natural medium for representing many inferences of ordinary discourse, more natural indeed than standard predicate logic. Though it has much less expressive power than predicate logic, being in its elementary form equivalent to monadic predicate calculus, it has much to recommend it from a pedagogical point of view, a fact recognised by Łukasiewicz, whose university textbook *Elements of Mathematical Logic* [1] augmented propositional calculus not with predicate calculus but with Aristotelian syllogistic.

The version of term logic we shall present owes much in inspiration to the logical reforms of Franz Brentano [2–4] with some influence from the logical writings of Lewis Carroll.

2. Language

2.1. Grammar

The grammar of our language will be categorial, with two basic categories: sentence (s) and term (n). (It is standard in categorial grammars to notate the nominal category by 'n' for 'name' rather than 't' for 'term'. We are following this tradition notationally, though we call the category by the older expression 'term'.) A functor category, the category of functor expressions taking arguments of categories $\beta_1, \ldots, \beta_n$ as arguments and forming an expression of category α, will be denoted as $\alpha \langle \beta_1 \ldots \beta_n \rangle$.

2.2. Basic Vocabulary

The Table 1 Basic Vocabulary below gives the basic expression used, together with their syntactic categories, categorial indices, and how we describe them.

Table 1. Basic Vocabulary.

Category	Index	Expressions	Description
Monadic Connective	s⟨s⟩	~	Sentential negation
Dyadic Connectives	s⟨ss⟩	$\wedge \vee \rightarrow \leftrightarrow$	[Standard]
Term Variables	n	$a, b, c, a_1, a_2, \ldots$	
Term Constant	n	Λ	Empty term
Monadic Term Functor	n⟨n⟩	'	Term negation
Dyadic Term Functor	n⟨nn⟩	[juxtaposition]	Term conjunction
Monadic Predicate	s⟨n⟩	N	Non-existence predicate
Dyadic Predicate	s⟨nn⟩	=	Identity predicate

The intended meanings of the term-logical constants are given in the Table 2 below:

Table 2. Meanings of Basic Term-Logical Constants.

Expression	Meaning	Example
Λ	Non-existing thing	
a'	non-a	non-animal
ab	a which is a b	doctor who is a musician
Na	there are no a	there are no unicorns
$a = b$	to be a is (the same thing as) to be b	to be a widow is to be a woman whose husband has died

2.3. Basic Syntax

In the interest of simplicity and brevity of expression, we delicately abuse the use/mention distinction and do not introduce special metavariables.

2.4. Terms

Any term variable or term constant is a term
If a is a term, so is $(a)'$
If a and b are terms, so is (ab)
Nothing else is a term except as allowed by definitions.

2.5. Sentences

If a and b are terms, $a = b$ is a sentence

If a is a term, Na is a sentence
If p is a sentence, so is $\sim(p)$
If p and q are sentences, so are $(p \wedge q)$, $(p \vee q)$, $(p \rightarrow q)$ and $(p \leftrightarrow q)$
Nothing else is a sentence except as allowed by definitions.

We will omit parentheses where no ambiguity results. Propositional connectives are assumed to bind in the order negation, conjunction, disjunction, implication, equivalence.

3. Axioms

3.1. Propositional Logic Background

We presuppose without mention axioms sufficient for classical bivalent propositional logic, with substitution and modus ponens as inference rules.

3.2. Term-Logical Axioms

3.2.1. Intensional

for =
ID $a = a$ (Identity)
LEIB $a = b \rightarrow (p[a] \rightarrow p[b])$ (Leibniz)
where $p[x]$ is any sentential context containing the term x.

Justification

Self-identity and Leibniz's Law are standardly characteristic of identity.
for = and term conjunction
IDEM $aa = a$ (Idempotence)
COMM $ab = ba$ (Commutativity)
ASSOC $a(bc) = (ab)c$ (Associativity)

Justification

For idempotence: to be an a which is an a is the same thing as to be an a
For commutativity: to be an a which is a b is the same thing as to be a b which is an a
For associativity: to be an a which is a (b which is a c) is the same thing as to be an (a which is a b) which is a c.

for = and '
DN $a'' = a$ (Term Double Negation)

Justification

To be a non-non-a is the same thing as to be an a.
for =, ' and term conjunction
DIST $a(bc)' = ((ab')'(ac')')'$ (Distribution)

Justification

This is the least self-evident of our axioms. It can be made more evident by considering an empty Venn or Carroll diagram for three terms and their negations: the three out of eight cells indicated by the left-hand side of the identity are the same as those picked out by the right-hand side, namely the cells for abc', $ab'c$ and $ab'c'$. Given the De Morgan definition of Term Addition ADD below, it permits the derivation of the desirable distribution laws DIST1 and DIST2 below.

3.2.2. Extensional

for N and $\wedge$
HEID   N$\wedge$ (Heidegger's Law)

Justification

> There is no thing which does not exist.
> for N and term conjunction

NWK $Na \to Nab$ (N-Weakening)

Justification

> If there are no as, there are no as which are bs.
> for N, term conjunction and '

TNC Naa' (Term Non-Contradiction)

Justification

> There are no as which are non-as. This is the term-logical version of the Principle of (Excluded) Contradiction, going back to Aristotle.

NEXH $Nab \wedge Nab' \to Na$ (N-Exhaustion)

Justification

> If there are no as which are bs, and no as which are non-bs, then there are no as (at all).

3.3. Definitions

The Table 3. below gives definitions of the constant expressions we define using the basic ones. The form is either an identity for terms or a propositional equivalence for functors.

Table 3. Definitions.

Name	Definition	Description	Reading
UN	$V = \Lambda'$	Universal Term	thing; object
ADD	$a + b = (a'b')'$	Term Addition	a or b
EX	$Ea \leftrightarrow \sim Na$	Existence	There are a; a exist
NO	$a \mid b \leftrightarrow Nab$	Universal Negative	No a are b
ALL	$a \subset b \leftrightarrow Nab'$	Universal Positive	All a are b
SOM	$a \triangle b \leftrightarrow Eab$	Particular Positive	Some a are b
AEQ	$a \equiv b \leftrightarrow Nab' \wedge Nba'$	Term Equivalence	The a are the b

These readings should be self-evident. The definitions NO, ALL and SOM are due to Brentano [2] (p.121) [4,5].

4. A Few Theorems

IDAEQ $a = b \to a \equiv b$ NC, LEIB, aeq AEQ
(Identity entails Equivalence)
EWK $Eab \to Ea$ NWK, contrap., EX
(Existential Weakening)
EEXH $Ea \to Eab \vee Eab'$ NWK, contrap., EX
(Existential Exhaustion)
TDS $Ea \wedge Nab \to Eab'$ NEXH, contrap., EX
(Term Disjunctive Syllogism)
DIST1 $a(b + c) = ab + ac$
(Distribution, First Form)

Proof.

$a(b + c) = a(b'c')'$ ADD
$= ((ab'')'(ac'')')'$ DIST
$= ((ab)'(ac)')'$ DN
$= ab + ac$ ADD
DIST2 $\quad a + bc = (a + b)(a + c)$
(Distribution, Second Form) □

Proof.

$a + bc = (a'(bc)')$ ADD
$= (((a'b')'(a'c')')')'$ DIST
$= ((a'b')'(a'c')')''$ rewrite
$= (a'b')'(a'c')'$ DN
$= (a + b)(a + c)$ ADD
EXCL N$(ab)' \leftrightarrow$ N$a' \wedge$ Nb'
(Exclusion) □

Proof.

1.	N$(ab)'$	A for CP (assumption for conditional proof)
2.	N$(ab)' \to$ N$a'(ab)'$	nwk
3.	N$a'(ab)$	nc, nwk, assoc, comm
4.	N$a'(ab)'$	1, 2, MP
5.	Na'	3, 4, NEXH
6.	Nb'	similiter
7.	N$a' \wedge$ Nb'	5, 6
8.	N$(ab)' \to$ N$a' \wedge$ Nb'	1–7 CP
9.	N$a' \wedge$ Nb'	A for CP
10.	E$(ab)'$	A for RAA (assumption for reductio ad absurdum)
11.	E$a(ab)' \vee$ E$a'(ab)'$	10, EEXH
12.	E$a(ab)'$	2nd disjunct incompatible with Na' from 9
13.	E$ab(ab)' \vee$ E$ab'(ab)'$	12, EEXH, term shuffling
14.	Contradiction: first disjunct by TNC, second contradicts Nb' from 9	
15.	N$(ab)'$	10, 14, *reductio*
16.	N$a' \wedge$ N$b' \to$ N$(ab)'$	9–15 CP
17.	N$(ab)' \leftrightarrow$ N$a' \wedge$ Nb'	8, 16 □

Corollary. N$(a + b) \leftrightarrow$ N$a \wedge$ Nb

Some Sample Syllogisms BARBARA $b \subset c, a \subset b \vdash a \subset c$

Proof.

1.	$b \subset c$	A
2.	$a \subset b$	A
3.	Nbc'	1, ALL
4.	Nab'	2, ALL
5.	Nabc'	3, NWK
6.	N$ab'c'$	4, nwk
7.	Nac'	5, 6, NEXH
8.	$a \subset c$	7, all

DARII $b \subset c, a \triangle b \vdash a \triangle c$ □

Proof.

1.	$b \subset c$	A
2.	$a \mathbin{\Delta} b$	A
3.	Nbc'	1, all
4.	Eab	2, SOM
5.	$Nabc'$	3, NWK
6.	$Eabc$	4, 5, TDS, DN
7.	Eac	6, EWK
8.	$a \mathbin{\Delta} c$	7, som

DARAPTI $Eb, b \subset c, b \subset a \vdash a \mathbin{\Delta} c$ □

Proof.

1.	Eb	A
2.	$b \subset c$	A
3.	$b \subset a$	A
4.	Nbc'	2, ALL
5.	Nba'	3, ALL
6.	Eab	1, 5, tds, dn, comm
7.	$Nabc'$	4, NWK
8.	$Eabc$	6, 7, TDS, DN
9.	Eac	8, EWK
10.	$a \mathbin{\Delta} c$	9, SOM □

DARAPTI is one of those syllogisms whose validity is dependent on existential import of the subject term of the two premises: this is made explicit as the first premise.

In fact, every valid categorical syllogism has one of just three forms. We let * be a toggle operator taking positive terms to negative terms and vice versa, that is, if a is positive $a^* = a'$, while if a is negative, $a = b'$, $a^* = b$. Then every syllogism has as its core one of the three valid inference forms

POSITIVE $Eab, Nbc^* \vdash Eac$ (cf. DARII)
NEGATIVE $Nab^*, Nbc \vdash Nac$ (cf. CELARENT)
IMPORT $Ea, Nab^*, Nbc^* \vdash Eac$ (cf. BARBARI)

All can be derived from one of these by choosing b or c to be positive or negative, relabelling, swapping the order of premises, and applying commutativity ($ab = ba$) to obtain simple conversion. Furthermore, either POSITIVE or NEGATIVE is derivable from the other via partial contraposition and relabelling, so in the end Aristotelian categorical syllogistic owes its validity to just two forms of syllogistic inference, with a little propositional help.

Before concluding this section, a word about the ironic designation 'Heidegger's Law'. The basic non-existence predicate 'N' is best read as "Nothing is (a)", and the definitionally empty term 'Λ' can often be read as 'nothing'. The axiomatic formula 'NΛ' can then be read as 'Nothing is (a) nothing', or, with a little linguistic chivvying, 'Nothing noths' or *Das Nichts nichtet*. Of course, Heidegger did not intend to say anything so straightforward or trivial, but it does refute Carnap's claim that the sentence has to be nonsense. *Au contraire*: suitably understood, it is a logical law.

It may seem a little perverse to have based this logic on the negative idea of non-existence rather than the positive one of existence. Of course, it is possible to do it the other way around, but in general the axioms for N are more satisfyingly elegant than those for E.

5. Intension and Extension

One of the standard principles of the Boolean algebra that emerged from Boole's and others' work on the algebra of terms in the nineteenth century is that all empty terms are identical: we have, e.g., that $aa' = bb'$, $Na \wedge Nb \rightarrow a = b$, $Na \rightarrow a = \Lambda$. These are *not* theorems of our system and it is important to see why. Their analogues with equivalence '≡' replacing identity '=' *are* theorems, and if we were to add an axiom of extensionality

EXT $a \equiv b \rightarrow a = b$

they would be theorems, and there would be no distinction between identity and equivalence. Most nineteenth century algebraic logicians understood their logic extensionally, so would be happy with this simplification. However, Brentano was not, and nor am I. The axioms involving identity = and those involving non-existence N are distinct in intent. Existence and non-existence have to do, for the most part, with contingent facts: there are narwals; there are no unicorns. There are some non-contingent principles involving N, obviously, our axioms such as term non-contradiction, but the premises in syllogisms and the antecedents in NWK and NEXH are typically contingent in application to actual propositions and actual inferences.

As will be seen more clearly when we consider diagrammatic representation, the axioms governing identity have to do not with contingent propositions but with the framework of discourse within which propositions and inference are employed. In any of the syllogisms considered, we are looking at three terms, their negations and conjunctions. For three terms, there are eight maximally specific combinations of conjunction and negation, for example $ab'c$, and the question may then arise whether N or E is true of this term. The axioms governing identity (and conjunction and negation) are formal synonymies, there to tell us, in advance of any statements about what does or does not exist, when term expressions relate to the same possibilities. Of these, the most obvious perhaps is $a = a''$, term double negation. No contingent facts have any bearing on these two expressions' relating to the same possibility of existence or non-existence. For this reason, I call the principles governing '=' *intensional* and those governing 'N' *extensional*. That does not mean I here endorse a modal logic or possible worlds, simply that the role played by framework description is different from and prior to that played by questions of existence and non-existence.

6. Consistency

The system is consistent. In the empty universe, every term is empty, and the extensional axioms are trivially true. Interpreting identity as equivalence, so are the intensional axioms. The empty universe is expressly not ruled out by the system: the dual to Heidegger's Law, namely.

EV There is something (rather than nothing) is not a theorem, because it is false for the empty universe.

7. Decidability

It is well known that first-order monadic predicate logic is decidable [6]. We may interpret the term logic in monadic predicate logic by associating each term a with a monadic predicate A

$$a \mapsto Ax$$

the term Λ with a necessarily empty predicate, for example

$$\Lambda \mapsto \sim(x = x)$$

with complex terms as follows

$$a' \mapsto \sim(Ax)$$

$$ab \mapsto Ax \wedge Bx$$

and the predicates as follows

$$a = b \mapsto \forall x(Ax \leftrightarrow Bx)$$

$$Na \mapsto \sim\exists x(Ax).$$

In this way, each formula of the term logic is correlated with a formula of monadic predicate logic. The interpretation validates extensionality. It can be seen that all the axioms of the term logic system are valid formulas of monadic first-order predicate logic, and the validity of a formula or inference with finitely many premises containing n term variables may be decided on a domain of no more than 2^n individuals.

8. Tree Proof Techniques

Formulas and inferences with finitely many premises may be tested for validity or invalidity using tree techniques. The basic ideas were presented earlier for a slightly simpler system, so we can be brief [7]. We assume that all rules for trees for propositional logic are available, and we confine attention to term formulas using only the basic vocabulary of term variables, Λ, =, ', conjunction, and N, as well as propositional connectives. Any defined constants are eliminated first as per their definitions. The counterexample set of an inference to be tested consists of the premises together with the negation of the conclusion, or the negation of a formula if that formula's validity is to be tested. A tree starts with the counterexample set. It may then be extended according to the following rules:

1. In a formula, terms may be replaced by terms identical to them according to the intensional axioms.
2. Any sentence N(ab)' may be replaced by the two sentences Na', Nb' by EXCL.
3. These and DN may be used to drive term negations inwards so they only occur singly and modify term letters only.
4. A branch containing ~Nab may be extended by ~Na, or ~Nb, or both. (In the next two rules, b is a term occurring in the premises but not occurring in a.)
5. A branch containing Na may be extended by Nab and by Nab'.
6. A branch containing ~Na splits and continues with ~Nab in one branch and ~Nab' in the other.
7. Open branches are extended until all variables from the premises occur in any remaining branch, with term negations inmost, i.e., modifying a single term letter or constant term. Branches close under the following conditions:
8. The branch contains two contradictory formulas, for example E$ab'c$ and N$ab'c$
9. The branch contains a formula ~NΛ
10. The branch contains a formula ~Naa'.

If all branches close, the formula or inference is valid; if any branch remains open, the formulas along it may all be true and constitute a counterexample.

9. Diagram Techniques

Diagrams for deciding the validity of logical inferences go back centuries, but the first effective ones are due to John Venn [8]. The idea, as applied to term logic, is to start with a diagram consisting of as many areas, or *cells*, as there are conjunctions of all simple terms and their negations contained in an inference. If there are n simple terms, that will be 2^n cells. Venn's own curvilinear diagrams are inferior to the rectilinear ones proposed by Lewis Carroll, who ingeniously constructed diagrams for up to eight different simple terms, and indicated how to extend these further [9] (p. 245 ff.: "My Method of Diagrams". Carroll was incidentally the first to use trees as an aid for solving logic problems: *ibid.*, 279 ff. Since one of his problems ("Froggy's Problem", *ibid.*, 338 ff.) is a sorites in 18 terms, which would require a diagram with 262,144 cells, taxing human capacity to solve, further aids were clearly needed.) The method for term logic as for syllogistic is to shade out those cells corresponding to N propositions, and indicate by crosses those cells corresponding to E propositions. The chief difficulty is that an E proposition whose term is not a maximal compound of simple terms and their negations must straddle several cells disjunctively, a problem compounded in any term-logical formula or inference employing disjunction or its equivalent. For this reason, diagrams are practicable only for relatively small and straightforward problems. Trees branch easily, but the only way to branch a diagram is to treat several diagrams disjunctively.

An unfilled diagram for n term variables, with its 2^n cells, represents the framework within which N and E propositions employing these variables are to be represented, and is neutral with respect to such propositions. The axioms for identity are then to be understood as indicating different but formally equivalent ways in which cells or groups of cells are indicated. This is why they play a different role in the logic from the N and E propositions.

10. Quantifiers

It is natural to extend the term logic employed to date with variable-binding quantifiers. One reason is simply to enhance the representative scope of the system. Quantifiers binding term variables do not affect the decidability of the resulting system, (Ackermann, *loc.cit.*) but they bring greater expressive power. We take the universal quantifier as primitive and add the following axiom schemes, where A and B are sentences:

$$\text{QDIST} \qquad \forall a(A \to B) \to (A \to \forall a(B))$$

where a is any term variable which is not free in A;

$$\text{QINST} \qquad \forall a(A) \to A[t/a]$$

where t is a term expression (variable, constant or compound), $A[t/a]$ is the result of substituting t for all free occurrences of a in A, and no free occurrence of a in A is in a well-formed part of A of the form $\forall t(B)$ [10] (p. 172).

The particular quantifier may then be introduced in the standard way as dual of the universal:

$$\text{FORSOME} \qquad \exists a(A) \leftrightarrow \sim\forall a(\sim A)$$

It should be noted that the particular quantifier does not in this system carry existential import. Since it is a theorem that $\sim E\Lambda$, it follows that $\exists a(\sim Ea)$, so the quantifier cannot very well mean 'there exists', but must mean, neutrally, 'for some'. If we wish to talk about existing things, we have the predicate 'E' to hand.

One of the ways in which quantifiers introduce greater expressive power is that they facilitate expressions of number. Hitherto, expressions of the form Ea only said that there is some a. This is compatible with there being one, two, ... any number of as, and this is why the cells of any diagram for finitely many propositions need only be finite in number. Indeed, the terms need not denote individuals or pluralities of individuals at all: they could denote numberless stuffs, as do mass terms in ordinary language. The syllogism in Darii

All morphine is highly addictive
Some pain medication is morphine:

therefore Some pain medication is highly addictive

is no less valid for being about stuffs ("substances") rather than individuals. If we wish to introduce numbers, to count individuals or also consignments of stuff, we need quantifiers. Here is how to define 'at least two':

$$\geq 2 \qquad E_{\geq 2}a \leftrightarrow \exists x(Eax \wedge Eax')$$

So we can define 'exactly one' as

$$=1 \qquad E_{=1}a \leftrightarrow Ea \wedge \sim E_{\geq 2}a$$

11. Pegagogical Advantages of Term Logic

For students coming to logic with little or no background except in propositional calculus, term logic is quite natural and easy to understand. It is close to natural language (no bound variables, quantifiers as phrases not operators, logical and grammatical form closely similar); by comparison with predicate logic, there is minimal paraphrasing required; it has a straightforward and intuitive denotational semantics requiring no set theory, and is easy to do without special symbols. It can be treated by methods building easily on those of propositional logic: semantic diagrams, natural deduction proofs, axioms and trees. It has an accessible metalogic: it is sound, complete, and decidable, uses finitistic methods, affords a variety of approaches and good illustration of basic concepts. In difficulty, it is only slightly more complex than truth-tables and natural deduction for propositional logic, and is readily scriptable should one wish to write suitable computer programs. For an introduction to the history of logic, it allows much greater scope for comparison than post-Fregean predicate logic. It allows a variety of bases, apart from the one we have chosen. Alternative bases are equational (Leibniz, Boole, Jevons), subsumptional (Leibniz, Peirce, Schröder), existential (Leibniz, Brentano, Carroll), traditional (Aristotle, Łukasiewicz), or based on the singular copula (Leśniewski, Słupecki). It admits of various extensions, most obviously by introducing predicates, especially relational predicates, but also modal [11], towards Leśniewskian logic [3,12], and introducing higher types, up to and including full simple type theory.

On the negative side, it delays students' encounter with ∀ and ∃, with relations and multiple generality, has rather few links to modern mathematics, and being now unorthodox, suffers from a modern textbook gap. Most textbooks highlighting term logic (Łukasiewicz excepted) are antiquated attempts to keep pre-Fregean logic alive, often for non-logical reasons.

Nevertheless, I hope enough has been shown in this paper to suggest that term logic, whether done this way or in some other way, remains worthy of the attention of logicians and teachers of logic.

Funding: This research received no external funding.

Conflicts of Interest: No conflict of interest.

References

1. Łukasiewicz, J. *Elements of Mathematical Logic*; Wojtasiewicz, O., Translator; Pergamon: Oxford, UK, 1963.
2. Brentano, F. *Die Lehre vom richtigen Urteil*; Mayer-Hillebrand, F., Ed.; Francke: Bern, Switzerland, 1956.
3. Simons, P. A Brentanian Basis for Leśniewskian Logic. *Logique et Analyse* **1984**, *27*, 297–307.
4. Simons, P. Judging Correctly: Brentano and the Reform of Elementary Logic. In *The Cambridge Companion to Brentano*; Jacquette, D., Ed.; Cambridge University Press: Cambridge, UK, 2003; pp. 45–65.
5. Simons, P. Brentano's Reform of Logic. *Topoi* **1987**, *6*, 23–63. [CrossRef]
6. Ackermann, W. *Solvable Cases of the Decision Problem*; North-Holland: Amsterdam, The Netherlands, 1954.
7. Simons, P. Tree Proof for Syllogistic. *Studia Logica* **1989**, *48*, 539–554. [CrossRef]
8. Venn, J. On the diagrammatic and mechanical representation of propositions and reasonings. *Philos. Mag.* **1880**, *59*, 1–18. [CrossRef]
9. Carroll, L.; Dodgson, C.L. *Lewis Carroll's Symbolic Logic*; Bartley, W.W., III, Eds.; Potter: New York, NY, USA, 1977.
10. Church, A. *An Introduction to Mathematical Logic*; Princeton University Press: Princeton, NJ, USA, 1956.
11. Simons, P. Calculi of Names: Free and Modal. In *New Essays in Free Logic, In Honour of Karel Lambert*; Morscher, E., Hieke, A., Eds.; Kluwer: Dordrecht, The Netherlands, 2001; pp. 49–65.
12. Słupecki, J.S. Leśniewski's Calculus of Names. *Studia Logica* **1955**, *3*, 7–72. [CrossRef]

© 2020 by the author. Licensee MDPI, Basel, Switzerland. This article is an open access article distributed under the terms and conditions of the Creative Commons Attribution (CC BY) license (http://creativecommons.org/licenses/by/4.0/).

Article
Distribution Tableaux, Distribution Models

J.-Martín Castro-Manzano

Faculty of Philosophy, UPAEP University, Puebla 72410, Mexico; josemartin.castro@upaep.mx

Received: 3 December 2019; Accepted: 26 March 2020; Published: 17 April 2020

Abstract: The concept of distribution is a concept within traditional logic that has been fundamental for the syntactic development of Sommers and Englebretsen's term functor logic, a logic that recovers the term syntax of traditional logic. The issue here, however, is that the semantic counterpart of distribution for this logic is still in the making. Consequently, given this disparity between syntax and semantics, in this contribution we adapt some ideas of term functor logic tableaux to develop models of distribution, thus providing some alternative formal semantics to help close this breach.

Keywords: semantic tree; term logic; distribution

1. Introduction

The concept of distribution is a concept within traditional logic that is applied when a term appears under the scope of a universal quantifier. Thus, for example, in the proposition "All men are mortal" we say the term "men" is distributed whereas the term "mortal" is not. In spite of its simplicity, the importance of this concept is far from being overstated, given that it allows us to define rules essential to syllogistic, a term logic at the core of traditional logic; however, specially since the objections raised by Geach [1], distribution—as traditional logic—is seemingly out of favor.

Nevertheless, distribution serves a logical purpose (cf. [2–5]) that has been fundamental for the syntactic development of Sommers and Englebretsen's term functor logic, a logic that recovers the term syntax of traditional logic [4,6,7]. The issue here, however, is that the formal semantics counterpart of distribution for this logic is still in the making. Of course, it goes without saying that Sommers and Englebretsen have already provided semantic standards for their logic, but such semantics are philosophical rather than formal, so to speak (cf. [6,8]). Consequently, given this disparity between syntax and semantics, in this contribution we adapt some ideas of term functor logic tableaux [9] to develop models of distribution, thus providing some alternative formal semantics to help close this breach.

In order to reach this goal we proceed in the following way. First we expound some preliminaries about syllogistic, distribution, and term functor logic and its tableaux; then we develop our main contribution by exploring some links between the rules of term functor logic and the concept of the model.

2. Preliminaries

2.1. Syllogistic

Syllogistic is a logic term at the core of traditional logic that deals with inference between categorical propositions. A *categorical proposition* is a proposition composed by two terms, a quantity, and a quality. The subject and the predicate of a proposition are called *terms*: the term-schema S denotes the subject term of the proposition and the term-schema P denotes the predicate. The *quantity* may be either universal (*All*) or particular (*Some*), and the *quality* may be either affirmative (*is*) or negative (*is not*). These categorical propositions have a *type* denoted by a label (either a (universal affirmative, SaP), e (universal negative, SeP), i (particular affirmative, SiP), or o (particular negative, SoP)) that allows us to determine a *mood*, that is, a sequence of three categorical propositions ordered in such a way that the first two propositions are premises (major and minor) and the last one is a conclusion. A *categorical syllogism*, then, is a mood with three terms one of which appears in both premises but not in the conclusion. This particular term, usually denoted with the term-schema M, works as a link between the remaining terms and is known as the middle term. According to the position of this middle term, four *figures* can be set up in order to encode the valid syllogistic moods (Table 1) (for sake of brevity, but without loss of generality, we omit the syllogisms that require existential import).

Table 1. Valid syllogistic moods.

First Figure	Second Figure	Third Figure	Fourth Figure
aaa	eae	iai	aee
eae	aee	aii	iai
aii	eio	oao	eio
eio	aoo	eio	

2.2. Distribution

Now, in order to determine the validity of the previous moods we can invoke the concept of distribution since, as we mentioned before, it allows us to define rules instrumental for showing the validity of such inferences. These rules state (i) that a syllogistic inference is valid if and only if for every term A, if A is distributed in the conclusion, then A is distributed in the premises as well; and (ii) that a syllogistic inference is valid if and only the middle term M is distributed once in the premises. Thus, for instance, the following inference (Table 2) is valid since the term "dogs" is distributed in the conclusion and in the minor premise; and the middle term, "mammals," is distributed once in the premises (at the major premise).

Table 2. A valid syllogism: aaa-1.

	Proposition
1.	All mammals are animals.
2.	All dogs are mammals.
⊢	All dogs are animals.

Previously we claimed that the concept of distribution is applied when a term appears under the scope of a universal quantifier, but the very concept of distribution is far from being clear, given that there are several accounts of it. For example, according to some medieval theories of supposition a term is said to be distributed in a proposition when such a term refers to all that it means, as when a term is under the scope of a universal quantifier. In this sense we can understand, for instance, the notion of Peter of Spain [10], who talks of distribution as *multiplicatio termini communis per signum universale facta* (i.e., [distribution is] the multiplication of the common term made by a *universal* sign (translation and emphasis are ours)) or the notion advanced by the Port-Royal tradition [11] that claims that to say that a term is distributed in a proposition is to say that such a term *doit être pris universellement* (i.e., must be taken *universally* (translation and emphasis are ours)).

Nevertheless, the notion of distribution developed by Keynes [12]—and well defended by Sommers [3] and Wilson [5], in our opinion—is probably worth mentioning, since this is the one that Geach [1] discusses. Keynes' idea is that a term is said to be distributed when reference is made to all the items denoted by it; and it is said to be undistributed when they are only referred to partially. More formally, and following [3], we assume that a term A is distributed in a proposition p in the next sense:

- A is distributed in p if and only if p entails a proposition of the form "every A is ..."

- A is not distributed in p if and only if A is distributed in the contradictory of p.

Hence for syllogistic we say that, given its quantity and its quality, a categorical proposition distributes its terms if and only if they have a universal quantity or a negative quality: this will be formalized in the next section with the help of term functor logic.

2.3. Term Functor Logic

Term Functor Logic (TFL, for short) is a plus-minus algebra [4,6,7,13–15] that employs terms rather than first order language elements such as individual variables or quantifiers (cf. [4,16–19]). According to this algebra—or "logibra", as Sommers dubbed it—, the four categorical propositions can be represented by the following syntax [6], where "−" indicates that a term is distributed, and "+" that a term is not distributed:

- $SaP := -S + P = -S - (-P) = -(-P) - S = -(-P) - (+S)$
- $SeP := -S - P = -S - (+P) = -P - S = -P - (+S)$
- $SiP := +S + P = +S - (-P) = +P + S = +P - (-S)$
- $SoP := +S - P = +S - (+P) = +(-P) + S = +(-P) - (-S)$

Given this representation, this plus-minus algebra provides a simple rule for syllogistic inference: a conclusion follows validly from a set of premises if and only if (i) the sum of the premises is algebraically equal to the conclusion and (ii) the number of conclusions with particular quantity (*viz.*, zero or one) is the same as the number of premises with particular quantity ([6] p. 167). This rule is an algebraic rendition of the *dictum de omni et nullo* (this is the principle that states that everything that is affirmed (or denied) of a whole can be affirmed (or denied) of a part (cf. [4,6,7,14,20]))—hence its name, *DON*—and can be formally deployed as follows:

$$\Gamma \vdash \phi \; iff \; \begin{cases} (i) \; \sum_{alg}(\Gamma) = \sum_{alg}(\phi), \; and \\ (ii) \; |particulars(\Gamma)| = |particulars(\phi)| \end{cases}$$

where Γ stands for a (possibly empty) set of premises; ϕ stands for a conclusion; $\sum_{alg}$ for an algebraic sum; and $|particulars|$ for a function that returns the number of particular propositions in a set of terms.

Thus, for instance, if we consider a valid syllogism, say the mood aaa from the first figure (i.e., aaa-1), we can see how the application of *DON* produces the right conclusion (Table 3).

Table 3. A valid syllogism: aaa-1.

	Proposition	TFL
1.	All mammals are animals.	$-M + A$
2.	All dogs are mammals.	$-D + M$
⊢	All dogs are animals.	$-D + A$

Indeed, in the previous example we can clearly see how the rule works: (i) if we add up the premises we obtain the algebraic expression $(-M + A) + (-D + M) = -M + A - D + M = -D + A$, so that the sum of the premises is algebraically equal to the conclusion and the conclusion is $-D + A$, rather than $+A - D$, because (ii) the number of conclusions with particular quantity (zero in this case) is the same as the number of premises with particular quantity (zero in this case). Additionally, we must mention that this algebraic approach is not only capable of representing syllogistic, since it can also represent relational, singular, and compound propositions with ease and clarity while preserving its main idea, namely, that inference is a logical procedure between terms [14].

2.4. TFL Tableaux

Now, since alternative proof methods for TFL are still in the making, [9] has developed tableaux for it. So, following [21,22] we say a *tableau* is an acyclic connected graph determined by nodes and vertices, or more precisely, a labeled directed rooted tree. The node at the top is called the *root*. The nodes at the bottom are called *tips*. Any path from the root down a series of vertices is a *branch*. To test an inference for validity we construct a tableau which begins with a single branch at whose nodes occur the premises and the rejection of the conclusion: this is the *initial list*. We then apply the rules that allow us to extend the initial list (Scheme 1).

$$
\begin{array}{cc}
-A \pm B & +A \pm B \\
\diagup\diagdown & | \\
-A^i \quad \pm B^i & +A^i \\
 & | \\
 & \pm B^i
\end{array}
$$

Scheme 1. Term Functor Logic (TFL) tableaux rules.

In Scheme 1, from left to right, the first rule is the rule for a (e) propositions, and the second rule is the rule for i (o) propositions. Notice that, after applying a rule, we introduce some index $i \in \{1, 2, 3, \ldots\}$. For propositions a and e, the index may be any number; for propositions i and o, the index has to be a new number if they do not already have an index. In addition, following TFL tenets, we assume the followings rules of rejection: $-(\pm A) = \mp A$, $-(\pm A \pm B) = \mp A \mp B$, and $-(--A--A) = +(-A)+(-A)$.

As usual, a tableau is *complete* if and only if every rule that can be applied has been applied. A branch is *closed* if and only if there are terms of the form $\pm A^i$ and $\mp A^i$ on two of its nodes; otherwise it is *open*. A closed branch is indicated by writing a $\bot$ at the end of it; an open branch is indicated by writing ∞. A tableau is *closed* if and only if every branch is closed; otherwise it is *open*. So, again as usual, A is a logical consequence of the set of terms Γ (i.e., $\Gamma \vdash A$) if and only if there is a complete closed tableau whose initial list includes the terms of Γ and the rejection of A (i.e., $\Gamma \cup \{-A\} \vdash \bot$). Accordingly, up next we provide proofs of the valid syllogistic moods of the first figure (Scheme 2).

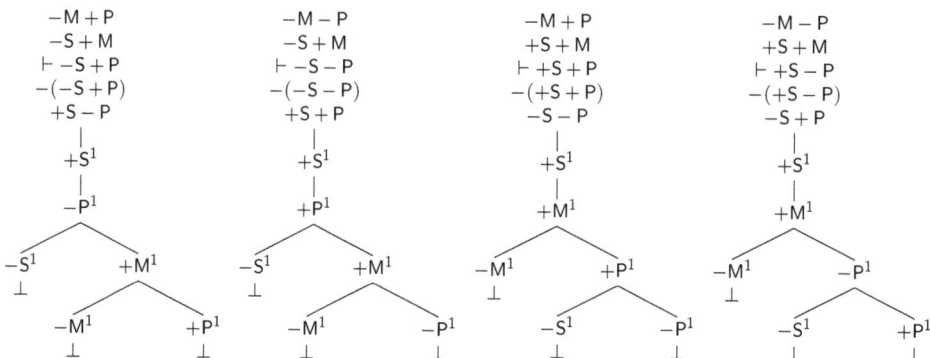

Scheme 2. Moods aaa-1, eae-1, aii-1, and eio-1.

To describe the process we follow to unfold each tableaux consider the first one. The first three lines are the premises (major and minor) and the conclusion, and the fourth line is the rejection of the conclusion: all these lines but the conclusion define the initial list. Then the fifth line is the result of applying a rule of rejection to the conclusion. Then the next couple of lines is the result of applying the rule for an i proposition to the fifth line, picking index 1. Then the first split results from applying the rule for an a proposition to the second line (i.e., the minor premise), also picking index 1, since we want the indexes to unify. This split produces two branches, one of which (the leftmost) includes terms $+S^1$ and $-S^1$ on two of its nodes, and hence is closed; the remaining branch is not closed yet, so we continue with the same process: we split the last available premise (the major one) to obtain, again, a couple of branches, one of which (the leftmost) includes terms $-M^1$ and $+M^1$ on two of its nodes, and hence is closed; and the other (the rightmost) that contains terms $+P^1$ and $-P^1$ on two of its nodes, and hence is closed as well.

More precisely, we can state this general procedure as follows (cf. [23]): given a syllogism $\Gamma \vdash \phi$, a tableau T for such a syllogism is defined as a tableau constructed by following the next steps: (1) Initiate the tableau for $\Phi = \Gamma \cup -\phi$ (in order to produce the initial list). (2) Let T be a tableau for Φ, let B be a branch of T, and let A be a term in $\Phi \cup B$. Take an arbitrary instance of a tableau rule in Scheme 1 with premise A and n extensions. Obtain the tree T' by extending B with n new subtrees whose nodes are the terms in the extensions of the rule instance. Then T' is a tableau for Φ (in order to expand the tableaux). (3) Let T be a tableau for Φ, B a branch of T, $\pm A$ and $\mp A$ terms in $B \cup \Phi$, and let σ be a substitution from terms to indexes such that $\sigma \pm A^i = \pm A^{\sigma i}$. If $\pm A^i$ and $\mp A^i$ are unifiable with a most general unifier σ, and T' is constructed by applying σ to all terms in T, then T' is a tableau for Φ (in order to close a tableau). As pointed out in [23], the most general unifiers are used instead of arbitrary substitutions as to avoid infinite or useless expansions.

As expected, this method is reliable in the sense that what can be proven using the inference rules of TFL produces closed complete tableaux, and vice versa (cf. [9]), but for the purposes of this study we need only the following:

Lemma 1. *An application of* DON *produces a closed complete tableau.*

Namely, that when we correctly apply *DON* to a syllogism we can also produce a closed complete tableau from said syllogism. We will refer to this result later.

3. Distribution Models

Broadly speaking, in modern predicate logic (MPL) we say an interpretation is a function that gives meaning to its symbols. Now, since, on the one hand, TFL avoids some important syntactic features of MPL but, on the other hand, TFL cannot escape the typical notions of syntax and semantics, and since TFL's formal semantics is not as developed as its philosophical semantics, here we adapt the usual notion of interpretation to build an analogous concept of interpretation for TFL. In short, the idea is that, since TFL tableaux make use of certain numbers, then maybe those numbers have an interpretation role in TFL.

Hence, with the previous background in mind, let us suggest that an interpretation in TFL is a triad $\mathfrak{J} = \langle I, T, v \rangle$ where $I = \{1, 2, 3, \ldots\}$ is a set of indexes, $T = \{\pm A, \pm B, \pm C, \ldots\}$ is a set of terms, and v is a function that assigns terms a finite set of indexes. To exemplify this notion of interpretation let us consider tableaux for some invalid inferences (Scheme 3) and let us focus on the open branches, since those branches, as usual, induce said interpretations.

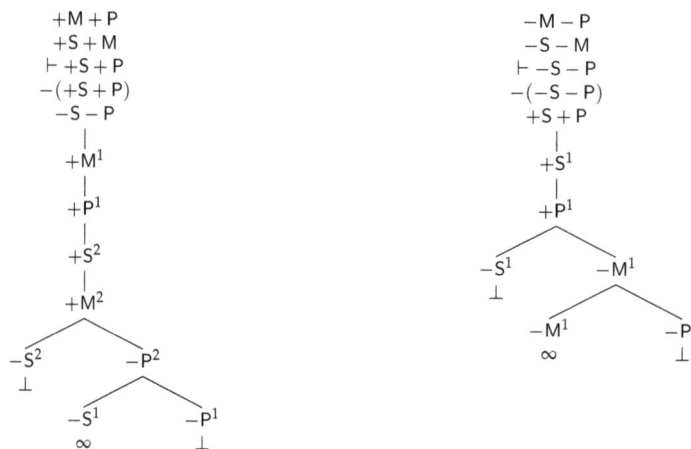

Scheme 3. Moods iii-1 and eee-1.

So, from left to right, the first tableau corresponds to the invalid inference iii-1 and thus has an open branch. The first five lines define the initial list. The next couple of lines results from applying the rule for an i proposition to the major premise; similarly, the next couple of lines results from applying the same rule but to the minor premise. Then the first split comes from applying the rule for an a proposition to the fifth line; and the second split is the result of the same process.

As we can see, after performing this process we obtain three branches, one of which will not close. This branch defines an interpretation $\mathfrak{I}$ such that $I = \{1,2\}$ (because those are the indexes associated to the open branch); and $v(+S) = \{2\}$, $v(+M) = \{1,2\}$, $v(+P) = \{1\}$, and $v(-P) = \{2\}$. Consequently, if we add up a subset of the values of this interpretation—say $v(+S) = \{2\}$, $v(+M) = \{1\}$, and $v(+P) = \{1\}$—we obtain the following arithmetic sum of the inference:

As we can observe in Table 4, the arithmetic sum of the premises, which amounts to 5, is not equal to the arithmetic sum of the conclusion, which is equal to 3.

Table 4. An interpretation of iii-1.

	Proposition	Arithmetic Sum
1.	+M + P	1 + 1 = 2
2.	+S + M	2 + 1 = 3
⊢	+S + P	2 + 1 = 3

By following a similar process, we find that the second tableau defines the interpretation $\mathfrak{I}$ such that $I = \{1\}$, $v(-S) = \{i|i \notin I\}$, say $v(-S) = \{2\}$ (since $v(+S) = \{1\}$), $v(-M) = \{1\}$, and $v(-P) = \{i|i \notin I\}$, say $v(-P) = \{2\}$ (since $v(+P) = \{1\}$). Consequently, if we add up the values of this interpretation we obtain the arithmetic sum shown in Table 5:

Table 5. An interpretation of eee-1.

	Proposition	Arithmetic Sum
1.	−M − P	1 + 2 = 3
2.	−S − M	2 + 1 = 3
⊢	−S − P	2 + 2 = 4

As in the previous case, the arithmetic sum of the premises in Table 5, which amounts to 6, is not equal to the arithmetic sum of the conclusion, which is equal to 4.

Since we can see a pattern in these cases we suggest that the next definition is worthy of exploring:

$$\Gamma \models \phi \; iff \; \sum_{arit}(\Gamma) = \sum_{arit}(\phi),$$

namely, that ϕ is a consequence of Γ if and only if the arithmetic sum (i.e., $\sum_{arit}$) of the premises is equal to the arithmetic sum of the conclusion, given an interpretation $\mathfrak{I}$. Since this notion would be the semantic counterpart of $\Gamma \vdash \phi$, we can imagine that there must be a relation between $\Gamma \models \phi$ and $\Gamma \vdash \phi$: let us entertain this relation by exploring the following propositions.

Proposition 1. *If $\Gamma \vdash \phi$ then $\Gamma \models \phi$.*

Proof. For *reductio*, suppose $\Gamma \vdash \phi$ but $\Gamma \not\models \phi$. If $\Gamma \vdash \phi$ then ϕ follows from Γ by an application of DON. Then, by Lemma 1, there is a closed complete tableau of $\Gamma \cup -\phi$ such that for every branch of the tableau there are two terms $+A$ and $-A$ whose indexes are equal. Now, if $\Gamma \not\models \phi$, then there is an interpretation $\mathfrak{I}$ that assigns a set of indexes in such a way that at least one arithmetic sum of the premises is not equal to the arithmetic sum of the conclusion, but if this is the case is because there is a branch in the tableau in which there is no pair of opposing terms or there is a pair of opposing terms that do not have equal indexes. In either case, there are no two terms $+A$ and $-A$ whose indexes are equal, in which case $\Gamma \cup -\phi$ would not produce a closed complete tableau, but this contradicts the initial assumption. □

Proposition 2. *If $\Gamma \models \phi$ then $\Gamma \vdash \phi$.*

Proof. Take the contrapositive of the original proposition: if $\Gamma \not\vdash \phi$ then $\Gamma \not\models \phi$. Now suppose, for *reductio*, that $\Gamma \not\vdash \phi$ but $\Gamma \models \phi$. If $\Gamma \not\vdash \phi$ then there is an open tableaux for $\Gamma \cup -\phi$ such that there are no two terms $+A$ and $-A$ whose indexes are equal. Now, we have to consider four cases in which this situation would occur, two in which there is a pair of opposing terms that do not have equal indexes, and two in which there is no pair of opposing terms.

For the first case consider the first tableau displayed in Scheme 4, from left to right, that induces the intepretation $\mathfrak{I}$ with $I = \{i\}$ and $v(+S) = v(\mp P) = \{i\}$, so that $v(-S) = v(\pm P) = \{j | j \neq i\}$ in such a way that $\sum_{arit}(\phi) = 2j$, but since $\Gamma \models \phi$, then $\sum_{arit}(\Gamma) = 2j$ as well. However, the only way this can be is if Γ includes something of the form $\{\ldots, -S \pm P, \ldots\}$ or $\{\ldots, -S + M, -M \pm P \ldots\}$, given that the intepretation $\mathfrak{I}$ would assign $v(-S) = v(\pm P) = \{j\}$, in which case $-S \in \Gamma$ and $\pm P \in \Gamma$, but that would contradict the assumption that there are no such terms in Γ.

Scheme 4. General cases.

Similarly, for the second case consider the second tableau that induces the intepretation $\mathfrak{I}$ with $I = \{i\}$ and $v(-S) = v(\mp P) = \{i\}$, so that $v(+S) = v(\pm P) = \{j | j \neq i\}$ in such a way that $\sum_{arit}(\phi) = 2j$, but since $\Gamma \models \phi$, then $\sum_{arit}(\Gamma) = 2j$. However, the only way this can be is if Γ includes something of the form $\{\ldots, +S \pm P, \ldots\}$, given that the intepretation $\mathfrak{I}$ would assign $v(+S) = v(\pm P) = \{j\}$, in which case $+S \in \Gamma$ and $\pm P \in \Gamma$, but again, that would contradict the assumption that there are no such terms in Γ.

For the remaining cases in which there is no pair of opposing terms consider that any interpretation $\mathfrak{I}$ would result from an open tableau, in which case there would never be two terms $+A$ and $-A$ whose indexes are equal, and that would render $\sum_{arit}(\Gamma) \neq \sum_{arit}(\phi)$ since there will be an extra index in Γ or an extra index in ϕ. □

As expected, the conjunction of these propositions suggests that $\Gamma \models \phi$ if and only if $\Gamma \vdash \phi$, that is to say, that the algebraic proof method of TFL (i.e., TFL's syntax), namely the algebraic sum, is related to an arithmetic interpretation, namely an arithmetic sum given an interpretation of (un)distributed terms.

4. Conclusions

To better explain what we have tried to accomplish in this contribution let us wrap up what we have done. TFL's syntax, which recovers the term syntax of syllogistic, relies on the notion of an algebraic sum that depends upon the concept of distribution (hence the use of plus and minus). In turn, the notion of an algebraic sum finds a reliable proxy in TFL tableaux. However, since TFL tableaux induce numerical interpretations, we have noticed that TFL's semantics may rely on arithmetic, rather than algebraic, sums. The idea is, then, that the syntactic notion of algebraic sum has a correspondent notion of arithmetic sum, but since these sums are dependant upon distribution, rather than truth values, we have called these interpretations distribution models.

We think that this result, albeit simple, is worthy of exploring since it promotes the revision and revival of term logics as tools that might be more interesting and powerful than once they seemed (*contra* [1,24,25])). Of course, we need to check if these results hold for more general cases, but if our intuitions are correct then we have found a link between the algebraic rules of term functor logic and a natural interpretation or model of TFL in terms of arithmetic sums. We believe, thus, that by assuming the traditional stance of distribution we still can obtain interesting results. We still are, as Sommers once said, friends of distribution.

Funding: This research was funded by an UPAEP Research Grant.

Acknowledgments: We would like to thank the referees for valuable comments and suggestions.

Conflicts of Interest: The author declares no conflict of interest.

References

1. Geach, P.T. *Reference and Generality: An Examination of Some Medieval and Modern Theories*; Contemporary Philosophy/Cornell University, Cornell University Press: Ithaca, NY, USA, 1962.
2. Williamson, C. Traditional Logic as a Logic Distribution-values. *Log. Et Anal.* **1971**, *14*, 729–746.
3. Sommers, F. Distribution Matters. *Mind* **1975**, *LXXXIV*, 27–46, doi:10.1093/mind/LXXXIV.1.27. [CrossRef]
4. Sommers, F. *The Logic of Natural Language*; Clarendon Library of Logic and Philosophy, Clarendon Press/Oxford University Press: Oxford, UK; New York, NY, USA, 1982.
5. Wilson, F. The distribution of terms: A defense of the traditional doctrine. *Notre Dame J. Form. Log.* **1987**, *28*, 439–454. [CrossRef]
6. Englebretsen, G. *Something to Reckon with: The Logic of Terms*; Canadian Electronic Library, Books Collection; University of Ottawa Press: Ottawa, ON, Canada, 1996.
7. Sommers, F.; Englebretsen, G. *An Invitation to Formal Reasoning: The Logic of Terms*; Ashgate: Farnham, UK, 2000.
8. Englebretsen, G. *Bare Facts and Naked Truths: A New Correspondence Theory of Truth*; Ashgate Pub. Limited: Farnham, UK, 2006.
9. Castro-Manzano, J.M.; Reyes-Cárdenas, P.O. Term Functor Logic Tableaux. *South Am. J. Log.* **2018**, *4*, 1–22.
10. de Rijk, L. *Peter of Spain (Petrus Hispanus Portugalensis): Tractatus: Called Afterwards Summule Logicales. First Critical ed. from the Manuscripts*; van Gorcum & Co.: Assen, The Netherlands, 1972.

11. Arnauld, A.; Nicole, P.; Buroker, J. *Antoine Arnauld and Pierre Nicole: Logic Or the Art of Thinking*; Cambridge Texts in the History of Philosophy; Cambridge University Press: Cambridge, UK, 1996.
12. Keynes, J. *Studies and Exercises in Formal Logic: Including a Generalisation of Logical Processes in Their Application to Complex Inferences*; Macmillan: London, UK 1887.
13. Sommers, F. On a Fregean Dogma. In *Problems in the Philosophy of Mathematics*; Lakatos, I., Ed.; Studies in Logic and the Foundations of Mathematics; Elsevier: Amsterdam, The Netherlands, 1967; Volume 47, pp. 47–81. [CrossRef]
14. Englebretsen, G. *The New Syllogistic*; Peter Lang: Bern, Switzerland 1987.
15. Englebretsen, G.; Sayward, C. *Philosophical Logic: An Introduction to Advanced Topics*; Bloomsbury Academic: London, UK, 2011.
16. Quine, W.V.O. Predicate Functor Logic. In *Proceedings of the Second Scandinavian Logic Symposium*; Fenstad, J.E., Ed.; North-Holland: Amsterdam, The Netherlands, 1971.
17. Noah, A. Predicate-functors and the limits of decidability in logic. *Notre Dame J. Form. Log.* **1980**, *21*, 701–707. [CrossRef]
18. Kuhn, S.T. An axiomatization of predicate functor logic. *Notre Dame J. Form. Log.* **1983**, *24*, 233–241. [CrossRef]
19. Sommers, F. Intelectual Autobiography. In *The Old New Logic: Essays on the Philosophy of Fred Sommers*; Oderberg, D.S., Ed.; Bradford Book: Cambridge, MA, USA, 2005; pp. 1–24.
20. Bastit, M. Jan Łukasiewicz contre le dictum de omni et de nullo. *Philos. Sci.* **2011**, *15*, 55–68. [CrossRef]
21. D'Agostino, M.; Gabbay, D.M.; Hähnle, R.; Posegga, J. *Handbook of Tableau Methods*; Springer: Berlin/Heidelberg, Germany, 1999.
22. Priest, G. *An Introduction to Non-Classical Logic: From If to Is*; Cambridge Introductions to Philosophy; Cambridge University Press: Cambridge, UK, 2008.
23. Hähnle, R. Tableaux and Related Methods. In *Handbook of Automated Reasoning (in 2 volumes)*; Robinson, J.A., Voronkov, A., Eds.; Elsevier: Amsterdam, The Netherlands; MIT Press: Cambridge, MA, USA, 2001; pp. 100–178. [CrossRef]
24. Carnap, R. Die alte und die neue Logik. *Erkenntnis* **1930**, *1*, 12–26. [CrossRef]
25. Geach, P.T. *Logic Matters*; University of California Press: Berkeley, CA, USA, 1980.

© 2020 by the authors. Licensee MDPI, Basel, Switzerland. This article is an open access article distributed under the terms and conditions of the Creative Commons Attribution (CC BY) license (http://creativecommons.org/licenses/by/4.0/).

Article

The *Zahl-Anzahl* Distinction in Gottlob Frege: Arithmetic of Natural Numbers with *Anzahl* as a Primitive Term

Eugeniusz Wojciechowski

Division of Philosophy of Nature, Hugo Kołłątaj Agriculture University of Cracow, 29 Listopada 46, 31-425 Cracow, Poland; eugeniusz.wojciechowski01@gmail.com

Received: 15 October 2019; Accepted: 24 December 2019; Published: 31 December 2019

Abstract: The starting point is Peano's expression of the axiomatics of natural numbers in the framework of Leśniewski's elementary ontology. The author enriches elementary ontology with the so-called Frege's predication scheme and goes on to propose the formulations of this axiomatic, in which the original natural number (*N*) term is replaced by the term *Anzahl* (*A*). The functor of the successor (*S*) is defined in it.

Keywords: Peano's axiomatics of natural numbers; Leśniewski's elementary ontology; Frege's predication scheme; Frege's *Zahl-Anzahl* distinction

1. Introduction

The term *Anzahl* functions in German in various numerical phrases, more or less defined. It appears in such contexts as "Anzahl der Äpfel" (i.e., "number of apples") or "neun ist Anzahl der Planeten des Sonnensystems" (i.e., "nine is the number of planets of the Solar System"). In numerical phrases of the indefinite type, we use the combination of the number name (*Zahl*), which is an abstract object, and the general name, which is the product of a general name (here, the word apple/apples) with a name present only implicitly, which somehow characterizes the objects falling within the extension of the first name (usually a general name referring to the place they occupy). However, number phrases of the definite type (v. gr. "neun ist Anzahl den Planeten des Sonnensystem"or "Anzahl... ist gleich dem Anzahl...") are discussed in Gottlob Frege's works. Number phrases of the definite type are generally given to us through elementary expressions of the type "*a* is *Ax*", where *a* is a number and *x* a (definite) general name. There is a specific functor *A* (from "Anzahl") of the *n/n* category. In Frege, "Anzahl" refers to a concept determined by a single-argument predicate (*Begriffswort*). Given these remarks, we enrich the framework of elementary ontology with the so-called Frege's predication scheme, to propose the formulations of these axiomatics in which the original natural number (*N*) term is replaced by the term Anzahl (*A*), and we present an original depiction of the arithmetic of natural numbers with *Anzahl* as a primitive term.

2. Preliminaries

2.1. Elementary Ontology

The specific axiom of elementary ontology (**OE**) is:

A0 $x\varepsilon y \leftrightarrow \Sigma z(z\varepsilon x) \wedge \Pi zu(z\varepsilon x \wedge u\varepsilon x \to z\varepsilon u) \wedge \Pi z(z\varepsilon x \to z\varepsilon y)$

The secondary rules, which are direct consequences of axiom A0 include:

R1 $x\varepsilon y / x\varepsilon x$
R2 $x\varepsilon y \wedge y\varepsilon z / x\varepsilon z$
R3 $x\varepsilon y \wedge y\varepsilon z / y\varepsilon x$

The nominal constants of object and contradictory object are defined as follows:

DV	$x \varepsilon V \leftrightarrow x \varepsilon x$	x is an object;
DΛ	$x \varepsilon \Lambda \leftrightarrow x \varepsilon x \wedge \sim x \varepsilon x$	x is a contradictory object.

The functors of existence, singularity, being an object, weak inclusion, strong inclusion, extension identity, identity, and negation are also introduced by definition:

Dex	$ex(x) \leftrightarrow \Sigma z(z \varepsilon x)$	x exists;
Dsol	$sol(x) \leftrightarrow \Pi zu(z \varepsilon x \wedge u \varepsilon x \to z \varepsilon u)$	at most one object is x;
Dob	$ob(x) \leftrightarrow x \varepsilon x$	is object x;
D⊂	$x \subset y \leftrightarrow \Pi z(z \varepsilon x \to z \varepsilon y)$	x is included in y (weak inclusion);
D⊏	$xy \leftrightarrow \Sigma z(z \varepsilon x) \wedge \Pi z(z \varepsilon x \to z \varepsilon y)$	x is included in y (strong inclusion);
D○	$xy \leftrightarrow \Pi z(z \varepsilon x \leftrightarrow z \varepsilon y)$	x is extension identical with y;
D=	$x = y \leftrightarrow x \varepsilon y \wedge y \varepsilon x$	x is identical with y;
Dn	$x \varepsilon n y \leftrightarrow x \varepsilon x \wedge \sim x \varepsilon y$	x is non y.

We also adopt the definition of the of satisfying functor:

Dstsf $x \varepsilon stsf(\phi) \leftrightarrow x \varepsilon x \wedge \phi(x)$ x is satisfying ϕ

The system of elementary ontology can be founded on the calculus of predicates without identity. Jerzy Słupecki's work [1] may serve as an introduction to elementary ontology.

2.2. Peano's Axiomatics

The system of arithmetic of natural numbers can be founded on the axiomatics given by Peano [2]. The formulation of these axiomatics in the framework of elementary ontology can be found in Leśniewski's works ([3], chapter 4, p. 129):

LA1	$1 \varepsilon N$
LA2	$a \varepsilon N \to Sa \varepsilon N$
LA3	$a \varepsilon N \to \sim Sa \varepsilon 1$
LA4	$a \varepsilon N \wedge b \varepsilon N \wedge Sa = Sb \to a = b$
LA5	$1 \varepsilon x \wedge \Pi b(b \varepsilon N \wedge b \varepsilon x \to Sb \varepsilon x) \wedge a \varepsilon N \to a \varepsilon x$

The last axiom is the axiom of mathematical induction. Axioms A1–A5, as depicted below in Section 4, are equivalent to the axioms given by Peano.

2.3. Elementary Ontology with Frege's Predication Scheme

Here, the system of elementary ontology is extended with the functor *sub* of the n/n category. The elementary expression "$x \varepsilon sub(y)$" is read here as: "x is subordinated to y". The specific axioms of this system (**OE**^{sub}) have the following shapes (see [4]):

SA1	$x \varepsilon sub(y) \to sub(x) \subset sub(y)$
SA2	$x \varepsilon sub(y) \to \sim y \subset sub(x)$
SA3	$x \varepsilon sub(y) \to y \varepsilon y$
SA4	$sub(x) sub(y) \to yy$

The functor of *concept* is introduced by definition:

DC $x \varepsilon Cy \leftrightarrow x \varepsilon x \wedge \Pi z(z \varepsilon y \leftrightarrow z \varepsilon sub(x))$ x is a concept y

The key theses with functors *sub* and *C* include:

$x \varepsilon sub(Cy) \to x \varepsilon y$ If x is subordinated to the concept y, then x is y,

and:

$Cx \varepsilon sub(Cy) \to x \subset y$ If the concept x is subordinated to the concept y, then x is included in y

3. Idea

The term "Anzahl" functions in German in various numerical phrases, more or less defined. The word *Anzahl* appears in such contexts as: *Anzahl der Äpfel* (*number of apples*), *Anzahl von Freuden* (*number of friends*), *fünf is Anzahl der Bäumen in meinem Garten* (*five is the number of trees in my garden*), or *neun ist Anzahl der Planeten des Sonnensystems* (*nine is the number of planets of the Solar System*).

3.1. An Indefinite Numerical Phrase with the Term Anzahl

In number phrases of the indefinite type we have to deal with the combination of the number name (*Zahl*) being an abstract object with the general name, which is the product of a general name (e.g., apple/apples) and a name present only implicitly, which somehow characterizes the objects falling within the extension of the first name (usually a general name referring to the place they occupy). For example, "five apples" occupy a certain place on the table.

A phrase of this type, which is a nominal expression, can be rendered as follows:

$$a \circ x,$$

where a is a number (*Anzahl*) and x is the name of object of a given sort, which, in combination with a hidden name characterizing these objects, gives us our universe of discourse. Since a and x are names, $a \circ x$ is also a name, then $\circ$ is a functor of the—n/nn category.

3.2. A Definite Numerical Phrase with the Term Anzahl

Numerical phrases of the definite type ("neun ist Anzahl den Planeten des Sonnensystem" or "Anzahl... ist gleich dem Anzahl...") are discussed in Gottlob Frege's works (see [5], p. 20 and [6], p. 88). In number phrases of the definite type, we use an elementary expression of the type:

$$a\varepsilon A x$$

where a is a number, ε is a functor (of the s/nn category), and x is a (definite) general name. A specific functor A (from Anzahl) of the n/n category appears here.

In Frege, "Anzahl" refers to a concept designated by a single-argument predicate (*Begriffswort*), but also a functor of the $n/(s/n)$ category. The elementary phrase with this functor could be expressed in the following way: $a\varepsilon A(P)$, where the (single-argument) predicate P is equivalent to the name x (from the phrase "$a\varepsilon Ax$"). Our expression existing in the framework of the calculus of names is closer to natural language, in which the category of names is understood broadly (individual and general names), as opposed to only individual names, as in Frege's language (*Namen=Eigennamen*).

We will continue to deal with definite number phrases with the term *Anzahl*. We shall return to indefinite phrases with this term in the final part of the present paper.

4. Arithmetic of Natural Numbers

As our starting point, we shall adopt the following five axioms given by Leśniewski, including 0 in natural numbers, which is in accordance with the contemporary formulations of Peano's axiomatics:

A1	$0\varepsilon N$
A2	$a\varepsilon N \to Sa\varepsilon N$
A3	$a\varepsilon N \to\, \sim Sa\varepsilon 0$
A4	$a\varepsilon N \wedge b\varepsilon N \wedge Sa = Sb \to a = b$
A5	$0\varepsilon x \wedge \Pi b(b\varepsilon N \wedge b\varepsilon x \to Sb\varepsilon x) \wedge a\varepsilon N \to a\varepsilon x$

Our initial system is $\mathbf{OE}^{sub}$ enriched with these axiomatics. The language of this system ($\mathbf{OE}^{sub}$[A1,A2,A3,A4,A5]) is extended with number nominal variables (*a*,*b*,*c*), referring to natural numbers. Apart from the standard rule of detachment (MP):

MP $\alpha, \alpha \to \beta / \beta$

we shall adopt the rule of substitution:

RS $\alpha / \alpha[x/t]$,

where t is a nominal variable/constant or (in particular) a number nominal variable/constant:

$\alpha / \alpha[a/t]$,

where t is a number nominal variable/constant.

Having adopted the definition of the natural number:

DN $a\varepsilon N \leftrightarrow a\varepsilon a \wedge \Sigma x(a\varepsilon Ax)$,

we can eliminate constant N by means of functor A:

B1 $0\varepsilon A\Lambda$
B2 $a\varepsilon Ax \to \Sigma z(Sa\varepsilon Az)$
B3 $a\varepsilon Ax \to \sim Sa\varepsilon A\Lambda$
B4 $a\varepsilon Ax \wedge b\varepsilon Ay \wedge Sa\varepsilon Sb \to a\varepsilon b$
B5 $0\varepsilon x \wedge \Pi by(b\varepsilon x \wedge b\varepsilon Ay \to Sb\varepsilon x) \wedge a\varepsilon Az \to a\varepsilon x$

Axioms A1–A5 are consequences of Axioms B1–B5:

(A1) $0\varepsilon N$ [B1,DN]
(A2) $a\varepsilon N \to Sa\varepsilon N$ [DN,B2]
(A3) $a\varepsilon N \to \sim Sa\varepsilon 0$ [DN,B1,R2,B3]
(A4) $a\varepsilon N \wedge b\varepsilon N \wedge Sa = Sb \to a = b$ [DN,D=,B4]
(A5) $0\varepsilon x \wedge \Pi b(b\varepsilon N \wedge b\varepsilon x \to Sb\varepsilon x) \wedge a\varepsilon N \to a\varepsilon x$ [DN,B5]

Next, we shall eliminate from axiomatics the functor of the successor (S) by introducing it by definition. We shall leave it only in the last axiom, for the purpose of shortening.

We shall adopt the axiomatics:

C1 $a\varepsilon a$
C2 $a\varepsilon Ax \to ex(nx)$
C3 $\sim ex(x) \to 0\varepsilon Ax$
C4 $a\varepsilon Ax \wedge b\varepsilon Ax \to a\varepsilon b$
C5 $0\varepsilon x \wedge \Pi by(b\varepsilon x \wedge b\varepsilon Ay \to Sb\varepsilon x) \wedge a\varepsilon Az \to a\varepsilon x$ (=B5)

We shall introduce the functor of the successor by definition:

DS $a\varepsilon Sb \leftrightarrow a\varepsilon nb \wedge \Pi xz(b\varepsilon Ax \wedge z\varepsilon nx \leftrightarrow a\varepsilon A(x \cup z) \wedge z\varepsilon nx) \wedge \Pi xz(0\varepsilon Ax \wedge z\varepsilon nx \to a\varepsilon Az) \wedge$
     $\Pi yz(a\varepsilon Az \wedge \sim ex(z) \to \sim b\varepsilon Ay) \wedge \Pi z(a\varepsilon Az \wedge z\varepsilon z \to b = 0)$

T1.1 $sol(Ax)$ [C4,Dsol]
T1.2 $A\Lambda = 0$ [OE,C3,T1.1,R3,D=]
T1.3 $a\varepsilon Ax \wedge b\varepsilon Ax \to a = b$ [C4,D=]
T1.4 $Sa\varepsilon Sa \leftrightarrow Sa\varepsilon na \wedge \Pi xz(a\varepsilon Ax \wedge z\varepsilon nx \leftrightarrow Sa\varepsilon A(x \cup z) \wedge z\varepsilon nx) \wedge \Pi xz(0\varepsilon Ax \wedge z\varepsilon nx \to Sa\varepsilon Az) \wedge \Pi yz(Sa\varepsilon Az \wedge \sim ex(z) \to \sim a\varepsilon Ay) \wedge \Pi z(Sa\varepsilon Az \wedge z\varepsilon z \to a = 0)$ [DS]
T1.5 $Sa\varepsilon A(x \cup z) \wedge z\varepsilon nx \to a\varepsilon Ax$ [R1,T1.4]
T1.6 $Sa\varepsilon Ax \wedge x\varepsilon x \to a = 0$ [R1,T1.4]
T1.7 $\sim Sa\varepsilon a$ [R1,T1.4.Dn]
T2.1 $0\varepsilon A\Lambda$ (=B1)   [OE,C3]
T2.2 $a\varepsilon Ax \to \Sigma z(Sa\varepsilon A)$ (=B2)   [C2,C1,Dex,T1.4]
T2.3 $a\varepsilon Ax \to \sim Sa\varepsilon A\Lambda$ (=B3)   [R1,T1.4,OE]
T2.4 $a\varepsilon Ax \wedge b\varepsilon Ay \wedge Sa\varepsilon Sb \to a\varepsilon b$ (=B4)   [OE,T1.4,C2,Dex,T1.5,T1.3]
T2.5 $\phi(0) \wedge \Pi b(b\varepsilon N \wedge \phi(b)\phi(Sb)) \wedge a\varepsilon N \to \phi(a)$ [C1,D$stsf$,DN,C5]

Of these, T2.5 is a typical formulation of an induction axiom.

5. Addition and Multiplication: Logical and Philosophical Analysis

Addition and multiplication as operations on numbers—with the numerical functor (Anzahl) as a primary functor—is, from a logical and philosophical point of view, an operation more complex than it usually appears.

5.1. Addition

The addition of numbers in the context of a numerical functor (*Anzahl*) establishes the number of objects of the same sort. For example, with five trees in the front garden and seven trees in the back garden, we reach the conclusion that we have 12 trees altogether. Using names y and z so that *y a tree in the front garden* and *z a tree in the back garden*, and taking into account the fact that the extensions of these names have no common elements ($\sim ex(y \cap z)$), we create in this case a common name *x a tree in the garden*, so that $xy \cup z$—in order to state the following: $12\varepsilon Ax$ with $5\varepsilon Ay$ and $7\varepsilon Az$, where between x, y, and z there are the connections described before. The operation of addition in this context can be generally defined as follows:

$$D + a\varepsilon b + c \leftrightarrow a\varepsilon a \wedge \Sigma xyz(xy \cup z \wedge \sim ex(y \cap z) \wedge a\varepsilon Ax \wedge b\varepsilon Ay \wedge c\varepsilon Az)$$

5.2. Multiplication

The multiplication of numbers in the context of the numerical functor (*Anzahl*) is more complicated. Let us take a similar example. Let us assume that in the garden we have trees which are grouped in four groups, each comprising five trees. There are two groups in front of the house (on the left and on the right) and two groups behind the house (one on the left and one on the right, too). We begin our operation by distinguishing four groups, which we treat as distributive classes determined by four names: y, z, u, and v, which are, respectively *a tree in the front garden on the left*, *a tree in the front garden on the right*, *a tree in the back garden on the left*, and *a tree in the back garden on the right*.

We shall mark these classes, respectively, as: Cy, Cz, Cu, and Cv. This distinction is accompanied by an ascertainment that for a certain name x with the extension—$xy \cup z \cup u \cup v$ (*x a tree in the garden*), these names are separate in terms of extension—$\sim ex(y \cap z)$, $\sim ex(z \cap u)$, and $\sim ex(u \cap v)$. Next, we state that there are four such groups/classes of trees—$4\varepsilon Aw$ $4\varepsilon Aw$, where w is a shortening of the name "*a group of trees in the garden*", and we finally state that each group of trees comprises five trees and each tree in the garden can be characterized as belonging strictly to one of these groups and as one of five trees belonging to this group. This last sentence can be shortly expressed as follows: each tree in the garden is characterized by a pair [*one of the groups, one of the trees of a given bank of trees assigned to this group*], to be eventually able to say that the number of trees in the garden is identical to the number of such pairs. Symbolically, it can be briefly expressed as follows:

$$D \odot a\varepsilon b \odot c \leftrightarrow a\varepsilon a \wedge \Sigma xyz(a\varepsilon Ax \wedge b\varepsilon Ay \wedge c\varepsilon Az \wedge \Pi u(u\varepsilon y \leftrightarrow \Sigma v(u\varepsilon Cv \wedge v \subset x)) \wedge x\varepsilon y \times z)$$

where $\times$ is a functor of Cartesian product of the n/nn category.

6. The New Formulation of the Arithmetic of Natural Numbers

We shall replace axiom C3 with a more intuitive one:

C3* $\quad xy \rightarrow AxAy$

The system of arithmetic of natural numbers in this formulation is: $\mathbf{OE}^{sub}$[C1,C2,C3*,C4,C5]. In the formulation of axiom C5, the functor of successor occurs, which we are introducing similarly, by means of the already given definition DS.

Here, the term C from $\mathbf{OE}^{sub}$ is interpreted in terms of class (see [7]):

DC $x\varepsilon Cy \leftrightarrow x\varepsilon x \wedge \Pi z(z\varepsilon y \leftrightarrow z\varepsilon sub(x))$ x is a distributive class (of objects which are) y.

In addition to the rules of substitution and detachment (MP), we shall also adopt the rules of omission and introduction for the list operator in the form (see [8]):

OL $x\varepsilon[z_1,\ldots,z_n]/x\varepsilon z_1 \vee \ldots \vee x\varepsilon z_n$ $[x_1,\ldots,x_n]\varepsilon y/x_1\varepsilon y \wedge \ldots \wedge z_n\varepsilon y$ if y is not list

IL $x\varepsilon z_1 \vee \ldots \vee x\varepsilon z_n/x\varepsilon[z_1,\ldots,z_n]$ $x_1\varepsilon y \wedge \ldots \wedge z_n\varepsilon y/[x_1,\ldots,x_n]\varepsilon y$

and the rule:

RL $[x_1,\ldots,x_n]\varepsilon[z_1,\ldots,z_n]/x_1\varepsilon z_1 \wedge [x_2,\ldots,x_n]\varepsilon[z_2,\ldots,z_n]$ $[x]\varepsilon[z]/x\varepsilon z$

Thanks to this rule, we obtain the property for two-element lists, $[x,y] = [z,u] \rightarrow x = z \wedge y = u$, which is the equivalent of property for an orderly pair in the framework of the set theory.

Now, let us define the functor of Cartesian product (compare [9], p. 176):

Dx $x\varepsilon y \times z \leftrightarrow x\varepsilon x \wedge \Pi uvw(u\varepsilon[u,w] \wedge v\varepsilon y \wedge w\varepsilon z \leftrightarrow u\varepsilon sub(x))$

Now, we can—quite formally—adopt the definitions of the operations of addition (+) and multiplication ($\odot$) in accordance with their interpretation:

D+ $a\varepsilon b + c \leftrightarrow a\varepsilon a \wedge \Sigma xyz(xy \cup z\wedge \sim ex(y \cap z) \wedge a\varepsilon Ax \wedge b\varepsilon Ay \wedge c\varepsilon Az)$

D$\odot$ $a\varepsilon b \odot c \leftrightarrow a\varepsilon a \wedge \Sigma xyz(a\varepsilon Ax \wedge b\varepsilon Ay \wedge c\varepsilon Az \wedge \Pi u(u\varepsilon y \leftrightarrow \Sigma v(u\varepsilon Cv \wedge v \subset x)) \wedge x\varepsilon y \times z)$

According to D+, addition is a symmetrical operation: $a\varepsilon b + c \leftrightarrow c + b$ ($a + b = b + a$).

The operation of multiplication in the sense $\odot$, in accordance with the definition D$\odot$, is not symmetrical. We shall introduce multiplication as a symmetrical operation (·) by definition:

D· $a\varepsilon b \cdot c \leftrightarrow (a\varepsilon b \odot c) \vee a\varepsilon c \odot b)$

The theses which are consequences of definition D0 and axiom A3* include:

T2.6 $0A\wedge$ [D0,D$\odot$]
T2.7 $\sim ex(x) \rightarrow 0\varepsilon Ax$ (=C3)   [Dex,**OE**,Dd,D$\odot$,C3*,R1,T2.6,R2]

7. The Term *Anzahl* in the Indefinite Sense

We shall now deal with number phrases with the term "Anzahl" in the indefinite sense, where, "Numbers (Anzahlen) are always numbers (Anzahlen) of something. They can be added (five apples and two apples make seven apples). However, it is impossible to multiply numbers (Anzahlen) by numbers (Anzahlen)" (see [10], p. 7).

Compounds such as five apples or two dogs fall, in accordance with the previous arrangements, under the scheme $a \circ x$. We shall introduce the functor ∘ by definition:

D∘ $x\varepsilon a \circ y \leftrightarrow x\varepsilon x \wedge \Sigma z(a\varepsilon A(y \cap z))$

The numbers (Anzahlen) in this sense behave like the so-called *denominate numbers* (like, for example: 5 m or 2 kg) present in the so-called *dimensional analysis*, which appears in physics. They can be added, provided that the same unit is preserved. Addition in such contexts can be defined as follows:

$$D \bigoplus x\varepsilon(a \circ y) \bigoplus (b \circ y) \leftrightarrow x\varepsilon(a+b) \circ y$$

Denominate numbers (*benannte Zahlen*) are the subject of analysis in one of Hermann von Helmholtz's works (see [11], p. 12).

The numbers (Anzahlen) in the indefinite sense cannot be multiplied, as opposed to denominate numbers in dimensional analysis.

8. Conclusions

A new, original depiction of arithmetic of natural numbers with *Anzahl* as a primitive term has been presented. The basis of this depiction is the calculus of names, which is a certain extension of elementary ontology (OE^{sub}). The notion of a pair (list) was introduced by means of rules OL, IL, and RL, the last of them playing a significant role. I have recently noticed that it is possible to define an ordered pair in the framework of the OE^{sub} system.

Funding: This research received no external funding.

Acknowledgments: I would like to express my sincere gratitude and highest appreciation to all referees of this article for all their useful remarks, comments, and suggestions to the content of the article and also its Englishlanguage verification. Individual thanks are due, in particular, to Luna Shen and Hunter Jia for their precious support.

Conflicts of Interest: The authors declare no conflict of interest.

References

1. Słupecki, J.S. "Leśniewski Calculus of names". *Studia Log. Int. J. Symb. Log.* **1955**, *3*, 7–70.
2. Peano, G. *Arithmetices Principia Nova Methodo Exposito*; Bocca: Turin, Italy, 1889.
3. Leśniewski, S. *Lecture Notes in Logic*; Srzednicki, J.T.J., Stachniak, Z., Eds.; Kluwer Academic: Dordrecht, The Netherlands, 1988.
4. Wojciechowski, E. "Rachunek nazw i schemat predykacji z *Begriffschrift* Gottloba Fregego" (Calculus of Names and Predication Scheme from Gottlob Frege's *Begriffsschrift*). In *Predykacja, negacja i kwantyfikacja (Predication, Negation and Quantification)*; Wojciechowski, E., Ed.; Aureus: Kraków, Poland, 2019.
5. Frege, G. *Die Grundlagen der Arithmetik: Eine Logisch Mathematische Untersuchung über den Begriff der Zahl*; Verlag von Wilhelm Koebner: Breslau, Germany, 1884.
6. Patzig, G. Gottlob Frege und die logische Analyse der Sprache. In *Sprache Und Logik*, 2nd ed.; Patzig, G., Ed.; Vandenhoeck & Ruprecht: Göttingen, Germany, 1981; pp. 77–100.
7. Wojciechowski, E. "Klasy dystrybutywne i klasy kolektywne" (Distributive Classes and Collective Classes). In *Predykacja, negacja i kwantyfikacja (Predication, Negation and Quantification)*; Wojciechowski, E., Ed.; Aureus: Kraków, Poland, 2019.
8. Wojciechowski, E. "Rachunek nazw z listami" (The Calculus of Names with Lists). *Rocz. Filoz.* **2011**, *59*, 35–50.
9. Borkowski, L. *Logika Formalna (Formal Logic)*, 2nd ed.; PWN: Warszawa, Poland, 1977.
10. Grote, A. *Anzahl, Zahl und Menge. Die Phänomenologischen Grundlagen der Arithemitk*; Felix Meiner Verlag: Hamburg, Germany, 1983.
11. Von Helmholz, H. Zählen und Messen, erkenntnisstheoretisch betrachtet. In *Philosophische Aufsätze, Eduard Zeller Ze Seinem Fünfzigjährigen Doctorjubiläum Gewidmet*; Fues Verlag: Leipzig, Germany, 1887; pp. 17–52.

© 2019 by the author. Licensee MDPI, Basel, Switzerland. This article is an open access article distributed under the terms and conditions of the Creative Commons Attribution (CC BY) license (http://creativecommons.org/licenses/by/4.0/).

Article

Hybrid Deduction–Refutation Systems

Valentin Goranko [1,2]

[1] Department of Philosophy, Stockholm University, SE-10691 Stockholm, Sweden; valentin.goranko@philosophy.su.se
[2] Visiting professorship at Department of Mathematics, University of Johannesburg, Johannesburg 2006, South Africa

Received: 27 August 2019; Accepted: 10 October 2019; Published: 21 October 2019

Abstract: Hybrid deduction–refutation systems are deductive systems intended to derive both valid and non-valid, i.e., semantically refutable, formulae of a given logical system, by employing together separate derivability operators for each of these and combining 'hybrid derivation rules' that involve both deduction and refutation. The goal of this paper is to develop a basic theory and 'meta-proof' theory of hybrid deduction–refutation systems. I then illustrate the concept on a hybrid derivation system of natural deduction for classical propositional logic, for which I show soundness and completeness for both deductions and refutations.

Keywords: deductive refutability; refutation systems; hybrid deduction–refutation rules, derivative hybrid rules, soundness, completeness, natural deduction, meta-proof theory

1. Introduction

1.1. Semantic vs. Deductive Refutability

Consider a generic logical system L, comprising a formal logical language with a given semantics. The basic semantic notion is that of L-*validity*: an L-formula A is said to be L-*valid* in L, denoted $\models_L A$, iff it is true in every L-model. Respectively, an L-formula A is said to be *refutable* in L (L-*refutable*), or L-*falsifiable*, denoted $\not\models_L A$, iff there is an L-model falsifying A, i.e., if it is not L-valid. (Note that in the usual logical semantics "not L-valid" means the same as "L-falsifiable", but, in some non-classical logical systems, such as paraconsistent logic, the semantics may allow the same formula to be both true and false, hence it may turn out both valid and falsifiable. This leads, inter alia, to terminological complications. To avoid these, I will exclude from consideration such paraconsistent semantics here, but will briefly discuss that case in Remark 6.)

Now, consider a deductive system **D** for L, with a derivation relation $\vdash_D$. Then, the basic deductive notion associated with **D** is *provability in D* of a given L-formula A, denoted as usual by $\vdash_D A$. If **D** is sound and complete for L, then $\vdash_D A$ corresponds precisely to validity, i.e., $\models_L A$. In general, this may not be the case, but, still, provability is the intended syntactic counterpart of validity (and, more generally, of logical consequence).

Then, what is the precisely matching syntactic notion to semantic refutability? One can argue that it is *not* "non-provability" but is rather the notion of *"deductive refutability"*, i.e., existence of a formal derivation in a suitable *derivation system for L-refutable formulae*. Following Łukasiewicz, a new symbol, $\dashv_L$, can be introduced for that notion, where $\dashv_L A$ means *"A is deductively refutable in L"*, i.e., *"the non-validity of A in L is established deductively"*, or *"the refutation of A in L is formally derived/derivable"*.

Thus, the notion of *formal (deductive) refutation* arises. The idea goes far back in history, already to Aristotle, who essentially applied that idea to 'derive' some non-valid syllogisms from others, but it was not pursued further, until Łukasiewicz revived it in the early-mid 20th century and introduced the notion of *(deductive) refutation system*. For further details on the origins and history of refutation systems, see [1–3]. For a recent comprehensive overview on the research and literature on refutation systems, see [4].

It should, of course, be noted that the idea of formally proving unprovability in a given formal deductive system has been crucial for the development of Proof Theory since its very inception by Gentzen and others, and has been fundamental in Logic since Gödel's incompleteness theorems. Moreover, results in Structural Proof Theory (see [5]), such as normalisation results for systems of Natural Deduction (see [6,7]) make it possible to obtain precise mathematical proofs of unprovability in such system, which can be appropriately formalised. What the theory of refutation systems proposes further is to consider the concept of deductive refutability as a first-class citizen and treat it as an object of study in its own right.

1.2. Related Work and Main Contributions

The overall development of refutations systems so far has been driven by the ideas to employ and 'simulate' traditional deductive systems, rather than to interact with them. In particular, a commonly pursued goal has been to design *pure* refutations systems, involving *only* the relation of refutability, but not at all that of provability. Even in the cases of refutations systems involving both refutability and provability, the latter is typically used as an auxiliary, 'black box' operator, to enable the applications of some 'mixed' rules of refutation inference, such as Modus Tollens (see further).

An alternative philosophy, promoted in the present work, is *to treat both notions of provability and refutation on a par* and to seek to develop *combined* deductive systems where both notions not only coexist, but actually interact and cooperate with each other, for the sake of ultimately deriving the correct validity/non-validity status of the formula or logical consequence in question.

Related Work

The idea of combining deductions and refutations in common systems of derivations, here called *hybrid* (The use of the term "hybrid" in the context of this work is not related to deduction in the so-called "hybrid logic"). I hope that the use of that term here would not create terminological confusion in the literature) *deduction–refutation systems* (also, for short, *hybrid derivation systems*), can be traced back implicitly to some works of Łukasiewicz and Carnap. However, to my knowledge, it was first explicitly proposed in [3] but apparently not pursued further since then. Still, several similar or related ideas have been proposed and discussed (though not as follow-up works to [3]) in the meantime, including (chronologically):

- The idea of 'complementary systems' for sentential logic, suggested by Bonatti and Varzi in [8] is related in spirit, though technically different from the idea of hybrid refutation systems, as it considers the complementary systems, for deductions and for refutations, acting separately.
- Similarly, in [9], Skura studies 'symmetric inference systems', that is, pairs of essentially non-interacting inference systems, and shows how they can be used for characterizing maximal non-classical logic with certain properties. In particular, the method is applied there to paraconsistent logic.
- In [10], Wybraniec-Skardowska and Waldmajer explore the general theory of deductive systems employing the two dual consequence operators, the standard logical consequence, inferring validities, and the refutation consequence, inferring non-validities. Again, no interaction of these consequence operators is considered there.
- In [11], Caferra and Peltier, motivated by potential applications to automated reasoning, take a unifying perspective on deriving accepting or rejecting propositions from other, already accepted or rejected, propositions, thus considering separately each of the four consequence relations arising as combinations.
- In [12], Goré and Postniece combine derivations and refutations to obtain cut-free complete systems for bi-intuitionistic logic.

- In [13], Negri explores the duality of proofs and countermodels in labelled sequent calculi and develops a method for unifying proof search and countermodel construction for some modal and intuitionistic propositional logic over classes of Kripke frames with suitable frame conditions. In particular, for some of this logic, the method provides a decision procedure.
- In [14], Citkin considers essentially multiple-conclusion generalisations of hybrid inference rules studied here. Citkin discusses consequence relations and inference systems employing such rules and proposes a meta-logic for formalising propositional reasoning about such systems. Even though with different motivation and agenda, and with no technical results of the type pursued here, this work appears to be the closest in spirit to the idea of hybrid deduction–refutation systems studied in the present work.
- Likewise, in [15], Fiorentini and Ferrari explore the duality between unprovability and provability in forward proof-search for intuitionistic propositional logic and develop a refutation-complete sequent-based forward refutation calculus for it, following on their previous work [16].
- In [17], Rumfitt considers "reversals" of the rules of propositional Natural Deduction, to formalise derivations between "accepted" and "rejected" sentences. While the motivation is different from the one related to refutation systems, most (but not all!) resulting rules are essentially the same as the "hybrid refutation rules" obtained by contrapositive inversion of the rules of propositional Natural Deduction considered in Section 4. See Remark 7 on the distinction between the two types of rules.

Contributions and Structure of the Paper

The goal of this paper is to develop a basic proof theory and 'meta-proof' theory of hybrid deduction–refutation systems. After a brief description of refutation rules and systems in the preliminary Section 2, I present in Section 3 a basic theory of hybrid derivation rules and systems, including the notion of *inversion* of deduction and refutation rules and *canonical hybrid extensions* of deductive and refutation systems. In Section 4, I then illustrate these concepts on the natural deduction system for classical propositional logic $\mathbf{ND^{PL}}$, for which I develop a 'standard hybrid extension' $\mathcal{H}^s(\mathbf{ND^{PL}})$, for which I prove soundness and completeness with respect to both deductions and refutations in Section 5. Then, Section 6 discusses the 'meta-theory' of hybrid derivation systems. The paper ends with some concluding remarks on potential applications and further work on hybrid derivation systems in Section 7.

2. Preliminaries

I will assume that the reader is familiar with basic logical notation and terminology for proof systems for classical logic. If necessary, see, e.g., [5,7], or [18].

2.1. Refutation Rules and Systems: Basic Concepts

Let us consider and fix a logical system L, comprising a formal logical language with a given semantics, defining the notion of validity and, respectively, logical consequence. Here, I will introduce the basic concepts of (axiomatic) refutation systems, generally (but not fully) following notation and terminology from [3,19], to which the reader is referred for further details; see also [4] for a bibliographic overview on refutation systems.

A **pure rule of refutation inference** is a rule scheme of the type

$$\frac{\dashv \psi_1, \ldots, \dashv \psi_n}{\dashv \gamma},$$

where $\psi_1, \ldots, \psi_n, \gamma$ are propositional formulae. (This definition can be naturally extended to first-order languages. (Here, and further: the commas used to separate premises in the rules should not be considered as part of the formal syntax, but rather as typographical indication to separate these premisses. That way, derivations can be regarded as trees, as usual.) The intuitive meaning of that rule with respect to the logical system L is that, for any uniform substitution σ of formulae for the propositional variables occurring in the formulae $\psi_1, \ldots, \psi_n, \gamma$, if each of the formulae $\sigma(\psi_1), \ldots, \sigma(\psi_n)$, has been derived as non-valid (in L), then $\sigma(\gamma)$ is derived as non-valid (in L) too. In that sense, the rule is actually a *rule scheme*, and this will apply likewise for all propositional inference rules considered further in the paper.

A typical example of a pure rule of refutation inference is the **Disjunction rule**:

$$\frac{\dashv \varphi, \dashv \psi}{\dashv \varphi \vee \psi}.$$

Another important example is Łukasiewicz's **Reverse substitution rule scheme**:

$$\frac{\dashv \tau(\varphi)}{\dashv \varphi},$$

where τ is a uniform substitution. (Note that this is a substitutional scheme in two senses.)

Usually, pure refutation rules do not suffice to capture adequately semantic refutability in a refutation system, so we also consider a more general type of refutation rules, called **mixed refutation rules**, which are relativised to a given underlying deductive system **D** for the logical system L, as follows:

$$\frac{\vdash_\mathbf{D} \varphi_1, \ldots, \vdash_\mathbf{D} \varphi_m, \dashv \psi_1, \ldots, \dashv \psi_n}{\dashv \gamma},$$

where $\varphi_1, \ldots, \varphi_m, \psi_1, \ldots, \psi_n, \gamma$ are (here, propositional) formulae. The intuitive meaning of that rule with respect to the logical system L is that, for any uniform substitution σ of formulae for the propositional variables occurring in the formulae $\varphi_1, \ldots, \varphi_m, \psi_1, \ldots, \psi_n, \gamma$, if each of the formulae $\sigma(\varphi_1), \ldots, \sigma(\varphi_m)$ is derived by **D** (hence, assuming soundness of **D**, proved valid in L) and each $\sigma(\psi_1), \ldots, \sigma(\psi_n)$ has been derived as non-valid in L, then $\sigma(\gamma)$ is derived as non-valid in L too. A typical example is Łukasiewicz's rule **Reverse modus ponens** (aka, **Modus Tollens**):

$$\frac{\vdash_\mathbf{D} \varphi \to \psi, \dashv \psi}{\dashv \varphi}.$$

A **refutation system (associated with a given underlying deductive system D)** is a set $\mathcal{R}$ of (generally, mixed) refutation rules (where $\vdash$ is indexed with **D**). Refutation rules with no premises are called **structural refutation axioms**, and I will write them simply as $\dashv \theta$.

Remark 1. *Substitution closure of inference rules is a standard structurality condition in most non-classical propositional logic. However, in the case of mixed refutation rules, it has a rather non-trivial nature, as it makes an interesting connection with the notions of* unification and unifiability of propositional formulae; *see [20]. Indeed, a refutation axiom is sound under substitution instances precisely when it is* not unifiable. *In particular, this will make it necessary further to consider more general, not closed under substitutions' refutation axiom schemes, in addition to structural ones. Furthermore, unifiability of formulae is closely related to admissibility of rules. On the connections of these with refutation rules and systems, see [21].*

The issue of substitution closure of inferences will come up again later when hybrid inference rules are considered.

When defining refutation derivations in $\mathcal{R}$, one typically assumes that the necessary derivations in the underlying deductive system **D** are done separately, in advance or "on demand", whenever needed for the derivation of the target refutation, and as part of that derivation. In either case, the deductive system **D** is assumed to play only an auxiliary role for the functioning of the refutation system $\mathcal{R}$. Formally, a **refutation derivation in** $\mathcal{R}$, or just an $\mathcal{R}$-**derivation**, for a formula θ is a sequence $S_1, \ldots, S_t$, where S_t is $\dashv \theta$ and every S_i is either a refutation axiom, or is of the form $\vdash_\mathbf{D} \psi$

or is obtained from some already listed items in the sequence by applying a refutation rule from $\mathcal{R}$, by deriving the conclusion from suitable substitution instances of the premises. We now say that a formula θ is **refutable in** $\mathcal{R}$ (or, just $\mathcal{R}$**-refutable**) iff there is a refutation derivation for θ in $\mathcal{R}$.

Given a logical system L, we say that a refutation system $\mathcal{R}$ is:

- **refutation-sound**, or **Ł-sound, for** L, if *only* non-valid in L-formulae (more generally, logical consequences in L) are $\mathcal{R}$-refutable,
- **refutation-complete**, or **Ł-complete, for** L, if *all* non-valid in L-formulae (more generally, logical consequences in L) are $\mathcal{R}$-refutable.

2.2. Basic Refutation Systems for Classical Logic

The most common type of deductive systems are axiomatic (aka, Hilbert style) systems—respectively, the most common type of refutation systems are axiomatic refutation systems. Here is such a refutation system Ref^{CPC}, for any fixed sound and complete deductive system CPC for the Classical Propositional Logic PL, due to Łukasiewicz:

Refutation axiom: $\dashv \bot$.
Refutation rules:
Reverse Substitution **RS**:
$$\frac{\dashv \sigma(\varphi)}{\dashv \varphi}$$

for any uniform substitution σ.
Modus Tollens **MT**:
$$\frac{\vdash_{\text{CPC}} \varphi \to \psi,\ \dashv \psi}{\dashv \varphi}.$$

Remark 2. *Note that a refutation system can be Ł-complete for more than one logic. Indeed,* Ref^{CPC} *is Ł-complete not only for the* CPC, *but also for both maximal normal modal logic* $\mathbf{K} + \Box\bot$ *and* $\mathbf{K} + (p \leftrightarrow \Box p)$ *(see [3]), provided that the deductive system* CPC *in* $\vdash_{\text{CPC}}$ *in* **MT** *is replaced by one for the respective modal logic.*

Besides those in axiomatic style, some refutation systems have also been constructed for sequent calculi and in natural deduction style. In [22], Tiomkin constructed a sequent-style refutation calculus for FOL without function symbols and with the only logical connectives being $\lor, \neg, \forall$, and sketched a proof of its Ł-completeness for the formulae refutable in finite models. Independently, Goranko developed in [3] an Ł-complete sequent refutation calculi for the full language of PL, also extended there to some important normal modal logic. In [2], Tamminga developed a system of natural deduction for deriving the non-theorems of PL, proved there to be Ł-sound and Ł-complete.

3. Hybrid Derivation Systems: Basic Theory

3.1. Hybrid Deduction–Refutation Rules and Systems

Again, let us consider and fix a logical system L, comprising a formal propositional logical language with a given semantics defining the notion of L-validity and, more generally, logical consequence in L.

For greater generality and for the purposes of Section 4, the basic notions of hybrid deduction–refutation systems will be given here in terms of *sequents of formulae*, readily reducible to single formulae. By a **(single-conclusion) sequent**, we mean an expression of the type $\Gamma \bowtie \theta$, where Γ is a list (treated as a set) of formulae in L, θ is a formula in L, and $\bowtie \in \{\vdash, \dashv\}$. Sequents of the type $\Gamma \vdash \theta$ will be called **deductions**, while those of the type $\Gamma \dashv \theta$ will be called **refutations**. (From a general perspective, both deductions and refutations in our sense are treated syntactically as logical deductions, but we need a more differentiating and unambiguous terminology here.)

Semantically, $\Gamma \vdash \theta$ is meant to claim that the logical consequence $\Gamma \models \theta$ is valid in L, whereas $\Gamma \dashv \theta$ is meant to claim that $\Gamma \models \theta$ is falsifiable, hence not valid in L, i.e. that $\Gamma \not\models \theta$ holds in L. Thus, we say that a sequent $\Gamma \vdash \theta$ is **sound** in L if $\Gamma \models \theta$ is a valid logical consequence in L. Respectively, a sequent $\Gamma \dashv \theta$ is **sound** in L if $\Gamma \models \theta$ is a non-valid logical consequence in L.

Remark 3. *Note an important semantic distinction between deduction and refutation sequents: $\Gamma \models \theta$ is typically monotone by inclusion with respect to Γ, meaning that, if $\Gamma \models \theta$ and $\Gamma \subseteq \Gamma'$, then $\Gamma' \models \theta$; on the other hand, by contraposition, $\Gamma \not\models \theta$ is typically* anti-monotone *with respect to Γ, i.e., if $\Gamma \not\models \theta$ and $\Gamma' \subseteq \Gamma$, then $\Gamma' \not\models \theta$ but generally not vice versa. This will have the practical consequence that all inferences that involve refutation sequents $\Gamma \dashv \theta$ are sensitive, and generally intolerant, to adding extra formulae to Γ. See also Remark 1.*

Now, we will extend the refutation rules to *hybrid deduction–refutation rules*, also by adding premises as contexts, which now becomes essential in view of the remark above. These rules fall in two complementary types, defined below, where $\varphi_1, \ldots, \varphi_m, \psi_1, \ldots, \psi_n, \theta$ are (generally) schemes of formulae and $\Gamma, \Gamma_1, \ldots, \Gamma_m, \Delta, \Delta_1, \ldots, \Delta_n$ are sets of schemes of formulae in L. In the propositional case treated here, one can alternatively assume that all these are concrete formulae, but the rules employ uniform substitutions; see further. Because of the opposite monotonicity properties of the deduction and refutation sequents (see Remark 3), the hybrid rules generally have to employ sequents with different sets of premises.

A **hybrid deduction rule of inference** (based on a given deductive system **D**) is a rule of the type:

$$\text{HDR} \qquad \frac{\Gamma_1 \vdash \varphi_1, \ldots \Gamma_m \vdash \varphi_m, \Delta_1 \dashv \psi_1, \ldots, \Delta_n \dashv \psi_n}{\Gamma \vdash \theta}.$$

A **hybrid refutation rule of inference** (based on a given deductive system **D**) is a rule of the type:

$$\text{HRR} \qquad \frac{\Gamma_1 \vdash \varphi_1, \ldots \Gamma_m \vdash \varphi_m, \Delta_1 \dashv \psi_1, \ldots, \Delta_n \dashv \psi_n}{\Delta \dashv \theta}.$$

The two types of rules above will be called collectively **hybrid rules of inference**. Hybrid rules with no premises will be called respectively **deduction axioms** and **structural refutation axioms**, and we write them simply as sequents $\Gamma \vdash \theta$, respectively $\Delta \dashv \theta$.

These hybrid rules of inference will be regarded as *rule schemes under substitution*, like the earlier defined refutation rules, in the following sense. Every uniform substitution of formulae for the propositional variables occurring in the sequents in the rule creates an instance of the rule. For any such uniform substitution σ if all sequents resulting from applying σ to the premises of the rule, viz. $\sigma(\Gamma_1) \vdash \sigma(\varphi_1), \ldots, \sigma(\Gamma_m) \vdash \sigma(\varphi_m), \sigma(\Delta_1) \dashv \sigma(\psi_1), \ldots, \sigma(\Delta_n) \dashv \sigma(\psi_n)$ are derivable / have been derived, then the rule allows the derivation of the sequent resulting from applying σ to the conclusion, i.e., $\sigma(\Gamma) \vdash \sigma(\theta)$ in the case of **HDR**, resp. $\sigma(\Delta) \dashv \sigma(\theta)$ in the case of **HRR**. The respective semantic interpretation of the hybrid rules above in the case of propositional logical systems can be given as follows: for any uniform substitution σ, if each of the logical consequences $\sigma(\Gamma_1) \models \sigma(\varphi_1), \ldots, \sigma(\Gamma_m) \models \sigma(\varphi_m)$ is derived as valid and each of $\sigma(\Delta_1) \models \sigma(\psi_1), \ldots, \sigma(\Delta_n) \models \sigma(\psi_n)$ has been derived as non-valid, then $\sigma(\Gamma) \models \sigma(\theta)$ is derived as valid in the case of **HDR**, respectively $\sigma(\Delta) \models \sigma(\theta)$ derived as non-valid in the case of **HRR**, as defined above.

In addition (see Remark 1), we also need to allow more general, non-structural **refutation axiom schemes** of the type $\Gamma \dashv \theta$, where closure under substitution is not assumed, but syntactic constraints are imposed on Γ and θ. A simplest example is a scheme $p \dashv q$, where $p \neq q$. Clearly, allowing closure under substitution would produce unsound refutation sequents, such as $p \dashv p$. Of course, structural refutation axioms are special kinds of refutation axiom schemes, but it would be helpful to consider both types separately. Structural refutation axioms and refutation axiom schemes will be called collectively just **refutation axioms**. Note that, to make the general hybrid rules applicable, they must act in combination with some rules with no premises, i.e., deduction and refutation axioms, which provide an initial stock of derived sequents.

Remark 4. *Note that, unlike in standard refutation systems, in hybrid derivation systems, we will no longer assume that these rules act in the context of a separately pre-defined, purely deductive system **D**, which provides the initial stock of derived sequents $\Gamma \vdash \theta$ only, but rather that they define the notions of deduction derivations and refutation derivations on a par, by a mutual induction defined as expected, which I combine in one notion of hybrid derivation, defined further.*

A hybrid inference rule is **sound** for a given logical system L if it respects the intuitive interpretation above, i.e., whenever applied to sound premises in L, it produces a sound conclusion in L.

Here are some examples of hybrid rules:

- All standard deduction rules (in particular, axioms) are particular cases of hybrid deduction rules. In particular, such are all rules of sequent calculi and systems of natural deduction.
- The refutation rules defined in Section 2.1 are particular cases of hybrid refutation rules.
- In addition, suitable meta-properties of the given logical system L can be used to extract and justify specific new hybrid inference rules for it. An important example is the **Deductive consistency rule**

$$(\text{Cons}) \quad \frac{\vdash \varphi}{\dashv \neg \varphi},$$

which is justified ('sound') whenever the underlying deductive relation $\vdash$ is sound (hence consistent) for L. More generic examples will be given further.

A **hybrid deduction–refutation system**, or (for shorter) a **hybrid derivation system**, is a set $\mathcal{H}$ of hybrid rules of inference for a given logical language. A **hybrid derivation in** $\mathcal{H}$, or just an $\mathcal{H}$-**derivation**, for a sequent $\Gamma \bowtie \theta$ is a sequence of sequents $S_1, ..., S_t$, where S_t is $\Gamma \bowtie \theta$ and every S_i is either a deduction axiom or a refutation axiom, or is obtained from some already listed sequents in the sequence by applying a hybrid rule of inference from $\mathcal{H}$. Then, we say that the sequent $\Gamma \bowtie \theta$ is **derivable in** $\mathcal{H}$. Furthermore, we say that the logical consequence $\Gamma \models \theta$ is **deduced/deducible in** $\mathcal{H}$ if $\Gamma \vdash \theta$ is derivable in $\mathcal{H}$, and that $\Gamma \models \theta$ is **refuted/refutable in** $\mathcal{H}$ if $\Gamma \dashv \theta$ is derivable in $\mathcal{H}$.

In particular, $\mathcal{H}$ may contain all axioms and rules of a given traditional deduction system **D** (which can be an axiomatic system, a sequent calculus, a system of natural deduction, or a system of semantic tableaux). In such case, the derivations in $\mathcal{H}$ extend those in **D**, by enabling not only derivations of refutations based on **D**, but also possibly of some deductions not derivable in **D** (esp. in case **D** is incomplete).

Remark 5. *Note that hybrid derivation systems do not employ separate rules of uniform substitution, even to derived sequents $\Gamma \vdash \theta$ because of the non-preservation of refutations (that may have been used in the derivation) under such substitutions. (A similar remark is made in [14].) Still, uniform substitutions are used here for generating instances of the inference rules, as explained earlier..*

Some basic terminology will be needed in what follows. Recall (see footnote 1) the assumption that "not valid" and "non-valid" means "falsifiable" (but see also Remark 6). Given a logical system L, we say that a hybrid derivation system $\mathcal{H}$ is:

- **deductively sound for** L, or **D-sound for** L, if only logical consequences that are valid in L are $\mathcal{H}$-deducible.
- **refutationally sound for** L, or **R-sound for** L, if only logical consequences that are non-valid in L are $\mathcal{H}$-refutable.

- **Ł-sound for** L, if it is both D-sound and R-sound for L.
- **Ł-consistent**, if there is no Γ and θ such that both $\Gamma \vdash \theta$ and $\Gamma \dashv \theta$ are derivable in $\mathcal{H}$.
- **deductively complete for** L, or **D-complete for** L, if all logical consequences that are valid in L are $\mathcal{H}$-deducible.
- **refutationally complete for** L, or **R-complete for** L, if all logical consequences that are non-valid in L are $\mathcal{H}$-refutable.
- **Łukasiewicz-complete for** L, or **Ł-complete for** L, if it is both D-complete and R-complete for L.
- **Ł-saturated**, if for all Γ and θ, either $\Gamma \vdash \theta$ or $\Gamma \dashv \theta$ (possibly both) is derivable in $\mathcal{H}$.
- **Ł-adequate for** L, if it is both Ł-sound and Ł-complete for L.
- **Ł-balanced**, if it is both Ł-consistent and Ł-saturated.

Proposition 1. *Let L be a logical system and $\mathcal{H}$ a hybrid deduction–refutation system for L. Then:*

1. *If $\mathcal{H}$ is Ł-sound for L, then $\mathcal{H}$ is Ł-consistent.*
2. *If $\mathcal{H}$ is Ł-complete for L, then $\mathcal{H}$ is Ł-saturated.*
3. *If $\mathcal{H}$ is Ł-adequate for L, then $\mathcal{H}$ is Ł-balanced.*
4. *If $\mathcal{H}$ has a recursive set of rules and is Ł-adequate for L, then it provides a decision procedure for the valid logical consequences in L.*

Proof. Here, 1 and 2 are straightforward, since, for any Γ and θ, the logical consequence $\Gamma \models \theta$ is either valid or non-valid, but not both. Then, 3 follows immediately.

Likewise, 4 is immediate, as the recursiveness of $\mathcal{H}$ implies that all derived sequents in $\mathcal{H}$ can be recursively enumerated, the Ł-completeness of $\mathcal{H}$ means that, for every Γ and ϕ, either $\Gamma \vdash \phi$ or $\Gamma \dashv \phi$ (but not both) will eventually appear in that enumeration, and the Ł-soundness guarantees that, whatever the case is, it will correctly imply validity, resp. non-validity, of $\Gamma \models \phi$. □

Remark 6. *All notions defined above are meant to apply, in particular, to most general cases of derivation systems, which may possibly extend unsound, or even to paraconsistent deductive systems. (In a similar spirit, Citkin defines in [14] a more general motion of a logical system, as a pair consisting of a set of accepted and a set of rejected propositions, without assuming that these must be complementary, nor even disjoint.) However, in the case of paraconsistent semantics where a formula or logical consequence can be both valid and falsifiable, the term "non-valid" in the definitions of **R-soundness** and **R-completeness** should be replaced by "falsifiable", without assuming that the latter implies the former. Still, claims 1 and 3 in Proposition 1 will no longer hold for such semantics. (Thanks to the reviewer who pointed that out.) Still, note that, even if the deduction fragment of a hybrid derivation system may be D-unsound, or D-incomplete, for the given logical system, its refutation fragment may still be R-sound, or R-complete, and vice versa. An interesting example is the simple Ł-complete refutation system for Medvedev's logic of finite problems (for which no recursive axiomatization is known yet, but it has a co-r.e. set of validities) designed in [23], employing as the underlying deductive system the weaker Kreisel–Putnam's logic KP. Thus, the resulting hybrid system is D-incomplete but R-complete for Medvedev's logic.*

3.2. Inversion of Rules and Derivative Hybrid Rules

New hybrid rules can be defined in a uniform way as **derivative rules** from existing ones by using **inversion**: swapping one premise with the conclusion of the given rule and swapping $\vdash$ with $\dashv$ in both sequents. (The use of the term 'inversion' here is different from 'inversion principle' widely used in proof theory, see [5], but related to the term 'inversion' used in [2], when applied to single-premise rules. In addition, the idea of inverting inference rules was essentially used in the design and proof of completeness of the sequential refutation system for PL in [3].) For example, applying inversion to the rule Modus Ponens

$$\frac{\Gamma \vdash \phi, \; \Gamma \vdash \phi \to \psi}{\Gamma \vdash \psi}$$

produces the following derivative rules:

$$\frac{\Gamma \dashv \psi, \Gamma \vdash \phi \to \psi}{\Gamma \dashv \phi} \quad \text{and} \quad \frac{\Gamma \vdash \phi, \Gamma \dashv \psi}{\Gamma \dashv \phi \to \psi}.$$

Likewise, the Disjunction rule:

$$\frac{\Gamma \dashv \varphi, \Gamma \dashv \psi}{\Gamma \dashv \varphi \vee \psi}$$

produces the following derivative rules:

$$\frac{\Gamma \vdash \varphi \vee \psi, \Gamma \dashv \psi}{\Gamma \vdash \varphi} \quad \text{and} \quad \frac{\Gamma \dashv \varphi, \Gamma \vdash \varphi \vee \psi}{\Gamma \vdash \psi}.$$

The general definitions follow.

3.2.1. Inversion of Deduction Rules

The deduction rule

$$\frac{\Gamma_1 \vdash \varphi_1, \ldots, \Gamma_i \vdash \varphi_i, \ldots, \Gamma_m \vdash \varphi_m, \Delta_1 \dashv \psi_1, \ldots, \Delta_j \dashv \psi_j, \ldots, \Delta_n \dashv \psi_n}{\Gamma \vdash \theta}$$

produces each of the following derivative rules

$$\frac{\Gamma_1 \vdash \varphi_1, \ldots, \Gamma \dashv \theta, \ldots, \Gamma_m \vdash \varphi_m, \Delta_1 \dashv \psi_1, \ldots, \Delta_j \dashv \psi_j, \ldots, \Delta_n \dashv \psi_n}{\Gamma_i \dashv \varphi_i}$$

for each $i = 1, \ldots, m$, and

$$\frac{\Gamma_1 \vdash \varphi_1, \ldots, \Gamma_i \vdash \varphi_i, \ldots, \Gamma_m \vdash \varphi_m, \Delta_1 \dashv \psi_1, \ldots, \Gamma \dashv \theta, \ldots, \Delta_n \dashv \psi_n}{\Delta_j \vdash \psi_j}$$

for each $j = 1, \ldots, n$.

In particular, a deduction rule with no premises, i.e., a deduction axiom $\Gamma \vdash \theta$, will be regarded—without essential effect—as the rule

$$\frac{\vdash \top}{\Gamma \vdash \theta}.$$

Thus, it has one derivative rule:

$$\frac{\Gamma \dashv \theta}{\dashv \top}.$$

3.2.2. Inversion of Refutation Rules

Likewise, the refutation rule

$$\frac{\Gamma_1 \vdash \varphi_1, \ldots, \Gamma_i \vdash \varphi_i, \ldots, \Gamma_m \vdash \varphi_m, \Delta_1 \dashv \psi_1, \ldots, \Delta_j \dashv \psi_j, \ldots, \Delta_n \dashv \psi_n}{\Delta \dashv \theta}$$

produces each of the following derivative rules

$$\frac{\Gamma_1 \vdash \varphi_1, \ldots, \Delta \vdash \theta, \ldots, \Gamma_m \vdash \varphi_m, \Delta_1 \dashv \psi_1, \ldots, \Delta_j \dashv \psi_j, \ldots, \Delta_n \dashv \psi_n}{\Gamma_i \dashv \varphi_i}$$

for each $i = 1, \ldots, m$, and

$$\frac{\Gamma_1 \vdash \varphi_1, \ldots, \Gamma_i \vdash \varphi_i, \ldots, \Gamma_m \vdash \varphi_m, \Delta_1 \dashv \psi_1, \ldots, \Delta \vdash \theta, \ldots, \Delta_n \dashv \psi_n}{\Delta_j \vdash \psi_j}$$

for each $j = 1, ..., n$.

A refutation rule with no premises, i.e., a structural refutation axiom $\Gamma \dashv \theta$, will be regarded—again without essential effect—as the rule

$$\frac{\dashv \bot}{\Gamma \dashv \theta}.$$

Respectively, it also has one derivative rule

$$\frac{\Gamma \vdash \theta}{\vdash \bot}.$$

3.2.3. Soundness of Derivative Rules

Proposition 2. *Let L be a logical system and let R be a hybrid inference rule in the language of L, which is sound for L. Then, every derivative rule of R is sound for L too.*

Proof. Suppose first that R is a hybrid deduction rule

$$\frac{\Gamma_1 \vdash \varphi_1, \ldots, \Gamma_i \vdash \varphi_i, \ldots, \Gamma_m \vdash \varphi_m, \Delta_1 \dashv \psi_1, \ldots, \Delta_j \dashv \psi_j, \ldots, \Delta_n \dashv \psi_n}{\Gamma \vdash \theta}.$$

Consider the derivative refutation rule

$$\frac{\Gamma_1 \vdash \varphi_1, \ldots, \Gamma \dashv \theta, \ldots, \Gamma_m \vdash \varphi_m, \Delta_1 \dashv \psi_1, \ldots, \Delta_j \dashv \psi_j, \ldots, \Delta_n \dashv \psi_n}{\Gamma_i \dashv \varphi_i}$$

for $i \in \{1, ..., m\}$. To prove its soundness, consider any uniform substitution σ and suppose that all premises obtained after applying σ are sound, i.e., each of the logical consequences $\sigma(\Gamma_k) \models \sigma(\varphi_k)$, for $k = 1, ..., i-1, i+1, ..., m$, is valid and each of $\sigma(\Delta_k) \models \sigma(\psi_k)$, for $k = 1, ..., n$, as well as $\sigma(\Delta) \models \sigma(\theta)$, is non-valid. Then, $\sigma(\Gamma_i) \models \sigma(\varphi_i)$ must be non-valid too; otherwise, the soundness of R would imply the validity of $\sigma(\Delta) \models \sigma(\theta)$.

The argument for the soundness of derivative deduction rules is similar.

The proof when R is a hybrid refutation rule is completely analogous.
□

3.3. Canonical Hybrid Extensions of Deductive Systems

Given any deductive system **D**, its **canonical hybrid extension** $\mathcal{H}(\mathbf{D})$ is obtained by adding to **D** the derivative rules of all deduction rules (incl. axioms) of **D**.

Note that the sequent refutation systems proposed for PL and FOL in [3,22] are essentially constructed as (subsystems of) the canonical hybrid extensions of respective standard sequent deduction systems for these logic.

Proposition 2 implies that, if **D** is D-sound for a given logical system L, then $\mathcal{H}(\mathbf{D})$ is Ł-sound for L. If **D** is also D-complete for L, then $\mathcal{H}(\mathbf{D})$ cannot add more derivable deduction sequents, so it is D-complete too. In this case, $\mathcal{H}(\mathbf{D})$ extends **D** conservatively with respect to deductions, but it generally does add derivable refutation sequents. However, even then $\mathcal{H}(\mathbf{D})$ may generally not be R-complete, hence not Ł-complete, either. In particular, it *cannot* be R-complete if L is not decidable. The question of when $\mathcal{H}(\mathbf{D})$ is Ł-complete is one of the main questions of the general theory of hybrid derivation systems.

Likewise, given any refutation system $\mathcal{R}$, its **canonical hybrid extension** $\mathcal{H}(\mathcal{R})$ is obtained by adding to $\mathcal{R}$ the derivative rules of all refutation rules (incl. axioms) of $\mathcal{R}$. Again, by Proposition 2, if $\mathcal{R}$ is R-sound for a logical system L, then $\mathcal{H}(\mathcal{R})$ is Ł-sound for L. The question of Ł-completeness of $\mathcal{H}(\mathcal{R})$ is, again, generally open.

As for Ł-soundness, using Proposition 2, a straightforward induction on derivations proves the following.

Corollary 1. *Let **D** be a sound deductive system for a given logical system L. Then, $\mathcal{H}(D)$ is Ł-sound for L.*

4. Hybrid Extensions of the System of Natural Deduction for PL

I will illustrate here the concept of canonical hybrid extension, applied to the system of Natural Deduction for the classical propositional logic PL.

4.1. Hybrid Derivatives of the Rules for Natural Deduction for PL

Let us fix a standard version $\mathbf{ND}^{\mathsf{PL}}$ of a sound and complete system of Natural Deduction (ND) for PL (see [6], or [7], or [18]).

Every pure inference rule of $\mathbf{ND}^{\mathsf{PL}}$ produces one or two derivative hybrid rules. Note that the derivatives of introduction rules for $\vdash$ typically become hybrid elimination rules for $\dashv$ and vice versa.

Note also that the open assumptions must be explicitly listed in the rules because of the anti-monotonicity of the refutations (see Remark 3). For that reason and for better readability, the rules are presented further as rules over sequents.

4.2. Hybrid Derivatives of the Introduction Rules of $\mathbf{ND}^{\mathsf{PL}}$

For the record, here are the derivative rules produced from the introduction rules of $\mathbf{ND}^{\mathsf{PL}}$, where the arrows $\Rightarrow$ below indicate the respective transformations of deduction rules to their derivative hybrid rules:

$$(\wedge I) \frac{\Gamma \vdash \phi, \; \Gamma \vdash \psi}{\Gamma \vdash \phi \wedge \psi}$$

$$\Downarrow \qquad \Downarrow$$

$$(\wedge HE^l) \frac{\Gamma \dashv \phi \wedge \psi, \; \Gamma \vdash \psi}{\Gamma \dashv \phi} \quad (\wedge HE^r) \frac{\Gamma \vdash \phi, \; \Gamma \dashv \phi \wedge \psi}{\Gamma \dashv \psi},$$

$$(\vee I^l) \frac{\Gamma \vdash \phi}{\Gamma \vdash \phi \vee \psi} \qquad (\vee I^r) \frac{\Gamma \vdash \psi}{\Gamma \vdash \phi \vee \psi}$$

$$\Downarrow \qquad \Downarrow$$

$$(\vee HE^l) \frac{\Gamma \dashv \phi \vee \psi}{\Gamma \dashv \phi} \qquad (\vee HE^r) \frac{\Gamma \dashv \phi \vee \psi}{\Gamma \dashv \psi},$$

$$(\to I) \frac{\Gamma, \phi \vdash \psi}{\Gamma \vdash \phi \to \psi} \qquad (\neg I) \frac{\Gamma, \phi \vdash \bot}{\Gamma \vdash \neg \phi}$$

$$\Downarrow \qquad \Downarrow$$

$$(\to HE^1) \frac{\Gamma \dashv \phi \to \psi}{\Gamma, \phi \dashv \psi} \qquad (\neg HE^1) \frac{\Gamma \dashv \neg \phi}{\Gamma, \phi \dashv \bot}.$$

4.3. Hybrid Derivatives of the Elimination Rules of $\mathbf{ND}^{\mathsf{PL}}$

Here are the derivative rules produced from the elimination rules of $\mathbf{ND}^{\mathsf{PL}}$, where, again, the arrows $\Rightarrow$ below indicate the respective transformations of deduction rules to their derivative hybrid rules:

$$(\wedge E^l) \frac{\Gamma \vdash \phi \wedge \psi}{\Gamma \vdash \phi} \qquad (\wedge E^r) \frac{\Gamma \vdash \phi \wedge \psi}{\Gamma \vdash \psi}$$

$$\Downarrow \qquad \Downarrow$$

$$(\wedge\text{HI}^l) \frac{\Gamma \dashv \phi}{\Gamma \dashv \phi \wedge \psi} \qquad (\wedge\text{HI}^r) \frac{\Gamma \dashv \psi}{\Gamma \dashv \phi \wedge \psi},$$

$$(\vee\text{E}) \frac{\Gamma, \phi \vdash \theta, \ \Gamma, \psi \vdash \theta}{\Gamma, \phi \vee \psi \vdash \theta}$$

$$\Downarrow \qquad\qquad \Downarrow$$

$$(\vee\text{HI}^l) \frac{\Gamma, \phi \vee \psi \dashv \theta, \ \Gamma, \psi \vdash \theta}{\Gamma, \phi \dashv \theta}, \qquad (\vee\text{HI}^r) \frac{\Gamma, \phi \vdash \theta, \ \Gamma, \phi \vee \psi \dashv \theta}{\Gamma, \psi \dashv \theta},$$

$$(\to \text{E}) \frac{\Gamma \vdash \phi, \ \Gamma \vdash \phi \to \psi}{\Gamma \vdash \psi}$$

$$\Downarrow \qquad\qquad \Downarrow$$

$$(\to \text{HE}^2) \frac{\Gamma \dashv \psi, \ \Gamma \vdash \phi \to \psi}{\Gamma \dashv \phi}, \qquad (\to \text{HI}) \frac{\Gamma \vdash \phi, \ \Gamma \dashv \psi}{\Gamma \dashv \phi \to \psi},$$

$$(\neg\text{E}) \frac{\Gamma \vdash \phi, \ \Gamma \vdash \neg\phi}{\Gamma \vdash \bot}$$

$$\Downarrow \qquad\qquad \Downarrow$$

$$(\neg\text{HE}^2) \frac{\Gamma \dashv \bot, \ \Gamma \vdash \neg\phi}{\Gamma \dashv \phi}, \qquad (\neg\text{HI}) \frac{\Gamma \vdash \phi, \ \Gamma \dashv \bot}{\Gamma \dashv \neg\phi}.$$

4.4. Hybrid Derivatives of "Ex Falso" and "Reductio ad Absurdum"

The hybrid derivative of "Ex falso quodlibet" is produced as follows:

$$(\text{EFQ}) \frac{\Gamma \vdash \bot}{\Gamma \vdash \phi}$$

$$\Downarrow$$

$$(\text{HEQF}) \frac{\Gamma \dashv \phi}{\Gamma \dashv \bot}.$$

Respectively, here is the hybrid derivative of "Reductio ad absurdum":

$$(\text{RAA}) \frac{\Gamma, \neg\phi \vdash \bot}{\Gamma \vdash \phi}$$

$$\Downarrow$$

$$(\text{HRAA}) \frac{\Gamma \dashv \phi}{\Gamma, \neg\phi \dashv \bot}.$$

Note that this refutation rule is sound for PL, but not for the inuitionistic logic.

Remark 7. *Rumfitt considers in [17] (thanks to an anonymous reviewer for this reference) "reversals" of the rules of* **ND**^{PL} *to formalise derivations between "signed sentences" +A and −A used "to abbreviate Smiley's amalgams of questions with answers 'Is it the case that A? Yes' and 'Is it the case that A? No' " (ibid.). While the motivation is different from the one coming from refutation inference rules, most (but not all) resulting rules are essentially the same as the hybrid derivative rules for* **ND** *obtained here. However, there is an essential distinction between the meanings of the two types of rules, e.g.,: whereas rejection of a sentence implies acceptance of its negation, and deductive refutation of the validity of a sentence does not imply deduction of the validity of its negation. That distinction is manifested e.g., by the rules +¬I and −¬E in [17] as compared to the hybrid derivative rules for ¬ obtained and employed here.*

4.5. Atomic Refutations and Monotonicity Rules

The canonical extension $\mathcal{H}(\mathbf{ND}^{\mathsf{PL}})$ constructed above is easily seen to be too weak for deriving refutations, as it does not contain any refutation axioms nor hybrid refutation rules that only have deduction sequents as premises; hence, it cannot enable derivation of any refutation sequents yet. In order to compensate for that, we also need to add the following atomic **refutation axiom scheme** RefAx^{PL}:

$$\Gamma \dashv \phi,$$

where ϕ is a literal or $\bot$, all formulae in Γ are literals, Γ does not contain a complementary pair of literals, and $\phi \notin \Gamma$. Note that RefAx^{PL} is a non-structural refutation axiom scheme, i.e., not closed under uniform substitution.

In addition, the rules of $\mathcal{H}(\mathbf{ND}^{\mathsf{PL}})$ do not enable explicitly removing formulae from the left-hand side of a refutation sequent. To solve that deficiency and to streamline the hybrid derivation system, we also add the following two monotonicity rules:

- The rule Mon$^{\vdash}$: **Monotonicity of** $\vdash$

$$\frac{\Gamma \vdash \phi,\ \Gamma \subseteq \Gamma'}{\Gamma' \vdash \phi},$$

(Usually this rule is implicitly assumed in any traditional system of natural deduction.)

- The rule Mon$^{\dashv}$: **Anti-monotonicity of** $\dashv$

$$\frac{\Gamma \dashv \phi,\ \Gamma' \subseteq \Gamma}{\Gamma' \dashv \phi}.$$

Let us denote by $\mathcal{H}^s(\mathbf{ND}^{\mathsf{PL}})$ the extension of $\mathcal{H}(\mathbf{ND}^{\mathsf{PL}})$ obtained by adding the rules RefAx^{PL}, Mon$^{\vdash}$, and Mon$^{\dashv}$. The system $\mathcal{H}^s(\mathbf{ND}^{\mathsf{PL}})$ will be called the **standard hybrid extension of** $\mathbf{ND}^{\mathsf{PL}}$.

5. Some Results about the Standard Hybrid Extension of ND^{PL}

5.1. Soundness and Some Properties of $\mathcal{H}^s(\mathbf{ND}^{\mathsf{PL}})$

Proposition 3.

1. *Every rule of $\mathcal{H}^s(\mathbf{ND}^{\mathsf{PL}})$ is sound.*
2. *$\mathcal{H}^s(\mathbf{ND}^{\mathsf{PL}})$ is Ł-sound for PL and hence Ł-consistent.*
3. *If Γ is a satisfiable set of formulae, then $\Gamma \vdash \bot$ is not derivable in $\mathcal{H}^s(\mathbf{ND}^{\mathsf{PL}})$.*

Proof. 1. The soundness of all derivative rules for PL follows from the D-soundness of **ND** for PL and Proposition 2. Proving the soundness of RefAx^{PL}, Mon$^{\vdash}$, and Mon$^{\dashv}$ for PL is quite routine, and I leave out the details.

2. Now, the Ł-soundness of $\mathcal{H}^s(\mathbf{ND}^{\mathsf{PL}})$ for PL follows by a straightforward induction on hybrid derivations (Corollary 1). In particular, $\mathcal{H}^s(\mathbf{ND}^{\mathsf{PL}})$ extends conservatively $\mathbf{ND}^{\mathsf{PL}}$ with respect to deduction sequents.

3. Follows immediately from 2.
□

Lemma 1.

1. *If $\Gamma \dashv \phi$ is derivable in $\mathcal{H}^s(\mathbf{ND}^{\mathsf{PL}})$, then $\Gamma, \neg\phi \dashv \phi$ is derivable in $\mathcal{H}^s(\mathbf{ND}^{\mathsf{PL}})$.*
2. *If $\Gamma, \phi \dashv \psi$ is derivable in $\mathcal{H}^s(\mathbf{ND}^{\mathsf{PL}})$, then $\Gamma \dashv \phi \to \psi$ is derivable in $\mathcal{H}^s(\mathbf{ND}^{\mathsf{PL}})$.*

3. If $\Gamma \vdash \phi \to \psi$ is derivable in $\mathcal{H}^s(\mathbf{ND}^{\mathbf{PL}})$ and $\Gamma, \phi \dashv \theta$ is derivable in $\mathcal{H}^s(\mathbf{ND}^{\mathbf{PL}})$, then $\Gamma, \psi \dashv \theta$ is derivable in $\mathcal{H}^s(\mathbf{ND}^{\mathbf{PL}})$.

   Consequently, if $\Gamma \vdash \phi \leftrightarrow \psi$ is derivable in $\mathcal{H}^s(\mathbf{ND}^{\mathbf{PL}})$, then $\Gamma, \phi \dashv \theta$ is derivable in $\mathcal{H}^s(\mathbf{ND}^{\mathbf{PL}})$ iff $\Gamma, \psi \dashv \theta$ is derivable in $\mathcal{H}^s(\mathbf{ND}^{\mathbf{PL}})$.

4. If $\Gamma \vdash \phi \leftrightarrow \psi$ is derivable in $\mathcal{H}^s(\mathbf{ND}^{\mathbf{PL}})$, then $\Gamma, \theta \dashv \phi$ is derivable in $\mathcal{H}^s(\mathbf{ND}^{\mathbf{PL}})$ iff $\Gamma, \theta \dashv \psi$ is derivable in $\mathcal{H}^s(\mathbf{ND}^{\mathbf{PL}})$.

5. $\Gamma, \psi_1, ..., \psi_k \dashv \theta$ is derivable in $\mathcal{H}^s(\mathbf{ND}^{\mathbf{PL}})$ iff $\Gamma, \psi_1 \wedge ... \wedge \psi_k \dashv \theta$ is derivable in $\mathcal{H}^s(\mathbf{ND}^{\mathbf{PL}})$.

Proof.

1. Let $\Gamma \dashv \phi$ be derived in $\mathcal{H}^s(\mathbf{ND}^{\mathbf{PL}})$.

   Then, $\Gamma, \neg \phi \dashv \bot$ is derived in $\mathcal{H}^s(\mathbf{ND}^{\mathbf{PL}})$, by (HRAA).

   Hence, $\Gamma, \neg \phi \dashv \phi$ is derived in $\mathcal{H}^s(\mathbf{ND}^{\mathbf{PL}})$, by ($\neg HE^2$).

2. Suppose $\Gamma, \phi \dashv \psi$ is derivable in $\mathcal{H}^s(\mathbf{ND}^{\mathbf{PL}})$. Since $\Gamma, \phi \vdash \phi$ is derivable in $\mathcal{H}^s(\mathbf{ND}^{\mathbf{PL}})$, we derive $\Gamma, \phi \dashv \phi \to \psi$ by ($\to$ HI). Then, by the Anti-Monotonicity rule $\text{Mon}^{\dashv}$, $\Gamma \dashv \phi \to \psi$ is derived in $\mathcal{H}^s(\mathbf{ND}^{\mathbf{PL}})$.

3. Let $\Gamma \vdash \phi \to \psi$ be derivable in $\mathcal{H}^s(\mathbf{ND}^{\mathbf{PL}})$.

   Since $(\phi \to \psi) \to ((\psi \to \theta) \to (\phi \to \theta))$ is a classical tautology, $\Gamma \vdash (\phi \to \psi) \to ((\psi \to \theta) \to (\phi \to \theta))$ is derivable in $\mathcal{H}^s(\mathbf{ND}^{\mathbf{PL}})$.

   Hence, by Modus Ponens, $\Gamma \vdash (\psi \to \theta) \to (\phi \to \theta)$ is derivable in $\mathcal{H}^s(\mathbf{ND}^{\mathbf{PL}})$. (*)

   Now, suppose that $\Gamma, \phi \dashv \theta$ is derivable in $\mathcal{H}^s(\mathbf{ND}^{\mathbf{PL}})$.

   Then, by item 2, $\Gamma \dashv \phi \to \theta$ is derivable in $\mathcal{H}^s(\mathbf{ND}^{\mathbf{PL}})$.

   Therefore, $\Gamma \dashv \psi \to \theta$ is derivable in $\mathcal{H}^s(\mathbf{ND}^{\mathbf{PL}})$ by ($\to HE^2$) applied to the latter and (*). Then, finally, $\Gamma, \psi \dashv \theta$ is derivable in $\mathcal{H}^s(\mathbf{ND}^{\mathbf{PL}})$, by ($\to HE^1$).

4. Let $\Gamma \vdash \phi \leftrightarrow \psi$ be derivable in $\mathcal{H}^s(\mathbf{ND}^{\mathbf{PL}})$.

   Suppose that $\Gamma, \theta \dashv \phi$ is derivable in $\mathcal{H}^s(\mathbf{ND}^{\mathbf{PL}})$.

   Then, $\Gamma, \dashv \theta \to \phi$ is derivable in $\mathcal{H}^s(\mathbf{ND}^{\mathbf{PL}})$, by claim 2. (**)

   Since $(\phi \leftrightarrow \psi) \to ((\theta \to \psi) \to (\theta \to \phi))$ is a classical tautology, $\Gamma \vdash (\phi \leftrightarrow \psi) \to ((\theta \to \psi) \to (\theta \to \phi))$ is derivable in $\mathcal{H}^s(\mathbf{ND}^{\mathbf{PL}})$.

   Therefore, $\Gamma \vdash (\theta \to \psi) \to (\theta \to \phi)$ is derivable in $\mathcal{H}^s(\mathbf{ND}^{\mathbf{PL}})$.

   Hence, $\Gamma, \dashv \theta \to \psi$ is derivable in $\mathcal{H}^s(\mathbf{ND}^{\mathbf{PL}})$, by ($\to HE^2$) applied to the latter and (**).

   Then, finally, $\Gamma, \theta \dashv \psi$ is derivable in $\mathcal{H}^s(\mathbf{ND}^{\mathbf{PL}})$, by ($\to HE^1$).

5. It suffices to prove the claim when $k = 2$ and then apply a straightforward induction.

   Suppose $\Gamma, \psi_1, \psi_2 \dashv \theta$ is derivable in $\mathcal{H}^s(\mathbf{ND}^{\mathbf{PL}})$.

   Then, $\Gamma \dashv (\psi_1 \to (\psi_2 \to \theta))$ is derivable in $\mathcal{H}^s(\mathbf{ND}^{\mathbf{PL}})$, by applying claim 2 twice.

   Since $(\psi_1 \to (\psi_2 \to \theta)) \leftrightarrow ((\psi_1 \wedge \psi_2) \to \theta)$ is a classical tautology, $\Gamma \dashv (\psi_1 \wedge \psi_2) \to \theta$ is derivable in $\mathcal{H}^s(\mathbf{ND}^{\mathbf{PL}})$, by claim 4.

   Then, finally, $\Gamma, \psi_1 \wedge \psi_2 \dashv \theta$ is derivable in $\mathcal{H}^s(\mathbf{ND}^{\mathbf{PL}})$, by ($\to HE^1$).

   The converse direction is similar.

□

Given a truth assignment $\delta : \text{Prop} \to \{f, t\}$, for any propositional variable $p \in \text{Prop}$, let us define $p^\delta := p$ if $\delta(p) = t$, else $p^\delta := \neg p$.

Lemma 2. *Let Γ be a finite set of propositional formulae and let $\{p_1, ..., p_n\}$ contain all propositional variables occurring in formulae in Γ. Suppose δ is a truth assignment satisfying Γ and let $\Gamma^\delta = \Gamma \cup \{p_1^\delta, ..., p_n^\delta\}$. Then, $\Gamma^\delta \dashv \bot$ is derivable in $\mathcal{H}^s(\mathbf{ND}^{\mathsf{PL}})$.*

Proof. By items 3 and 5 of Lemma 1, it suffices to prove the claim assuming that all formulae in Γ are transformed to equivalent ones in CNF and then replaced by the list of elementary disjunctions occurring as conjuncts in that CNF. Thus, without loss of generality, we can assume that $\Gamma = \{\gamma_1, ..., \gamma_k\}$, where all γ_i are elementary disjunctions.

Take the satisfying assignment δ. By definition, δ also satisfies all literals in $\{p_1^\delta, ..., p_n^\delta\}$. Furthermore, $p_1^\delta, ..., p_n^\delta \dashv \bot$ is an atomic refutation axiom, hence derivable in $\mathcal{H}^s(\mathbf{ND}^{\mathsf{PL}})$.

Now, select from each γ_i in Γ a literal disjunct α_i that is satisfied by δ. Then, α_i must be in $\{p_1^\delta, ..., p_n^\delta\}$. Hence, $\{p_1^\delta, ..., p_n^\delta, \alpha_1, ..., \alpha_n\} = \{p_1^\delta, ..., p_n^\delta\}$.

Therefore, $p_1^\delta, ..., p_n^\delta, \alpha_1, ..., \alpha_n \dashv \bot$ is an atomic refutation axiom, hence derivable in $\mathcal{H}^s(\mathbf{ND}^{\mathsf{PL}})$. (*)

In addition, $\vdash \alpha_i \to \gamma_i$ is derivable in $\mathcal{H}^s(\mathbf{ND}^{\mathsf{PL}})$, for each $i = 1, ..., n$. Therefore, by applying repeatedly item 3 of Lemma 1, we can replace successively each α_i by γ_i in (*), thereby eventually proving the claim. □

By Anti-Monotonicity of $\dashv$, Lemma 2 immediately implies the following.

Corollary 2. *Let Γ be a finite satisfiable set of propositional formulae. Then, $\Gamma \dashv \bot$ is derivable in $\mathcal{H}^s(\mathbf{ND}^{\mathsf{PL}})$.*

5.2. Ł-Completeness and Ł-Adequacy of $\mathcal{H}^s(\mathbf{ND}^{\mathsf{PL}})$

Theorem 1. *The hybrid derivation system $\mathcal{H}^s(\mathbf{ND}^{\mathsf{PL}})$ is Ł-complete for the classical propositional logic* PL.

Proof. Due to the deductive completeness of $\mathbf{ND}^{\mathsf{PL}}$, of which $\mathcal{H}^s(\mathbf{ND}^{\mathsf{PL}})$ is a deductively conservative extension, it suffices to prove the R-completeness of $\mathcal{H}^s(\mathbf{ND}^{\mathsf{PL}})$, i.e., that the refutation of every non-valid in PL sequent is derivable there. Let $\Gamma \not\models \theta$. Then, there is a truth assignment δ satisfying Γ and falsifying θ. Therefore, δ satisfies $\Gamma \cup \{\neg\theta\}$. By Corollary 2, it follows that $\Gamma, \neg\theta \dashv \bot$ is derivable in $\mathcal{H}^s(\mathbf{ND}^{\mathsf{PL}})$. Then, by Rule ($\neg\mathsf{HE}^2$), $\Gamma, \neg\theta \dashv \theta$ is derivable in $\mathcal{H}^s(\mathbf{ND}^{\mathsf{PL}})$. Finally, by the Anti-Monotonicity Rule Mon$^\dashv$, we obtain that $\Gamma \dashv \theta$ is derivable in $\mathcal{H}^s(\mathbf{ND}^{\mathsf{PL}})$. QED. □

Proposition 3 and Theorem 1 together imply the following.

Corollary 3. *The hybrid derivation system $\mathcal{H}^s(\mathbf{ND}^{\mathsf{PL}})$ is Ł-adequate for* PL *and, therefore, it provides a syntactic decision procedure for* PL.

The system $\mathcal{H}^s(\mathbf{ND}^{\mathsf{PL}})$ and the ND-style refutation system developed in [2] are equivalent in terms of formal refutability, by virtue of the respective Ł-soundness and Ł-completeness results. Still, they are fairly different in style and it would be instructive to compare their proof-theoretic features, strengths and weaknesses, for the sake of possibly designing a better structured system of practical derivations based on $\mathcal{H}^s(\mathbf{ND}^{\mathsf{PL}})$.

Remark 8. *Note that only some of the derived hybrid refutation rules were used in the proofs of Ł-soundness and Ł-completeness, hence the others must be derivable, or at least admissible, in the reduction of $\mathcal{H}^s(\mathbf{ND}^{\mathsf{PL}})$ obtained by removing them. I leave the question of identifying a minimal Ł-complete subsystem of $\mathcal{H}^s(\mathbf{ND}^{\mathsf{PL}})$ to future investigation. In particular, however, the rule HRAA is used in the proof of Lemma 1, hence that proof is not applicable to the system $\mathcal{H}^s(\mathbf{ND}^{\mathsf{PL}})$ of Natural Deduction for the intuitionistic propositional logic IPL. Of course, it should not be applicable for IPL, e.g., because the refutation axiom $\dashv (p \vee \neg p)$ ought to be derivable there, while it is not in $\mathcal{H}^s(\mathbf{ND}^{\mathsf{PL}})$.*

6. Towards a Meta-Proof Theory of Hybrid Derivation Systems

Adding the relation $\dashv$ for syntactic refutation and building systems of formal derivations that involve it together with the standard provability relation $\vdash$ can be regarded as first steps towards internalising the notion of hybrid derivation into the logical language and then developing a theory for that notion that mirrors the proof theory of $\vdash$. In particular, derivability and refutability can now be treated on a par, as two related primitive concepts rather than as complementary ones where refutability is to be represented syntactically by non-provability. (Note, however, that, for any complete logic or theory, $\vdash$ and $\dashv$ applied to sequents of sentences are readily inter-reducible as complementary relations.) Thus, a proof theory of hybrid derivation systems emerges, extending and combining both the traditional proof theory and the theory of refutation systems.

Furthermore, the basic logical concepts of soundness, completeness, consistency, and satisfiability that relate syntax and semantics of a given logical system can now be all expressed and treated purely syntactically in terms of $\vdash$ and $\dashv$. Thus, a "meta-proof theory" of hybrid derivation systems now emerges too, studying the meta-logic of these concepts respective to the given logical system L. Here, I will only set the stage for development of such meta-proof theory and will raise some generic questions, but I leave its systematic study to future work.

To begin with, let us add a new meta-symbol **F**, for "absurd", "falsum", or "contradiction", to the meta-language of hybrid derivation systems. Now, new hybrid derivation rules can be added to the thus extended framework, in order to reflect basic meta-properties of the given hybrid derivation system:

▷ **Cons**, stating consistency:

$$\frac{\vdash \phi, \ \dashv \phi}{\mathbf{F}},$$

▷ "Ex (meta-)falso quodlibet", **EFQ**:

$$\frac{\mathbf{F}}{\vdash \phi}, \quad \frac{\mathbf{F}}{\dashv \phi},$$

▷ **Ł-Comp**: "Ł-completeness":

$$\frac{\begin{array}{c}[\vdash \phi]\\ \vdots \\ \mathbf{F}\end{array}}{\dashv \phi},$$

▷ **Ł-RAA**: "Ł-Reductio ad absurdum"

$$\frac{\begin{array}{c}[\dashv \phi]\\ \vdots \\ \mathbf{F}\end{array}}{\vdash \phi}.$$

Deductive completeness and Ł-completeness can now be *internalised* and stated as additional hybrid rules:

$$(\mathbf{Ded}) \ \frac{\begin{array}{cc}[\dashv \phi] & [\vdash \phi]\\ \vdots & \vdots \\ \vdash \psi & \vdash \psi\end{array}}{\vdash \psi} \qquad (\mathbf{Ref}) \ \frac{\begin{array}{cc}[\dashv \phi] & [\vdash \phi]\\ \vdots & \vdots \\ \dashv \psi & \dashv \psi\end{array}}{\dashv \psi}.$$

Some natural questions arise:

1. Can any of these meta-rules strengthen the deductive power of a given (not complete) hybrid derivation system?
2. In particular, can any of these bring about deductive completeness or Ł-completeness, when it does not hold without them?

A next natural step would be to strengthen the meta-language even further, to a full-fledged logical meta-language, involving meta-variables and quantification over derivable and refutable formulae (or, sequents). Then, for instance, the semantic relationship between $\vdash$ and $\dashv$ can be postulated in the meta-language as $\Gamma \vdash \phi \Leftrightarrow \sim \Gamma \dashv \phi$ (where $\sim$ is the meta-negation). (Some initial steps into studying propositional meta-theory of acceptance and rejection of formulae (sequents with empty lists of premises) in a similar spirit can be found in [14].) I leave the general study of the meta-proof theory of hybrid derivation system to future work.

Remark 9. *It should be noted that what I call here 'meta-proof theory' has essentially been studied in great depth for theories of the arithmetic in the context of Gödel's incompleteness theorems and, more generally, in the context of axiomatic theories of truth; see [24]. However, the general meta-proof theory proposed here makes no assumptions about the expressiveness of the object logic regarding definability of truth predicates in it, or in general, and consequently it has a much wider scope.*

7. Conclusions

7.1. Some Applications of Hybrid Derivation Systems

Arguably, hybrid derivation systems have a number of potential applications, both conceptual and technical, including:

- Hybrid derivation systems put proofs and refutations on equal footing and thus enable their comparative study and of the development of meta-proof theory, where the interaction of the concepts of deduction and syntactic refutation for a given logic is the object of study.
- Hybrid derivation systems can yield purely deductive decision procedures, as indicated in Proposition 1 and illustrated for PL in Section 5.
- Hybrid derivation systems can capture important classes of non-valid formulae in recursively axiomatizable but undecidable logic, such as FOL. They can also provide complete refutation systems for logical theories with co-r.e. validity. Typically, this is logic defined over a class of finite models, such as FOL in the finite or Medvedev's logic of finite problems (see respectively [23,25] for R-complete refutation systems for these).
- Hybrid derivation systems can *possibly* provide more succinct proof systems. This hypothesis is yet to be tried and tested.

7.2. Current and Future Work

Due to space and time limitations, this paper leaves many open ends and related questions, some of which have already been mentioned so far. In addition, here are some topics of current and follow-up work:

- Develop and understand the general meta-proof theory of hybrid derivation systems.
- Design Ł-complete hybrid derivation systems for the intuitionistic propositional logic and for some important modal logic (extending such results from [3]) and for other non-classical logic.
- Extend/modify $\mathcal{H}^s(\mathbf{ND}^{\mathrm{PL}})$ to hybrid derivation systems for classical and intuitionistic FOL that are R-complete for the non-validities in the finite. Characterise the set of refutable non-validities in these systems.

- Relate more explicitly hybrid derivation systems with tableaux systems. As the latter are designed to check satisfiability, i.e., non-validity of the negated input, they are naturally related to refutations and, hence, to hybrid derivation systems.
- Analyze the relation of the present work with Negri's work on proofs and countermodels in [13,26] and explore the interaction of these two approaches to develop systems combining proofs, refutations, and counter-model constructions for various non-classical logic.
- Another potentially interesting direction (suggested by an anonymous referee) for related further research is to explore the relation between hybrid derivation systems and methods for proof certification [27].
- Last but not least: a challenge worth pursuing in this area would be to obtain new decidability results by designing Ł-adequate hybrid deductive systems for logic that is not yet known to be decidable, such as Medvedev's logic.

Funding: This work was partly supported by research grant 2015-04388 of the Swedish Research Council.

Acknowledgments: I thank Sara Negri and Tom Skura, as well as the anonymous referees for careful reading, helpful suggestions, and some important corrections. I also thank the participants in the Refutation Symposium in Poznań 2018 and in the CLLAM seminar at the Philosophy Department of Stockholm University for some useful comments on earlier versions of this work presented at these events.

Conflicts of Interest: The author declares no conflicts of interest.

References

1. Wybraniec-Skardowska, U. Rejection in Łukasiewicz's and Słupecki's Sense. In *Lvov-Warsaw School. Past and Present*; Garrido, A., Wybraniec-Skardowska, U., Eds.; Birkhäuser, Basel, Switzerland, 2018; pp. 575–598.
2. Tamminga, A. Logics of Rejection: Two Systems of Natural Deduction. *Log. Anal.* **1994**, *146*, 169–208.
3. Goranko, V. Refutation systems in modal logic. *Stud. Log.* **1994**, *53*, 299–324. [CrossRef]
4. Goranko, V.; Pulcini, G.; Skura, T. Refutation systems: An overview and some applications to philosophical logic. **2019**, submitted.
5. Negri, S.; von Plato, J. *Structural Proof Theory*; Cambridge University Press, Cambridge, UK, 2001.
6. Prawitz, D. *Natural Deduction—A Proof-Theoretical Study*, 2nd ed.; Dover Publications, Mineola, NY, USA, 2006.
7. Von Plato, J. *Elements of Logical Reasoning*; Cambridge University Press, Cambridge, UK, 2014.
8. Bonatti, P.; Varzi, A.C. On the Meaning of Complementary Systems. In Proceedings of the 10th International Congress of Logic, Methodology and Philosophy of Science, Florence, Italy, 1995; Volume of Abstracts; Castagli, E., Konig, M., Eds.; International Union of History and Philosophy of Science: Jerusalem, Israel, 1995; p. 122.
9. Skura, T. Maximality and refutability. *Notre Dame J. Form. Log.* **2004**, *45*, 65–72. [CrossRef]
10. Wybraniec-Skardowska, U.; Waldmajer, J. On Pairs of Dual Consequence Operations. *Log. Univers.* **2011**, *5*, 177–203. [CrossRef]
11. Caferra, R.; Peltier, N. Accepting/rejecting propositions from accepted/rejected propositions: A unifying overview. *Int. J. Intell. Syst.* **2008**, *23*, 999–1020. [CrossRef]
12. Goré, R.; Postniece, L. Combining derivations and refutations for cut-free completeness in Bi-intuitionistic logic. *J. Log. Comput.* **2008**, *20*, 233–260. [CrossRef]
13. Negri, S. On the Duality of Proofs and Countermodels in Labelled Sequent Calculi. In *TABLEAUX 2013: Automated Reasoning with Analytic Tableaux and Related Methods*; Springer: Berlin/Heidelberg, Germany, 2013; pp. 5–9. [CrossRef]
14. Citkin, A. A Meta-Logic of Inference Rules: Syntax. *Log. Log. Philos.* **2015**, *24*, 313–337. [CrossRef]
15. Fiorentini, C.; Ferrari, M. Duality between unprovability and provability in forward proof-search for Intuitionistic Propositional Logic. *arXiv* **2018**, arXiv:1804.06689.
16. Fiorentini, C.; Ferrari, M. A Forward Unprovability Calculus for Intuitionistic Propositional Logic. In *TABLEAUX 2017: Automated Reasoning with Analytic Tableaux and Related Methods*; LNCS; Springer: Cham, Switzerland, 2017; Volume 10501, pp. 114–130.
17. Rumfitt, I. 'Yes and no'. *Mind* **2000**, *109*, 781–823. [CrossRef]
18. Goranko, V. *Logic as a Tool—A Guide to Formal Logical Reasoning*; Wiley, Chichester, UK, 2016.

19. Skura, T. Refutation systems in propositional logic. In *Handbook of Philosophical Logic*; Gabbay, D.M., Guenthner, F., Eds.; Springer: Dordrecht, The Netherlands, 2011; Volume 16, pp. 115–157.
20. Baader, F.; Ghilardi, S. Unification in modal and description logic. *Log. J. IGPL* **2011**, *19*, 705–730. [CrossRef]
21. Goudsmit, J.P. Admissibility and refutation: Some characterisations of intermediate logic. *Arch. Math. Log.* **2014**, *53*, 779–808. [CrossRef]
22. Tiomkin, M.L. Proving unprovability. In Proceedings of the LICS'88, Edinburgh, Scotland, UK, 5–8 July 1988; pp. 22–26. [CrossRef]
23. Skura, T. Refutation Calculi for Certain Intermediate Propositional Logics. *Notre Dame J. Form. Log.* **1992**, *33*, 552–560. [CrossRef]
24. Halbach, V. *Axiomatic Theories of Truth*, 2nd ed.; Cambridge University Press: Cambridge, UK, 2014.
25. Goranko, V.; Skura, T. Refutation Systems in the Finite. Available online: http://www.logic.ifil.uz.zgora.pl/refutation/files/goranko.skura.refutation.finite.pdf (accessed on 20 December 2018).
26. Negri, S. Proofs and Countermodels in Non-Classical Logics. *Log. Univ.* **2014**, *8*, 25–60. [CrossRef]
27. Chihani, Z.; Miller, D.; Renaud, F. Foundational Proof Certificates in First-Order Logic. In *Automated Deduction—CADE-24*; Springer: Berlin/Heidelberg, Germany, 2013; pp. 162–177. [CrossRef]

© 2019 by the author. Licensee MDPI, Basel, Switzerland. This article is an open access article distributed under the terms and conditions of the Creative Commons Attribution (CC BY) license (http://creativecommons.org/licenses/by/4.0/).

Article

A Kotas-Style Characterisation of Minimal Discussive Logic

Krystyna Mruczek-Nasieniewska *,† and Marek Nasieniewski *,†

Department of Logic, Nicolaus Copernicus University in Toruń, ul. Moniuszki 16/20, 87-100 Toruń, Poland
* Correspondence: mruczek@umk.pl (K.M.-N.); mnasien@umk.pl (M.N.)
† These authors contributed equally to this work.

Received: 26 August 2019; Accepted: 27 September 2019; Published: 1 October 2019

Abstract: In this paper, we discuss a version of discussive logic determined by a certain variant of Jaśkowski's original model of discussion. The obtained system can be treated as the minimal discussive logic. It is determined by frames with serial accessibility relation. As the smallest one, this logic can be treated as a basis which could be extended to richer discussive logics that are obtained by varying accessibility relation and resulting in a lattice of discussive logics. One has to remember that while formulating discussive logics there is no one-to-one determination of discussive logics by modal logics. For example, it is proved that Jaśkowski's logic D_2 can be expressed by other than **S5** modal logics. In this paper we consider a deductive system for the sketchily described minimal logic. While formulating the deductive system, we apply a method of Kotas that was used to axiomatize D_2. The obtained system determines a logic D_0 as a set of theses that is contained in D_2. Moreover, any discussive logic that would be expressed by means of the provided model of discussion would contain D_0, so it is the smallest discussive logic.

Keywords: discussive logics; the smallest discussive logic; discussive operators; seriality; accessibility relation; Kotas' method; modal logic

1. Introduction

Stanisław Jaśkowski's aim was to propose a calculus that would allow for explication of inconsistent theories by means of some consistent framework. As a result Jaśkowski developed a logic denoted as D_2 that was meant to be a basis for calculus that would not lead in general to overfull set of conclusions when applied to inconsistent set of premises. He used the scenario of a discussion as a model case. Intuitively, during discussions participants can contradict each other, but a possible external observer as well as particular participants would not conclude that everything follows from such discussion. (Some analysis on this matter can be found in [?].)

Interactions that take place between participants of a discussion are expressed formally by discussive counterparts of implication, conjunction and equivalence. Moreover, in Jaśkowski's intuitive model, operators take only auxiliary role and modal operators are not present in the object language of the discussive logic. Such a variant seems to be natural and has been considered in [?].

Our aim is to indicate the weakest logic that arises from a natural variant of Jaśkowski model of discussion and moreover, give an adequate deductive system for such a logic.

2. D_2 and the Minimal Variant of Discussive Logic

In the original formulation, Jaśkowski considered a situation in which there is no restriction on possible reactions of participants of a discussion, in other words he considered a model, where every two participants of the discussion are connected. It corresponds to the full accessibility relation that semantically allows to determine the logic **S5**. However, it is known that not every thesis is in fact used or needed to express discussive theses. What is only used, is the so-called M-counterpart of the logic **S5** (for investigations on this notion see [?]). In various papers it has been shown (see [? ? ? ? ?]) that to be able to formulate D_2, one can use various modal logics. (Jaśkowski's logic was meant to be a basis for a consequence relation and also in this case there can be given other systems than **S5** which also allow to express D_2-consequence relation (see [?]).) Moreover, one can introduce a general discussive consequence relation framework, in which D_2 would be the set of theses of one of its special cases (for details see [?]). However, this does not mean that any modal logic would be equally good to obtain D_2.

We will keep original Jaśkowski's meaning of discussive connectives of implication and conjunction. Jaśkowski's discussive implication denoted here as $\to_d$, is meant to be read as "if anyone states that p, then q" (see [?] p. 150, 1969), in modal terms: $\Diamond p \to q$. Discussive conjunction is usually interpreted as saying "p and someone said q", in the modal Jaśkowski's interpretation it is read as $p \wedge \Diamond q$. (The disjunction conjunction was introduced in the second Jaśkowski paper on discussive logic [?].) In both cases $\Diamond$ is originally interpreted as possibility that can refer to any participant of discussion. In our interpretation it will refer only to those participants, who are connected by the accessibility relation. In particular, it means that statements of participants, who are not connected to any disputant, make the whole discussion (since we are interested in expressing what logically follows, so we ought to consider each world — or in the nomenclature of the model of the discussion — each point of view) meaningless. As Jaśkowski says, every thesis of the discussive system during its interpretation ought to be preceded by the reservation: "in accordance with the opinion of one of the participants in the discourse", so "if a thesis is recorded in a discursive system, its intuitive sense ought to be interpreted so as if it were preceded by the symbol *Pos*" ([?] p. 149, 1969), which nowadays is denoted standardly by '$\Diamond$'. Taking into account what has been said, the minimal requirement for the considered model of discussion is that the 'outer' possibility ought to be ruled by a serial accessibility condition. From the formal point of view, the underlying modal logic that will be used for the formulation of the proposed variant of discussive logic, will be the deontic normal logic **D**. As it is known, it is semantically expressed by the class of frames with serial accessibility relation (where seriality means that for every world w there is a world v such that wRv).

To strictly formulate the presented idea we will need two formal languages: modal and discussive.

3. Modal and Discussive Languages

Throughout the paper we will use modal formulas that are formed in the standard way from propositional letters: 'p', 'q', 'r', 's', 't', 'p_0', 'p_1', 'p_2', ...; truth-value operators: '$\neg$', '$\vee$', '$\wedge$', '$\to$', and '$\leftrightarrow$' (connectives of negation, disjunction, conjunction, material implication and material equivalence, respectively); modal operators: the necessity and possibility operators '$\Box$' and '$\Diamond$'; and the brackets. Let For_m denote the set of all modal formulas. Of course, the set For_m includes the set of all classical formulas (without the use of '$\Box$' and '$\Diamond$'), in particular the set of all classical tautologies denoted as **CL**. The modal language plays only an auxiliary role in the formulation of discussive logic. Its object language is built again from propositional letters, truth-value operators '$\neg$' and '$\vee$' and discussive implication ($\to_d$), discussive conjunction ($\wedge_d$) and discussive equivalence ($\leftrightarrow_d$). The set of all discussive formulas is denoted by 'For_d'.

Basics of Normal Modal Logics

A normal modal logic is a set $M \subseteq \text{For}_m$ that fulfils the following conditions:

1. $\textbf{CL} \subseteq M$,

2. ***M*** is closed under *modus ponens* (??), uniform substitution (??) and necessitation rule (??):

$$\text{if } \varphi \text{ and } (\varphi \to \psi) \text{ belong of } M, \text{ so does } \psi. \tag{mp}$$
$$\text{if } \varphi \in M \text{ then } s(\varphi) \in M, \tag{us}$$
$$\text{if } \varphi \in M \text{ then } \Box\varphi \in M, \tag{rn}$$

3. ***M*** contains formulas (??) and (??)

$$\Diamond p \leftrightarrow \neg \Box \neg p \tag{df $\Diamond$}$$
$$\Box(p \to q) \to (\Box p \to \Box q) \tag{K}$$

As it is known, every normal modal logic contains the following formulas

$$\Box(p \to q) \to (\Diamond p \to \Diamond q) \tag{1}$$
$$(\Box p \to \Diamond q) \to \Diamond(p \to q) \tag{2}$$
$$\Diamond(p \to q) \to (\Box p \to \Diamond q) \tag{3}$$

D is the smallest normal logic containing (??):

$$\Box p \to \Diamond p \tag{D}$$

Standardly, **K** is the smallest normal modal logic and **S5** := **KT5**, that is, **S5** is the smallest normal modal logic containing (??) and (??), where

$$\Box p \to p \tag{T}$$
$$\Diamond p \to \Box \Diamond p \tag{5}$$

4. Discussive Logics

In the original formulation every two participants are connected one to another—in fact, for the explication of discussive implication one reads: 'if *anyone* states that p'. The same idea is applied for the modality that is corresponding to possibility expressing the point of view of an external observer. Hence, Jaśkowski's discussive logic D_2 is defined by means of **S5** as follows:

$$D_2 := \{\, A \in \text{For}_d : \Diamond i_1(A) \in \mathbf{S5} \,\},$$

where i_1 is a translation of discussive formulas into the modal language, that is, i_1 is a function from For_d into For_m such that:

1. $i_1(a) = a$, for any propositional letter a,
2. for any $A, B \in \text{For}_d$:
   (a) $i_1(\neg A) = \neg i_1(A)$,
   (b) $i_1(A \vee B) = i_1(A) \vee i_1(B)$,
   (c) $i_1(A \wedge_d B) = i_1(A) \wedge \Diamond i_1(B)$,
   (d) $i_1(A \to_d B) = \Diamond i_1(A) \to i_1(B)$,
   (e) $i_1(A \leftrightarrow_d B) = (\Diamond i_1(A) \to i_1(B)) \wedge \Diamond(\Diamond i_1(B) \to i_1(A))$.

One can also consider a more general case. Let **S** be any normal modal logic. Now, we can define

$$D_S := \{\, A \in \text{For}_d : \Diamond i_1(A) \in \mathbf{S} \,\}. \tag{4}$$

We easily see that in the case where there is no formula of the form $\Diamond(A)$ that would belong to a given modal logic **S**, then we have $D_S = \emptyset$. In particular, for any modal logic **S** that is determined by a class of Kripke frames, whose accessibility relation does not fulfil seriality condition, we obtain $D_S = \emptyset$.

(We use standard results from modal logic, for details see for example, References [? ?].) Of course, if for a given normal modal logic **S**, we have (??) $\in$ **S**, then **D** $\subseteq$ **S**. It is known (see Reference [?]) that one can consider various accessibility relations but the resulting discussive logic would be still the same. By definition, $D_{S5} = D_2$. We easily see that

Fact 1. *For any modal logics* $\mathbf{S}_1$ *and* $\mathbf{S}_2$, *if* $\mathbf{S}_1 \subseteq \mathbf{S}_2$, *then*

$$D_{\mathbf{S}_1} \subseteq D_{\mathbf{S}_2}$$

By induction on the complexity of a formula $\varphi \in \text{For}_m$, we can obtain:

Fact 2. *For every* $\varphi \in \text{For}_m$, *there is* $A \in \text{For}_d$ *such that* $\mathtt{i}_1(A) \leftrightarrow \varphi \in \mathbf{K}$.

In this paper we focus on the case where in the condition (??), **D** is taken as the modal system **S**. We denote the resulting system as D_0. Thus, by definition

$$D_0 := \{\, A \in \text{For}_d : \Diamond \mathtt{i}_1(A) \in \mathbf{D} \,\}. \tag{5}$$

In the context of the definition of D_0, first, let us observe that:

Lemma 1. *For any* $A, A_1, \ldots, A_n \in \text{For}_d$ *and any variables* $a_1, \ldots, a_n$,

$$(\mathtt{i}_1(A))(a_1/\mathtt{i}_1(A_1), \ldots, a_n/\mathtt{i}_1(A_n)) = \mathtt{i}_1(A(a_1/A_1, \ldots, a_n/A_n)) \in \mathbf{D}.$$

Proof. The proof goes by induction on the complexity of a formula A.
For the initial case, let $A = a_i$. We have: $(\mathtt{i}_1(a_i))(a_1/\mathtt{i}_1(A_1), \ldots, a_n/\mathtt{i}_1(A_n)) = \mathtt{i}_1(A_i) = \mathtt{i}_1(a_i(a_1/A_1, \ldots, a_n/A_n))$. If A is a variable that does not belong to $\{a_1, \ldots, a_n\}$, we have $(\mathtt{i}_1(A))(a_1/\mathtt{i}_1(A_1), \ldots, a_n/\mathtt{i}_1(A_n)) = A = \mathtt{i}_1(A(a_1/A_1, \ldots, a_n/A_n))$.
For the inductive step assume that inductive hypothesis holds for B and C. For the case of discussive conjunction observe that the following equations hold: $(\mathtt{i}_1(B \wedge_d C))(a_1/\mathtt{i}_1(A_1), \ldots, a_n/\mathtt{i}_1(A_n)) = (\mathtt{i}_1(B) \wedge \Diamond \mathtt{i}_1(C))(a_1/\mathtt{i}_1(A_1), \ldots, a_n/\mathtt{i}_1(A_n)) = (\mathtt{i}_1(B))(a_1/\mathtt{i}_1(A_1), \ldots, a_n/\mathtt{i}_1(A_n)) \wedge (\Diamond \mathtt{i}_1(C))(a_1/\mathtt{i}_1(A_1), \ldots, a_n/\mathtt{i}_1(A_n))$. Using inductive hypothesis and features of substitution, we have $(\mathtt{i}_1(B))(a_1/\mathtt{i}_1(A_1), \ldots, a_n/\mathtt{i}_1(A_n)) \wedge (\Diamond \mathtt{i}_1(C))(a_1/\mathtt{i}_1(A_1), \ldots, a_n/\mathtt{i}_1(A_n)) = (\mathtt{i}_1(B))(a_1/\mathtt{i}_1(A_1), \ldots, a_n/\mathtt{i}_1(A_n)) \wedge \Diamond((\mathtt{i}_1(C))(a_1/\mathtt{i}_1(A_1), \ldots, a_n/\mathtt{i}_1(A_n))) = \mathtt{i}_1(B(a_1/A_1, \ldots, a_n/A_n)) \wedge \Diamond \mathtt{i}_1(C(a_1/A_1, \ldots, a_n/A_n)) = \mathtt{i}_1((B(a_1/A_1, \ldots, a_n/A_n) \wedge_d C(a_1/A_1, \ldots, a_n/A_n))) = \mathtt{i}_1((B \wedge_d C)(a_1/A_1, \ldots, a_n/A_n))$. Similarly, also the proofs for $\neg$, $\vee$, $\rightarrow_d$ and $\leftrightarrow_d$ are straightforward. □

Fact 3. *The set* D_0 *is closed under substitution and modus ponens with respect to* $\rightarrow_d$.

Proof. Let $A \in D_0$ and $A \rightarrow B \in D_0$, that is $\Diamond \mathtt{i}_1(A) \in \mathbf{D}$ and $\Diamond \mathtt{i}_1(A \rightarrow B) \in \mathbf{D}$, so $\Diamond(\Diamond \mathtt{i}_1(A) \rightarrow \mathtt{i}_1(B))) \in \mathbf{D}$, by the distributivity of $\Diamond$ with respect to $\rightarrow$ $\square \Diamond \mathtt{i}_1(A) \rightarrow \Diamond \mathtt{i}_1(B) \in \mathbf{D}$, but by normality $\square \Diamond \mathtt{i}_1(A) \in \mathbf{D}$, hence $\Diamond \mathtt{i}_1(B))) \in \mathbf{D}$ and by definition (??), $B \in D_0$.
Assume now that $A \in D_0$, that is $\Diamond \mathtt{i}_1(A) \in \mathbf{D}$. Let us consider a result of uniform substitution $s(A)$ into A of formulas A_i for variables in A, where $1 \leq i \leq n$, for some n, that is, $s(A) = A(a_1/A_1, \ldots, a_n/A_n)$, where a_i are all variables in A. By Lemma ?? we know that $\mathtt{i}_1(s(A)) = (\mathtt{i}_1(A))(a_1/\mathtt{i}_1(A_1), \ldots, a_n/\mathtt{i}_1(A_n))$. Since **D** is a logic, so it is closed on substitution, so also $(\Diamond \mathtt{i}_1(A))(a_1/\mathtt{i}_1(A_1), \ldots, a_n/\mathtt{i}_1(A_n)) \in \mathbf{D}$. But the following equations hold $(\Diamond \mathtt{i}_1(A))(a_1/\mathtt{i}_1(A_1), \ldots, a_n/\mathtt{i}_1(A_n)) = \Diamond(\mathtt{i}_1(A)(a_1/\mathtt{i}_1(A_1), \ldots, a_n/\mathtt{i}_1(A_n))) = \Diamond \mathtt{i}_1(s(A))$, therefore $s(A) \in D_0$. □

A Comparison with Some Classical Theses

As one can easily see, none of the classical cases given below belong to D_0, although each of these formulas belong to D_2. To stress discussive interpretation we use the formulas from For_d.

$$p \to_d p$$
$$p \to_d p \vee q$$
$$p \wedge_d q \to_d p$$
$$p \to_d (q \to_d (p \wedge_d q))$$
$$(p \to_d (q \to_d r)) \to_d ((p \to_d q) \to_d (p \to_d r))$$
$$(p \to_d q) \to_d ((q \to_d r) \to_d (p \to_d r))$$
$$\neg\neg p \to_d p$$
$$p \to_d \neg\neg p$$

Each of these formulas can be rejected semantically. We use standard completeness theorem with respect to Kripke semantics for the logic **D**. As an example, let us consider the fifth formula, known as Frege syllogism. One can easily see that the respective translation:

$$\Diamond(\Diamond(\Diamond p \to (\Diamond q \to r)) \to (\Diamond(\Diamond p \to q) \to (\Diamond p \to r)))$$

is not a thesis of **D**, so the formula $(p \to_d (q \to_d r)) \to_d ((p \to_d q) \to_d (p \to_d r))$ does not belong to D_0. Similarly one falsifies the other cases.

5. A Kotas Style Deductive System for the Smallest Discussive Logic

We will characterise a discussive logic being a result of the given variant of the discussive model; that is in the case that the relation is serial and no other condition is assumed as regards accessibility relation. We will give an adequacy result for the given deductive system.

We will adopt a method of Kotas (see Reference [?]) that was used for indicating the way in which D_2 could be axiomatized. (There are other axiomatisations of D_2. In Reference [?] there is an axiomatisation of discussive logic but in a version with left discussive conjunction. For not so straight history of axiomatisation of D_2 see References [? ?].) The same method was used *inter alia* in Reference [?] to axiomatize a variant of D_2 with modal operators.

Let us use the following notation:

$$(\Box Ai) \text{ denotes } \Box\phi, \text{ for } (Ai) \text{ denoting } \phi \tag{6}$$

$$\Diamond\text{-}\mathbf{S} = \{A \in \text{For}_m : \Diamond A \in \mathbf{S}\} \tag{7}$$

$\Diamond$-**S** is called an M-counterpart of **S** (see Reference [?] (p. 70)). By definitions, for any normal logic $\mathbf{S} \supseteq \mathbf{D}$:

$$\mathbf{S} \subseteq \Diamond\text{-}\mathbf{S}$$

It is known that (see Reference [?] (p. 68)):

Fact 4.

$$\Diamond\text{-}\mathbf{D} = \mathbf{D} \tag{8}$$

Consider the following axiomatisation of **CL**:

$$p \to (q \to p) \tag{A1}$$
$$(p \to (q \to r)) \to ((p \to q) \to (p \to r)) \tag{A2}$$
$$p \land q \to p \tag{A3}$$
$$p \land q \to q \tag{A4}$$
$$p \to (q \to p \land q) \tag{A5}$$
$$p \to p \lor q \tag{A6}$$
$$q \to p \lor q \tag{A7}$$
$$(p \to q) \to ((r \to q) \to (p \lor r \to q)) \tag{A8}$$
$$(p \leftrightarrow q) \to (p \to q) \tag{A9}$$
$$(p \leftrightarrow q) \to (q \to p) \tag{A10}$$
$$(p \to q) \to ((q \to p) \to (p \leftrightarrow q)) \tag{A11}$$
$$(\neg p \to \neg q) \to (q \to p) \tag{A12}$$

and formulas

$$\Box(\Diamond p \leftrightarrow \neg\Box\neg p) \tag{$\Box$df$\Diamond$}$$
$$\Box(\Box(p \to q) \to (\Box p \to \Box q)) \tag{$\Box$K}$$
$$\Box(\Box p \to \Diamond p) \tag{$\Box$D}$$
$$\Box(\Box p \to p) \tag{$\Box$T}$$
$$\Box(\Diamond p \to \Box\Diamond p) \tag{$\Box$5}$$

Let $\Omega := \{(\Box Ai) : 1 \leq i \leq 12\} \cup \{(??), (??), (??)\}$ and let $\Omega_1 := \{(\Box Ai) : 1 \leq i \leq 12\} \cup \{(??), (??), (??), (??)\}$

Let us recall a theorem that allows to formulate **S5** syntactically with the use of the above mentioned formulas and rules.

Fact 5 ([?]).

1. **S5** is the smallest set including Ω_1 and closed under (??) and the following rules:

$$\frac{\Box\varphi, \ \Box(\varphi \to \psi)}{\Box\psi} \tag{$\Box$mp}$$

$$\frac{\Box\varphi}{\Box\Box\varphi} \tag{$\Box$rn}$$

$$\frac{\Box\varphi}{\varphi} \tag{rn$_\Leftarrow$}$$

2. $\Box$**S5** is the smallest set including Ω_1 and closed under (??), (??) and (??).
3. $\Diamond$-**S5** is the smallest set including Ω_1 and closed under (??), (??), (??), (??) and (??):

$$\frac{\Diamond\varphi}{\varphi} \tag{rp$_\Leftarrow$}$$

But in a quite similar way, one can formulate the logic **D**. Let **D**$^\vdash$ denote the smallest set including Ω and closed under substitution, (??), (??), (??) and (??)

$$\frac{\varphi, \ \Box(\varphi \to \psi)}{\psi} \tag{$\Box$mp$_-$}$$

We will follow a custom used for modal logics of calling elements of $\mathbf{D}^{\Vdash}$ *theses of a deductive system*. So $\mathbf{D}^{\Vdash}$ is the set of theses with respect to a deductive system $\Vdash$ determined by the given axioms $(\Box Ai) : 1 \leqslant i \leqslant 12$ and rules (??), (??), (??), (??) and an substitution.

Lemma 2. $\mathbf{D} = \mathbf{D}^{\Vdash}$.

Proof. We show that $\mathbf{D}^{\Vdash} \subseteq \mathbf{D}$. First, by the standard formulation of $\mathbf{D}$ and (??) we see that $\Omega \subseteq \mathbf{D}$—it is enough to use necessitation for respective axioms of $\mathbf{D}$. Besides by (??), $\mathbf{D}$ is closed on (??). We will prove that $\mathbf{D}$ is closed on (??). Assume that $\varphi, \Box(\varphi \to \psi) \in \mathbf{D}$. By (??), $\Box\varphi \in \mathbf{D}$, while by (??), we have $\Diamond(\varphi \to \psi) \in \mathbf{D}$ hence using (??) we obtain $\Box\varphi \to \Diamond\psi \in \mathbf{D}$. Thus, by modus ponens $\Diamond\psi \in \mathbf{D}$ and by (??), $\psi \in \mathbf{D}$. The fact that $\mathbf{D}$ is closed on (??) follows by axiom (??) and modus ponens. Finally, $\mathbf{D}$ is closed on (??) by necessitation.

For the reverse direction let us assume that $\varphi \in \mathbf{D}$. We can consider a proof $\varphi_1, \ldots, \varphi_k = \varphi$ of φ in the standard axiomatisation of $\mathbf{D}$. First observe that by (??) and $\Box Ai$, where, $1 \leqslant i \leqslant 12$ we obtain $\Box\psi$, for any classical tautology ψ — it is enough to consider a prove of ψ and the basis of the system with $\{Ai : 1 \leqslant i \leqslant 12\}$ as axioms, with modus ponens and substitution as rules of inference and next precede every element of the proof by $\Box$ and observe that the obtained sequence $\Box\varphi_1, \ldots, \Box\varphi_k$ is a proof of $\Box\psi$ on the basis of the given system of $\mathbf{D}^{\Vdash}$. Second, we see that any other axiom of $\mathbf{D}$ preceded by $\Box$ becomes an axiom of $\mathbf{D}^{\Vdash}$; besides, rules of (??) and (??) correspond respectively to necessitation and modes ponens in the original proof of φ. Hence, using induction on the length of the proof we see that $\Box\varphi$ has a mentioned proof in $\mathbf{D}^{\Vdash}$. We extend the sequence $\Box\varphi_1, \ldots, \Box\varphi_k = \Box\varphi$ to infer φ. By $(\Box??)$ we have $\Box(\Box\varphi \to \Diamond\varphi)$ and so using (??) and $\Box\varphi$ we get $\Diamond\varphi$, hence by (??) we infer φ. Let us finally add that both sets are closed in substitution. $\square$

In Reference [?] two translations were considered. $\mathtt{i}_1$ is a natural version of the first one adjusted to the considered here language. The translation $\mathtt{i}_2 \colon \mathrm{For}_m \longrightarrow \mathrm{For}_d$ given below is a version of $\mathtt{i}_2$ defined in Reference [?] where the case of $\Diamond$ is added:

1. $\mathtt{i}_2(a) = a$, for any propositional letter a,
2. for any $\varphi, \psi \in \mathrm{For}_m$:

    (a) $\mathtt{i}_2(\neg\varphi) = \neg \mathtt{i}_2(\varphi)$,
    (b) $\mathtt{i}_2(\Box\varphi) = \neg((\neg p \vee p) \wedge_d \neg \mathtt{i}_2(\varphi))$,
    (c) $\mathtt{i}_2(\Diamond\varphi) = (\neg p \vee p) \wedge_d \mathtt{i}_2(\varphi)$,
    (d) $\mathtt{i}_2(\varphi \vee \psi) = \mathtt{i}_2(\varphi) \vee \mathtt{i}_2(\psi)$,
    (e) $\mathtt{i}_2(\varphi \wedge \psi) = \neg(\neg \mathtt{i}_2(\varphi) \vee \neg \mathtt{i}_2(\psi))$,
    (f) $\mathtt{i}_2(\varphi \to \psi) = \neg \mathtt{i}_2(\varphi) \vee \mathtt{i}_2(\psi)$,
    (g) $\mathtt{i}_2(\varphi \leftrightarrow \psi) = \neg(\neg(\neg \mathtt{i}_2(\varphi) \vee \mathtt{i}_2(\psi)) \vee \neg(\neg \mathtt{i}_2(\psi) \vee \mathtt{i}_2(\varphi)))$.

The below Lemma is being proved similarly as Lemma 4.2 in Reference [?].

Lemma 3. *For any $\varphi \in \mathrm{For}_m$, $\mathtt{i}_1(\mathtt{i}_2(\varphi)) \leftrightarrow \varphi \in \mathbf{D}$.*

Proof. The proof goes by induction on the complexity of a given formula.
If φ is a variable, we have $\mathtt{i}_1(\mathtt{i}_2(\varphi)) = \varphi$, thus the thesis holds trivially.
Assume that the inductive thesis holds for formulas simpler than a given formula. Firstly, we have $\mathtt{i}_1(\mathtt{i}_2(\varphi \vee \psi)) = \mathtt{i}_1(\mathtt{i}_2(\varphi) \vee \mathtt{i}_2(\psi)) = \mathtt{i}_1(\mathtt{i}_2(\varphi)) \vee \mathtt{i}_1(\mathtt{i}_2(\psi))$. Thus, $\mathtt{i}_1(\mathtt{i}_2(\varphi \vee \psi)) \leftrightarrow (\varphi \vee \psi) \in \mathbf{D}$, by **CL**. Similarly by definition, $\mathtt{i}_1(\mathtt{i}_2(\neg\varphi)) = \mathtt{i}_1(\neg \mathtt{i}_2(\varphi)) = \neg \mathtt{i}_1(\mathtt{i}_2(\varphi))$, so the required equivalence holds by **CL**.
For the case of $\Box$, we have: $\mathtt{i}_1(\mathtt{i}_2(\Box\varphi)) = \mathtt{i}_1(\neg((\neg p \vee p) \wedge_d \neg \mathtt{i}_2(\varphi))) = \neg((\neg p \vee p) \wedge \Diamond \neg \mathtt{i}_1(\mathtt{i}_2(\varphi)))$, so the required equivalence holds by normality, in particular, by **CL**, extensionality and due to the fact that $\neg \Diamond \neg p \leftrightarrow \Box p \in \mathbf{K}$.
For the case of '$\Diamond$', we have: $\mathtt{i}_1(\mathtt{i}_2(\Diamond\varphi)) = \mathtt{i}_1((\neg p \vee p) \wedge_d \mathtt{i}_2(\varphi)) = (\neg p \vee p) \wedge \Diamond \mathtt{i}_1(\mathtt{i}_2(\varphi))$ so again, the required equivalence holds by normality.

Secondly, by **CL** and due to the following equations and equivalences, the inductive thesis holds for '$\wedge$', '$\rightarrow$' and '$\leftrightarrow$'.

For '$\wedge$': $i_1(i_2(\varphi \wedge \psi)) = i_1(\neg(\neg i_2(\varphi) \vee \neg i_2(\psi))) = \neg(\neg i_1(i_2(\varphi)) \vee \neg i_1(i_2(\psi)))$. But, of course, $\neg(\neg i_1(i_2(\varphi)) \vee \neg i_1(i_2(\psi))) \leftrightarrow (i_1(i_2(\varphi)) \wedge i_1(i_2(\psi))) \in \mathbf{D}$.
For the case of '$\rightarrow$': $i_1(i_2(\varphi \rightarrow \psi)) = i_1(\neg i_2(\varphi) \vee i_2(\psi)) = \neg i_1(i_2(\varphi)) \vee i_1(i_2(\psi))$. But $\neg i_1(i_2(\varphi)) \vee i_1(i_2(\psi)) \leftrightarrow i_1(i_2(\varphi)) \rightarrow i_1(i_2(\psi)) \in \mathbf{D}$.
For '$\leftrightarrow$': $i_1(i_2(\varphi \leftrightarrow \psi)) = i_1(\neg(\neg(\neg i_2(\varphi) \vee i_2(\psi)) \vee \neg(\neg i_2(\psi) \vee i_2(\varphi)))) = \neg(\neg(\neg i_1(i_2(\varphi)) \vee i_1(i_2(\psi))) \vee \neg(\neg i_1(i_2(\psi)) \vee i_1(i_2(\varphi))))$. However, $\neg(\neg(\neg i_1(i_2(\varphi)) \vee i_1(i_2(\psi))) \vee \neg(\neg i_1(i_2(\psi)) \vee i_1(i_2(\varphi)))) \leftrightarrow (i_1(i_2(\varphi)) \leftrightarrow i_1(i_2(\psi))) \in \mathbf{D}$. □

Due to Fact **??**, we also have a connection similar to the relation between $\mathbf{D}_2$ and $\Diamond$-**S5**:

Lemma 4.

1. For any $A \in \mathrm{For}_d$: $A \in \mathbf{D}_0$ iff $i_1(A) \in \mathbf{D}$.
2. For any $\varphi \in \mathbf{D}$, we have $i_2(\varphi) \in \mathbf{D}_0$.
3. If $\varphi \in \mathbf{D}$ and $\varphi \leftrightarrow \psi \in \mathbf{D}$, then $\psi \in \mathbf{D}$.

Proof. *Ad* 1. Let $A \in \mathbf{D}_0$. By the definition of $\mathbf{D}_0$ it means that $\Diamond i_1(A) \in \mathbf{D}$, that is, $i_1(A) \in \Diamond$-**D**. Thus, by Fact **??**, $i_1(A) \in \mathbf{D}$.
If $i_1(A) \in \mathbf{D}$, then by necessitation and (**??**), $\Diamond i_1(A) \in \mathbf{D}$, so $A \in \mathbf{D}_0$.
Ad 2. Let $\varphi \in \mathbf{D}$, that is, $\Diamond \varphi \in \mathbf{D}$. Then, by Lemma **??** and extensionality for **D**, we have $\Diamond i_1(i_2(\varphi)) \in \mathbf{D}$. Thus, $i_2(A) \in \mathbf{D}_0$.
Point 3 is obvious. □

To make the following consideration easier to follow, for $A, B \in \mathrm{For}_d$, let us denote the formula $\neg A \vee B$ as $A \rightarrow_c B$, and $\neg((\neg p \vee p) \wedge_d \neg A)$ as $\square^d A$, $(\neg p \vee p) \wedge_d A$ as $\Diamond^d A$, and $\square^d(A \rightarrow_c B))$, that is $\neg((\neg p \vee p) \wedge_d \neg(\neg A \vee B))$ as $A \twoheadrightarrow_d B$.

Let $\vdash_{\mathbf{D}_0}$ be the consequence relation determined by the set $i_2(\Omega)$ and the following formulas:

$$i_2(i_1(q \wedge_d r)) \twoheadrightarrow_d (q \wedge_d r) \tag{B1}$$

$$(i_2(i_1(q \rightarrow_d r)) \twoheadrightarrow_d (q \rightarrow_d r)) \tag{B2}$$

$$(i_2(i_1(q \leftrightarrow_d r)) \twoheadrightarrow_d (q \leftrightarrow_d r)) \tag{B3}$$

$$((q \wedge_d r) \twoheadrightarrow_d i_2(i_1(q \wedge_d r))) \tag{C1}$$

$$((q \rightarrow_d r) \twoheadrightarrow_d i_2(i_1(q \rightarrow_d r))) \tag{C2}$$

$$((q \leftrightarrow_d r) \twoheadrightarrow_d i_2(i_1(q \leftrightarrow_d r))) \tag{C3}$$

as axioms together with substitution and the following rules:

$$\frac{\square^d A \quad (A \twoheadrightarrow_d B)}{\square^d B} \tag{$\square^d\ \mathrm{mp}_{\mathrm{str}}$}$$

$$\frac{A \quad (A \twoheadrightarrow_d B)}{B} \tag{$\square^d\ \mathrm{mp}'_{\mathrm{str}}$}$$

$$\frac{\square^d A}{\square^d \square^d A} \tag{$\square^d\ \mathrm{rn}$}$$

$$\frac{\Diamond^d A}{A} \tag{$\mathrm{rp}^d_{\Leftarrow}$}$$

The proof of the following lemma is straightforward by induction on the complexity of a modal formula φ.

Lemma 5. *Let $\varphi, \psi_1, \ldots, \psi_n$ be modal formulas and ψ be a result of substitution of $\psi_1, \ldots, \psi_n$ respectively for atoms $a_1, \ldots, a_n$ in φ. Then the formula $\mathtt{i}_2(\psi)$ equals the result of substitution of $\mathtt{i}_2(\psi_1), \ldots, \mathtt{i}_2(\psi_n)$ for atoms $a_1, \ldots, a_n$ in $\mathtt{i}_2(\varphi)$.*

Directly from definitions and used notation, we have:

Fact 6. *For any $\varphi, \psi \in \text{For}_m$*

1. $\mathtt{i}_2(\Box(\varphi)) = \Box^d(\mathtt{i}_2(\varphi))$,
2. $\mathtt{i}_2(\Diamond(\varphi)) = \Diamond^d(\mathtt{i}_2(\varphi))$,
3. $\mathtt{i}_2(\Box(\varphi \to \psi)) = \mathtt{i}_2(\varphi) \twoheadrightarrow_d \mathtt{i}_2(\psi)$.

Lemma 6.

1. *For any $\varphi \in \mathbf{CL}$, we have $\vdash_{\mathbf{D}_0} \Box^d \mathtt{i}_2(\varphi)$.*
2. *For any $\varphi \in \mathbf{D}$, we have $\vdash_{\mathbf{D}_0} \mathtt{i}_2(\varphi)$.*

Proof. *Ad* 1. Assume that $\varphi \in \mathbf{CL}$. Then there is a proof $\chi_1, \ldots, \chi_n$, of $\varphi = \chi_n$ on the basis of (??)–(??), (??) and substitution. Consider the sequence $\Box^d \mathtt{i}_2(\chi_1), \ldots, \Box^d \mathtt{i}_2(\chi_n)$. By induction on the length of the sequence one can see that it is a proof on the basis of $\vdash_{\mathbf{D}_0}$, since its elements are either elements of $\mathtt{i}_2(\Omega)$ or arise by the application of (??) or substitution. For the case of substitution it is enough to use Lemma ?? and apply the substitution of $\mathtt{i}_2(\psi_1), \ldots, \mathtt{i}_2(\psi_m)$ for $a_1, \ldots, a_m$ in $\Box^d \mathtt{i}_2(\chi_i)$, if in the initial proof a substitution of formulas $\psi_1, \ldots, \psi_m$ for atoms $a_1, \ldots, a_m$ in χ_i was applied.

Ad 2. Assume that $\varphi \in \mathbf{D}$. Then there is a proof $\chi_1, \ldots, \chi_n$ of φ, in the sense of the consequence relation $\Vdash$. Consider $\mathtt{i}_2(\Box\chi_1), \ldots, \mathtt{i}_2(\Box\chi_n), (\Box^d \mathtt{i}_2(\chi_n) \twoheadrightarrow_d \Diamond^d \mathtt{i}_2(\chi_n)), \Diamond^d \mathtt{i}_2(\chi_n), \mathtt{i}_2(\chi_n) = \mathtt{i}_2(\varphi)$. By induction on $1 \leqslant i \leqslant n$, by the point 1 of this lemma and Fact ??, we can easily show that $\vdash_{\mathbf{D}_0} \mathtt{i}_2(\Box\chi_i)$, that is, $\vdash_{\mathbf{D}_0} \Box^d \mathtt{i}_2(\chi_i)$, while $(\Box^d \mathtt{i}_2(\chi_n) \twoheadrightarrow_d \Diamond^d \mathtt{i}_2(\chi_n))$ is an instance of $\mathtt{i}_2(??)$, $\Diamond^d \mathtt{i}_2(\chi_n)$ follows from $\mathtt{i}_2(\Box\chi_n)$ and $(\Box^d \mathtt{i}_2(\chi_n) \twoheadrightarrow_d \Diamond^d \mathtt{i}_2(\chi_n))$ by (??); and $\mathtt{i}_2(\varphi)$ follows from $\Diamond^d \mathtt{i}_2(\chi_n)$ by (??). □

Thus, we obtain that:

Fact 7. *The following formulas are theses with respect to $\vdash_{\mathbf{D}_0}$:*

$$(\mathtt{i}_2(\mathtt{i}_1(\neg p)) \twoheadrightarrow_d \neg p) \tag{9}$$

$$(\mathtt{i}_2(\mathtt{i}_1(p \lor q)) \twoheadrightarrow_d p \lor q) \tag{10}$$

Proof. For (??), observe that $(\mathtt{i}_2(\mathtt{i}_1(\neg p)) \twoheadrightarrow_d \neg p) = \Box^d(\mathtt{i}_2(\mathtt{i}_1(\neg p)) \to_c \neg p) = \Box^d(\neg p \to_c \neg p) = \Box^d \mathtt{i}_2(\neg p \to_c \neg p)$. So (??) follows by Lemma ??.??.
For (??), one can see that $(\mathtt{i}_2(\mathtt{i}_1(p \lor q)) \twoheadrightarrow_d p \lor q) = \Box^d(\mathtt{i}_2(\mathtt{i}_1(p) \lor \mathtt{i}_1(q)) \to_c p \lor q) = \Box^d(\mathtt{i}_2(\mathtt{i}_1(p)) \lor \mathtt{i}_2(\mathtt{i}_1(q)) \to_c p \lor q) = \Box^d(p \lor q \to_c p \lor q)$. So again, the condition is obtained by Lemma ??.??. □

The above fact can be extended for the case of any formulas used instead of p and q. Moreover, it can be generalised to any compound formula which results in the below Lemma. Its proof goes similarly as the proof of Reference [?] (Fact 4.6). However, there are essential changes for the case of discussive operators, so we present it for the sake of completeness of considerations.

Fact 8. *For any $A \in \text{For}_d$,*

$$\vdash_{\mathbf{D}_0} \Box^d(\mathtt{i}_2(\mathtt{i}_1(A)) \to_c A) \tag{11}$$

$$\vdash_{\mathbf{D}_0} \Box^d(A \to_c \mathtt{i}_2(\mathtt{i}_1(A))). \tag{12}$$

Proof. The proof goes simultaneously for both theses by induction on the complexity of a formula A. The case of atoms follows by Lemma ??.??.

Assume that the inductive hypothesis holds for any formula of complexity smaller than the complexity of A. That is, for the following cases $A = \neg B$, $A = B \vee C$, $A = B \wedge_d C$, $A = B \rightarrow_d C$ and $A = B \leftrightarrow_d C$, where $B, C \in \text{For}_d$ we assume that:

$$\vdash_{D_0} (i_2(i_1(B)) \dashv_d B),$$
$$\vdash_{D_0} (i_2(i_1(C)) \dashv_d C),$$
$$\vdash_{D_0} (B \dashv_d i_2(i_1(B))),$$
$$\vdash_{D_0} (C \dashv_d i_2(i_1(C))).$$

in other words

$$\vdash_{D_0} \Box^d(i_2(i_1(B)) \rightarrow_c B), \tag{13}$$
$$\vdash_{D_0} \Box^d(i_2(i_1(C)) \rightarrow_c C), \tag{14}$$
$$\vdash_{D_0} \Box^d(B \rightarrow_c i_2(i_1(B))), \tag{15}$$
$$\vdash_{D_0} \Box^d(C \rightarrow_c i_2(i_1(C))). \tag{16}$$

For the case of '$\neg$' let us notice that $\vdash_{D_0} \Box^d i_2((p \rightarrow q) \rightarrow (\neg q \rightarrow \neg p))$, by Lemma ??(??), that is, $\vdash_{D_0} ((p \rightarrow_c q) \dashv_d (\neg q \rightarrow_c \neg p))$. So, since $\neg i_2(i_1(B)) = i_2(i_1(\neg B))$, using the substitution p/B, $q/i_2(i_1(B))$, (??) and (??), we get:

$$\vdash_{D_0} \Box^d(i_2(i_1(\neg B)) \rightarrow_c \neg B).$$

Similarly, using the substitution $p/i_2(i_1(B))$, q/B, (??) and (??) we obtain:

$$\vdash_{D_0} \Box^d(\neg B \rightarrow_c i_2(i_1(\neg B))).$$

For the case of '$\vee$' notice that $\vdash_{D_0} \Box^d(i_2((p \rightarrow q) \rightarrow ((r \rightarrow s) \rightarrow ((p \vee r) \rightarrow (q \vee s)))))$, by Lemma ??(??), that is, $\vdash_{D_0} \Box^d((p \rightarrow_c q) \rightarrow_c ((r \rightarrow_c s) \rightarrow_c ((p \vee r) \rightarrow_c (q \vee s))))$ or equivalently

$$(p \rightarrow_c q) \dashv_d ((r \rightarrow_c s) \rightarrow_c ((p \vee r) \rightarrow_c (q \vee s))) \tag{17}$$

is a thesis with respect to $\vdash_{D_0}$. Thus, since $i_2(i_1(B)) \vee i_2(i_1(B)) = i_2(i_1(B \vee C))$, using the substitution $p/i_2(i_1(B))$, q/B, $r/i_2(i_1(C))$, s/C into (??), by (??) and (??), we get:

$$\vdash_{D_0} \Box^d((i_2(i_1(C)) \rightarrow_c C) \rightarrow_c ((i_2(i_1(B)) \vee i_2(i_1(C))) \rightarrow_c (B \vee C)))$$

that is,

$$\vdash_{D_0} (i_2(i_1(C)) \rightarrow_c C) \dashv_d ((i_2(i_1(B)) \vee i_2(i_1(C))) \rightarrow_c (B \vee C))$$

And again, by (??) and (??), we obtain:

$$\vdash_{D_0} \Box^d(i_2(i_1(B \vee C)) \rightarrow_c (B \vee C)).$$

Similarly, using the substitution p/B, $q/i_2(i_1(B))$, r/C, $s/i_2(i_1(C))$, to (??), we obtain:

$$(B \rightarrow_c i_2(i_1(B))) \dashv_d$$
$$\dashv_d ((C \rightarrow_c i_2(i_1(C))) \rightarrow_c ((B \vee C) \rightarrow_c (i_2(i_1(B)) \vee i_2(i_1(C)))))$$

hence, by (??) and (??), and again by (??) and (??) we get:

$$\vdash_{D_0} \Box^d((B \vee C) \to_c i_2(i_1(B \vee C))).$$

For the case of '$\wedge_d$' first let us observe that

$$i_2(i_1(B \wedge_d C)) = i_2(i_1(B) \wedge \Diamond i_1(C)) =$$
$$\neg(\neg i_2(i_1(B)) \vee \neg i_2(\Diamond i_1(C))) = \quad (18)$$
$$\neg(\neg i_2(i_1(B)) \vee \neg((\neg p \vee p) \wedge_d i_2(i_1(C)))).$$

Hence by (??) and definitions of i_1 and i_2 we get $\vdash_{D_0} \neg(\neg q \vee \neg((\neg p \vee p) \wedge_d r)) \mathrel{\exists_d} (q \wedge_d r)$. While by (??), similarly we obtain: $\vdash_{D_0} (q \wedge_d r) \mathrel{\exists_d} \neg(\neg q \vee \neg((\neg p \vee p) \wedge_d r))$. Therefore, by substitution we obtain

$$\vdash_{D_0} \neg(\neg B \vee \neg((\neg p \vee p) \wedge_d C)) \mathrel{\exists_d} (B \wedge_d C) \quad (19)$$

and

$$\vdash_{D_0} (B \wedge_d C) \mathrel{\exists_d} \neg(\neg B \vee \neg((\neg p \vee p) \wedge_d C)) \quad (20)$$

Notice that $\Box((t \to q) \to (\Box(r \to s) \to (t \wedge \Diamond r \to q \wedge \Diamond s))) \in \mathbf{D}$, so by Lemma ??(??) and Fact ??, $\vdash_{D_0} \Box^d(i_2((p \to q) \to (\Box(r \to s) \to (t \wedge \Diamond r \to q \wedge \Diamond s))))$, that is,

$$\vdash_{D_0} (t \to_c q) \mathrel{\exists_d}$$
$$((r \mathrel{\exists_d} s) \to_c (\neg(\neg t \vee \neg((\neg p \vee p) \wedge_d r)) \to_c \neg(\neg q \vee \neg((\neg p \vee p) \wedge_d s)))). \quad (21)$$

By substitution: $t/i_2(i_1(B))$, q/B, $r/i_2(i_1(C))$, s/C, (??) and (??), we get:

$$\Box^d((i_2(i_1(C)) \mathrel{\exists_d} C) \to_c$$
$$(\neg(\neg i_2(i_1(B)) \vee \neg((\neg p \vee p) \wedge_d i_2(i_1(C)))) \to_c \neg(\neg B \vee \neg((\neg p \vee p) \wedge_d C))))$$

But by (??) applied to (??) we have

$$\vdash_{D_0} \Box^d(i_2(i_1(C)) \mathrel{\exists_d} C)$$

Therefore, again by (??):

$$\Box^d(\neg(\neg i_2(i_1(B)) \vee \neg((\neg p \vee p) \wedge_d i_2(i_1(C)))) \to_c \neg(\neg B \vee \neg((\neg p \vee p) \wedge_d C)))$$

But by Lemma ??(??) we have

$$\vdash_{D_0} (p \to_c q) \mathrel{\exists_d} ((q \to_c r) \to_c (p \to_c r)), \quad (22)$$

so by substitution $p/\neg(\neg i_2(i_1(B)) \vee \neg((\neg p \vee p) \wedge_d i_2(i_1(C))))$, $q/\neg(\neg B \vee \neg((\neg p \vee p) \wedge_d C))$ and $r/(B \wedge_d C)$, using (??) we obtain:

$$\Box^d((\neg(\neg B \vee \neg((\neg p \vee p) \wedge_d C)) \to_c (B \wedge_d C)) \to_c$$
$$(\neg(\neg i_2(i_1(B)) \vee \neg((\neg p \vee p) \wedge_d i_2(i_1(C)))) \to_c (B \wedge_d C)))$$

that is,

$$(\neg(\neg B \vee \neg((\neg p \vee p) \wedge_d C)) \to_c (B \wedge_d C)) \mathrel{\exists_d}$$
$$(\neg(\neg i_2(i_1(B)) \vee \neg((\neg p \vee p) \wedge_d i_2(i_1(C)))) \to_c (B \wedge_d C))$$

Thus, from (??) using (??) we infer

$$\Box^d(\neg(\neg i_2(i_1(B)) \vee \neg((\neg p \vee p) \wedge_d i_2(i_1(C)))) \to_c (B \wedge_d C))$$

what is the required formula due to the observation (??).

For the reverse implication, using (??) but substituting $t/B, q/i_2(i_1(B)), r/C, s/i_2(i_1(C))$ and next appyling (??) and (??) we infer

$$\Box^d((C \mathbin{\dashv_d} i_2(i_1(C))) \to_c$$
$$(\neg(\neg B \vee \neg((\neg p \vee p) \wedge_d C)) \to_c \neg(\neg i_2(i_1(B)) \vee \neg((\neg p \vee p) \wedge_d i_2(i_1(C))))))$$

While applying (??) for (??) we obtain

$$\vdash_{D_0} \Box^d(C \mathbin{\dashv_d} i_2(i_1(C)))$$

Thus, by (??):

$$\Box^d(\neg(\neg B \vee \neg((\neg p \vee p) \wedge_d C)) \to_c \neg(\neg i_2(i_1(B)) \vee \neg((\neg p \vee p) \wedge_d i_2(i_1(C)))))$$

Using an instance of (??), where $p/(B \wedge_d C)$, $q/\neg(\neg B \vee \neg((\neg p \vee p) \wedge_d C))$ and $r/\neg(\neg i_2(i_1(B)) \vee \neg((\neg p \vee p) \wedge_d i_2(i_1(C))))$, by (??) and (??), and then thanks to the above formula and again (??) we obtain

$$\Box^d((B \wedge_d C) \to_c \neg(\neg i_2(i_1(B)) \vee \neg((\neg p \vee p) \wedge_d i_2(i_1(C)))))$$

which is the required formula by the observation (??).

The case of '$\to_d$'. By definitions, we have:

$$\begin{aligned} i_2(i_1(B \to_d C)) &= i_2(\Diamond i_1(B) \to i_1(C)) = \\ &\neg((\neg p \vee p) \wedge_d i_2(i_1(B))) \vee i_2(i_1(C)) \end{aligned} \qquad (23)$$

Thus, by (??) and (??) we have

$$\neg((\neg p \vee p) \wedge_d q) \vee r \mathbin{\dashv_d} (q \to_d r)$$

$$(q \to_d r) \mathbin{\dashv_d} \neg((\neg p \vee p) \wedge_d q) \vee r$$

So by substitution we have

$$\neg((\neg p \vee p) \wedge_d B) \vee C \mathbin{\dashv_d} (B \to_d C) \qquad (24)$$

$$(B \to_d C) \mathbin{\dashv_d} \neg((\neg p \vee p) \wedge_d B) \vee C \qquad (25)$$

Let us notice that $\Box(\Box\Box(q \to r) \to \Box(\Diamond q \to \Diamond r)) \in \mathbf{K} \subseteq \mathbf{D}$. Hence, by Lemma ??(??): $\vdash_{D_0} i_2(\Box(\Box\Box(q \to r) \to \Box(\Diamond q \to \Diamond r)))$, that is, by definition of i_2 and by Fact ??:

$$\vdash_{D_0} \Box^d(\Box^d \Box^d(q \to_c r) \to_c \Box^d(\Diamond^d q \to_c \Diamond^d r)).$$

or equivalently

$$\vdash_{D_0} \Box^d \Box^d(q \to_c r) \mathbin{\dashv_d} \Box^d(((\neg p \vee p) \wedge_d q) \to_c ((\neg p \vee p) \wedge_d r)). \qquad (26)$$

so by substitution $q/B, r/\mathtt{i}_2(\mathtt{i}_1(B))$, we obtain that

$$\Box^d \Box^d (B \to_c \mathtt{i}_2(\mathtt{i}_1(B))) \dashv_d \quad (27)$$
$$\Box^d(((\neg p \vee p) \wedge_d B) \to_c ((\neg p \vee p) \wedge_d \mathtt{i}_2(\mathtt{i}_1(B)))).$$

is a thesis of D_0. Applying (??) to (??) we get $\Box^d \Box^d (B \to_c \mathtt{i}_2(\mathtt{i}_1(B)))$, hence by (??) we infer

$$\Box^d(((\neg p \vee p) \wedge_d B) \to_c ((\neg p \vee p) \wedge_d \mathtt{i}_2(\mathtt{i}_1(B)))).$$

By a substitution $q/C, r/\mathtt{i}_2(\mathtt{i}_1(C))$ in (??) and by (??) applied for (??) instead of (??) and finally by (??) we obtain:

$$\Box^d(((\neg p \vee p) \wedge_d C) \to_c ((\neg p \vee p) \wedge_d \mathtt{i}_2(\mathtt{i}_1(C)))). \quad (28)$$

As one can see, $\Box((\Diamond q \to \Diamond t) \to ((r \to s) \to ((\Diamond t \to r) \to (\Diamond q \to s)))) \in \mathbf{K} \subseteq \mathbf{D}$. Hence by Lemma ??(??) we obtain

$$(((\neg p \vee p) \wedge_d q) \to_c ((\neg p \vee p) \wedge_d t)) \dashv_d \quad (29)$$
$$((r \to_c s) \to_c ((((\neg p \vee p) \wedge_d t) \to_c r) \to_c (((\neg p \vee p) \wedge_d q) \to_c s)))$$

Now, we use the substitution: $q/B, t/\mathtt{i}_2(\mathtt{i}_1(B)), r/\mathtt{i}_2(\mathtt{i}_1(C)), s/C$ and applying (??) to (??) we infer

$$(\mathtt{i}_2(\mathtt{i}_1(C)) \to_c C) \dashv_d$$
$$((((\neg p \vee p) \wedge_d \mathtt{i}_2(\mathtt{i}_1(B))) \to_c \mathtt{i}_2(\mathtt{i}_1(C))) \to_c (((\neg p \vee p) \wedge_d B) \to_c C))$$

and next by (??) applied to (??) we have

$$(((\neg p \vee p) \wedge_d \mathtt{i}_2(\mathtt{i}_1(B))) \to_c \mathtt{i}_2(\mathtt{i}_1(C))) \dashv_d (((\neg p \vee p) \wedge_d B) \to_c C)$$

But by (??) it means that

$$\vdash_{D_0} \mathtt{i}_2(\mathtt{i}_1(B \to_d C)) \dashv_d (((\neg p \vee p) \wedge_d B) \to_c C) \quad (30)$$

Now, we use (??), (??) and the above formula and act similarly as in the case of '$\wedge_d$', we conclude that:

$$\vdash_{D_0} (\mathtt{i}_2(\mathtt{i}_1(B \to_d C)) \dashv_d (B \to_d C)).$$

For the reverse implication we apply the following substitution to (??): $q/\mathtt{i}_2(\mathtt{i}_1(B)), t/B, r/C, s/\mathtt{i}_2(\mathtt{i}_1(C))$ and we receive:

$$(((\neg p \vee p) \wedge_d \mathtt{i}_2(\mathtt{i}_1(B))) \to_c (\neg p \vee p) \wedge_d B)) \dashv_d$$
$$\dashv_d ((C \to_c \mathtt{i}_2(\mathtt{i}_1(C))) \to_c \quad (31)$$
$$((((\neg p \vee p) \wedge_d B) \to_c C) \to_c (((\neg p \vee p) \wedge_d \mathtt{i}_2(\mathtt{i}_1(B))) \to_c \mathtt{i}_2(\mathtt{i}_1(C)))))$$

Similarly as (??) we obtain $\Box^d(((\neg p \vee p) \wedge_d \mathtt{i}_2(\mathtt{i}_1(B))) \to_c ((\neg p \vee p) \wedge_d B))$. Hence, applying (??) to (??) we infer

$$(C \to_c \mathtt{i}_2(\mathtt{i}_1(C))) \dashv_d$$
$$((((\neg p \vee p) \wedge_d B) \to_c C) \to_c (((\neg p \vee p) \wedge_d \mathtt{i}_2(\mathtt{i}_1(B))) \to_c \mathtt{i}_2(\mathtt{i}_1(C))))$$

So again by (??) and (??) we obtain

$$(((\neg p \vee p) \wedge_d B) \rightarrow_c C) \dashv_d (((\neg p \vee p) \wedge_d i_2(i_1(B))) \rightarrow_c i_2(i_1(C)))$$

So acting as previously, by (??), (??), (??) and (??) we obtain

$$\vdash_{D_0} \Box^d((B \rightarrow_d C) \rightarrow_c i_2(i_1(B \rightarrow_d C))).$$

For the case of '$\leftrightarrow_d$' notice that by Lemma ??(??), $\vdash_{D_0} \Box^d(i_2((p \rightarrow q) \rightarrow ((q \rightarrow p) \rightarrow (p \leftrightarrow q))))$, that is, $\vdash_{D_0} \Box^d((p \rightarrow_c q) \rightarrow_c ((q \rightarrow_c p) \rightarrow_c \neg(\neg(p \rightarrow_c q) \vee \neg(q \rightarrow_c p))))$. Thus, by substitutions, inductive hypotheses (??), (??), (??), (??) and applying (??) and definition of i_2 we get:

$$\Box^d \neg(\neg(B \rightarrow_c i_2(i_1(B))) \vee \neg(i_2(i_1(B)) \rightarrow_c B)) \tag{32}$$

$$\Box^d \neg(\neg(C \rightarrow_c i_2(i_1(C))) \vee \neg(i_2(i_1(C)) \rightarrow_c C)) \tag{33}$$

Hence, by the use of the following thesis of **K** $\subseteq$ **D**:

$$\Box((r \leftrightarrow s) \rightarrow (\Box(t \leftrightarrow q) \rightarrow (\Box\Box(r \leftrightarrow s) \rightarrow (((\Diamond q \rightarrow s) \wedge \Diamond(\Diamond s \rightarrow q))$$
$$\rightarrow ((\Diamond t \rightarrow r) \wedge \Diamond(\Diamond r \rightarrow t)))))),$$

by Lemma ??(??), Fact ??, (??), (??) and substitution $r/C, s/i_2(i_1(C)), t/B, q/i_2(i_1(B))$ we get:

$$\Box^d(\Box^d \neg(\neg(B \rightarrow_c i_2(i_1(B))) \vee \neg(i_2(i_1(B)) \rightarrow_c B)) \rightarrow_c$$
$$(\Box^d \Box^d \neg(\neg(\neg C \vee i_2(i_1(C))) \vee \neg(\neg i_2(i_1(C)) \vee C))$$
$$\rightarrow_c (\neg((\neg \Diamond^d i_2(i_1(B)) \rightarrow_c i_2(i_1(C))) \vee \neg \Diamond^d(\Diamond^d i_2(i_1(C)) \rightarrow_c i_2(i_1(B))))$$
$$\rightarrow_c \neg(\neg(\Diamond^d B \rightarrow_c C) \vee \neg \Diamond^d(\Diamond^d C \rightarrow_c B)))))$$

By the result of application of (??) to (??) and again (??) we infer

$$\Box^d(\Box^d \Box^d \neg(\neg(\neg C \vee i_2(i_1(C))) \vee \neg(\neg i_2(i_1(C)) \vee C))$$
$$\rightarrow_c (\neg((\neg \Diamond^d i_2(i_1(B)) \rightarrow_c i_2(i_1(C))) \vee \neg \Diamond^d(\Diamond^d i_2(i_1(C)) \rightarrow_c i_2(i_1(B))))$$
$$\rightarrow_c \neg(\neg(\Diamond^d B \rightarrow_c C) \vee \neg \Diamond^d(\Diamond^d C \rightarrow_c B))))$$

Now we twice apply (??) to (??) and again use (??), as a result we get:

$$\Box^d(\neg(\neg(\Diamond^d i_2(i_1(B)) \rightarrow_c i_2(i_1(C))) \vee \neg \Diamond^d(\Diamond^d i_2(i_1(C)) \rightarrow_c i_2(i_1(B)))) \rightarrow_c$$
$$\neg(\neg(\Diamond^d B \rightarrow_c C) \vee \neg \Diamond^d(\Diamond^d C \rightarrow_c B))).$$

Thus, since by definitions of the functions i_1 and i_2, $\neg(\neg(\Diamond^d i_2(i_1(B)) \rightarrow_c i_2(i_1(C))) \vee \neg \Diamond^d(\Diamond^d i_2(i_1(C)) \rightarrow_c i_2(i_1(B)))) = i_2((\Diamond i_1(B) \rightarrow i_1(C)) \wedge \Diamond(\Diamond i_1(C) \rightarrow i_1(B))) = i_2(i_1(B \leftrightarrow_d C))$, we get:

$$\Box^d(i_2(i_1(B \leftrightarrow_d C)) \rightarrow_c \neg(\neg(\Diamond^d B \rightarrow_c C) \vee \neg \Diamond^d(\Diamond^d C \rightarrow_c B))).$$

From axiom (??) we obtain:

$$\Box^d(\neg(\neg(\Diamond^d B \rightarrow_c C) \vee \neg \Diamond^d(\Diamond^d C \rightarrow_c B)) \rightarrow_c (B \leftrightarrow_d C)).$$

Thus, again by a substitution to $i_2(\Box((p \rightarrow q) \rightarrow ((q \rightarrow r) \rightarrow (p \rightarrow r))))$ and the application of (??), we get:

$$\vdash_{D_0} \Box^d(i_2(i_1(B \leftrightarrow_d C)) \rightarrow_c (B \leftrightarrow_d C)).$$

Similarly we show:
$$\vdash_{D_0} \Box^d((B \leftrightarrow_d C) \rightarrow_c i_2(i_1(B \leftrightarrow_d C))).$$

By induction on the complexity of formulas, we obtain (??) and (??). □

Theorem 1. *The set of theses with respect to the consequence relation $\vdash_{D_0}$ equals D_0.*

Proof. Let $A \in D_0$. By Lemma ??, $i_1(A) \in D$. Thus, by Lemma ??, there is a proof $\chi_1, \ldots, \chi_n = i_1(A)$ such that for each $i \in \{1, \ldots, n\}$, χ_i is one of the formulas $(\Box??)$–$(\Box??)$, $(\Box??)$, $(\Box??)$, $(\Box??)$ or χ_i is a result of the application of substitution, (??), (??), (??), (??), for some formulas preceding χ_i in the sequence. Let us consider the following sequence of formulas $i_2(\chi_1), \ldots, i_2(\chi_{n-1}), i_2(i_1(A))$, $\Box^d(i_2(i_1(A)) \rightarrow_c A)$, A. By induction on i we see that the fragment of the sequence $i_2(\chi_1), \ldots, i_2(\chi_n) = i_2(i_1(A))$ is a proof in the given axiomatic system, while the formula $\Box^d(i_2(i_1(A)) \rightarrow_c A)$ is a thesis of the system, by Fact ??. So the formula A completes the proof as a result of application of the rule (??) to two preceding formulas. Thus, $\vdash_{D_0} A$

Now, for the reverse direction, assume that for a given formula A there is a proof $C_1, \ldots, C_n = A$ of the formula A within the considered deductive system. By the definition of D_0, it is enough to prove that for any $i \in \{1, \ldots, n\}$, $\lozenge i_1(C_i) \in D$ or, equivalently, $i_1(C_i) \in D$.

Firstly, let us consider the case of axioms. Suppose that $C_i \in i_2(\Omega)$. For the case $(\Box??)$–$(\Box??)$ that is, when $C_i = i_2(\Box\varphi)$, for some $\varphi \in \mathbf{CL} \subseteq D$—it is enough to apply Lemma ??(??). Then $i_1(i_2(\Box\varphi)) \leftrightarrow \Box\varphi \in D$, by Lemma ??. Therefore, by Lemma ??(3), $i_1(i_2(\Box\varphi)) = i_1(C_i) \in D$. In the case of other axioms from the set $i_2(\Omega)$, we act similarly.

Ad(??): $i_1(\Box^d(i_2(i_1(q \wedge_d r)) \rightarrow_c (q \wedge_d r))) = i_1(\Box^d(i_2(q \wedge \lozenge r) \rightarrow_c (q \wedge_d r))) = i_1(\neg((\neg p \vee p) \wedge_d \neg(\neg(\neg q \vee \neg((\neg p \vee p) \wedge_d r)) \rightarrow_c (q \wedge_d r)))) = \neg((\neg p \vee p) \wedge \lozenge \neg(\neg(\neg q \vee \neg((\neg p \vee p) \wedge \lozenge r)) \rightarrow_c (q \wedge \lozenge r)))$. But $\neg((\neg p \vee p) \wedge \lozenge \neg(\neg(\neg q \vee \neg((\neg p \vee p) \wedge \lozenge r)) \rightarrow_c (q \wedge \lozenge r))) \leftrightarrow \Box((q \wedge \lozenge r) \rightarrow_c (q \wedge \lozenge r))$ belongs to $\mathbf{D}$. Thus, (??) $\in D_0$, thanks to Lemma ??.

Ad (??): $i_1(\Box^d(i_2(i_1(q \rightarrow_d r)) \rightarrow_c (q \rightarrow_d r))) = i_1(\neg((\neg p \vee p) \wedge_d \neg(((\neg p \vee p) \wedge_d q \rightarrow_c r) \rightarrow_c (q \rightarrow_d r)))) = \neg((\neg p \vee p) \wedge \lozenge \neg(((\neg p \vee p) \wedge \lozenge q \rightarrow_c r) \rightarrow_c (\lozenge q \rightarrow r)))$. However, $\neg((\neg p \vee p) \wedge \lozenge \neg(((\neg p \vee p) \wedge \lozenge q \rightarrow_c r) \rightarrow_c (\lozenge q \rightarrow r))) \leftrightarrow \Box((\lozenge q \rightarrow_c r) \rightarrow_c (\lozenge q \rightarrow r))$ is a thesis of $\mathbf{D}$. Thus, (??) $\in D_0$, by Lemma ??.

Ad (??): $i_1(\Box^d(i_2(i_1(q \leftrightarrow_d r)) \rightarrow_c (q \leftrightarrow_d r))) = i_1(\Box^d(i_2((\lozenge q \rightarrow r) \wedge \lozenge(\lozenge r \rightarrow q)) \rightarrow_c (q \leftrightarrow_d r))) = i_1(\neg((\neg p \vee p) \wedge_d \neg(\neg(\neg((\neg p \vee p) \wedge_d q \rightarrow_c r) \vee \neg((\neg p \vee p) \wedge ((\neg p \vee p) \wedge_d r \rightarrow_c q))) \rightarrow_c (q \leftrightarrow_d r)))) = \neg((\neg p \vee p) \wedge \lozenge \neg(\neg(\neg((\neg p \vee p) \wedge \lozenge q \rightarrow_c r) \vee \neg((\neg p \vee p) \wedge \lozenge((\neg p \vee p) \wedge \lozenge r \rightarrow_c q))) \rightarrow_c ((\lozenge q \rightarrow r) \wedge \lozenge(\lozenge r \rightarrow q))))$. Notice that $\neg((\neg p \vee p) \wedge \lozenge \neg(\neg(\neg((\neg p \vee p) \wedge \lozenge q \rightarrow_c r) \vee \neg((\neg p \vee p) \wedge \lozenge((\neg p \vee p) \wedge \lozenge r \rightarrow_c q))) \rightarrow_c ((\lozenge q \rightarrow r) \wedge \lozenge(\lozenge r \rightarrow q)))) \leftrightarrow \Box((\lozenge q \rightarrow_c r) \wedge \lozenge(\lozenge r \rightarrow_c q) \rightarrow_c ((\lozenge q \rightarrow r) \wedge \lozenge(\lozenge r \rightarrow q)))$. Thus, (??) belongs to D_0, similarly, by Lemma ??.

Ad (??): $i_1(\Box^d((q \wedge_d r) \rightarrow_c i_2(i_1(q \wedge_d r)))) = \neg((\neg p \vee p) \wedge \lozenge \neg((q \wedge \lozenge r) \rightarrow_c \neg(\neg q \vee \neg((\neg p \vee p) \wedge \lozenge r))))$. Thus again, (??) $\in D_0$.

Ad (??): $i_1((q \rightarrow_d r) \dashv_d i_2(i_1(q \rightarrow_d r))) = \neg((\neg p \vee p) \wedge \lozenge \neg((\lozenge q \rightarrow r) \rightarrow_c ((\neg p \vee p) \wedge \lozenge q \rightarrow_c r)))$. Hence (??) $\in D_0$.

Ad (??): $i_1((q \leftrightarrow_d r) \dashv_d i_2(i_1(q \leftrightarrow_d r))) = \neg((\neg p \vee p) \wedge \lozenge \neg(((\lozenge q \rightarrow r) \wedge \lozenge(\lozenge r \rightarrow q)) \rightarrow_c \neg(\neg((\neg p \vee p) \wedge \lozenge q \rightarrow_c r) \vee \neg((\neg p \vee p) \wedge \lozenge((\neg p \vee p) \wedge \lozenge r \rightarrow_c q)))))$. As one can see (??) $\in D_0$.

Ad (??). Assume that $\Box^d A$ and $\Box^d(A \rightarrow_c B)$ belong to D_0, that is, $\lozenge i_1(\neg((\neg p \vee p) \wedge_d \neg A))$ and $\lozenge i_1(\neg((\neg p \vee p) \wedge_d \neg(A \rightarrow_c B)))$ are theses of $\mathbf{D}$. By Fact ?? equivalently, $(\neg((\neg p \vee p) \wedge \lozenge \neg i_1(A))) \in D$ and $(\neg((\neg p \vee p) \wedge \lozenge \neg(i_1(A) \rightarrow_c i_1(B)))) \in D$, in other words $\Box i_1(A) \in D$ and $\Box(i_1(A) \rightarrow_c i_1(B)) \in D$, hence of course $\Box i_1(B) \in D$. Equivalently, $\neg \lozenge \neg i_1(B) \in D$ and $\neg((\neg p \vee p) \wedge \lozenge \neg i_1(B)) \in D$, but by definition of i_1, $\neg((\neg p \vee p) \wedge \lozenge \neg i_1(B)) = i_1(\neg((\neg p \vee p) \wedge_d \neg B))$, hence also $i_1(\neg((\neg p \vee p) \wedge_d \neg B)) \in D$, therefore $\Box^d B \in D_0$.

Ad (??). Suppose that A and $\Box^d(A \rightarrow_c B)$ belong to D_0, that is, $\lozenge i_1(A)$ and $\lozenge i_1(\Box^d(A \rightarrow_c B))$ belong to $\mathbf{D}$; and so $\lozenge(\neg((\neg p \vee p) \wedge \lozenge \neg(\neg i_1(A) \vee i_1(B)))) \in D$. Therefore, $\lozenge \Box(i_1(A) \rightarrow i_1(B)) \in D$ and by Fact ??: $\Box(i_1(A) \rightarrow i_1(B)) \in D$. Hence $\lozenge i_1(B) \in D$, so $B \in D_0$. by definition of $\mathbf{D}$.

Ad (??): Let $\Box^d A \in D_0$, that is, $\Diamond \mathtt{i}_1(\neg((\neg p \vee p) \wedge_d \neg A)) \in \mathbf{D}$; and in other words $\Diamond(\neg((\neg p \vee p) \wedge \Diamond \neg \mathtt{i}_1(A))) \in \mathbf{D}$. But this means that $\Diamond \Box \mathtt{i}_1(A) \in \mathbf{D}$ and by Fact ??, also $\Box \mathtt{i}_1(A) \in \mathbf{D}$. Hence by necessitation $\Box \Box \mathtt{i}_1(A) \in \mathbf{D}$, so again by necessitation and (??) also $\Diamond \Box \Box \mathtt{i}_1(A) \in \mathbf{D}$. Equivalently, $\Diamond \neg \Diamond \neg \neg \Diamond \neg \mathtt{i}_1(A) \in \mathbf{D}$. This can be rewritten as $\Diamond \neg ((\neg p \vee p) \wedge \Diamond \neg \neg ((\neg p \vee p) \wedge \Diamond \neg \mathtt{i}_1(A))) \in \mathbf{D}$. That is $\Diamond \mathtt{i}_1(\neg((\neg p \vee p) \wedge_d \neg \neg((\neg p \vee p) \wedge_d \neg A))) \in \mathbf{D}$. Hence $\Diamond \mathtt{i}_1(\Box^d \Box^d A) \in \mathbf{D}$, that is, $\Box^d \Box^d A \in D_0$.

Ad (??): Let $\Diamond^d A \in D_0$, that is, $\Diamond \mathtt{i}_1((\neg p \vee p) \wedge_d A) \in \mathbf{D}$, equivalently $\Diamond(\neg p \vee p) \wedge \Diamond \mathtt{i}_1(A) \in \mathbf{D}$,; and so $\Diamond \Diamond \mathtt{i}_1(A) \in \mathbf{D}$. Then by Fact ??, also $\Diamond \mathtt{i}_1(A) \in \mathbf{D}$; and so $A \in D_0$.

Finally, we consider the case of substitution. Assume that A belongs to D_0, that is, $\Diamond \mathtt{i}_1(A)$ is a thesis of $\mathbf{D}$. Now, let us consider a substitution of formulas $A_1, \ldots, A_n$ for variables $a_1, \ldots, a_n$ in A. Let us also consider $\mathtt{i}_1(A)(a_1/\mathtt{i}_1(A_1), \ldots, a_n/\mathtt{i}_1(A_n))$ the result of substitution of $\mathtt{i}_1(A_1), \ldots, \mathtt{i}_n/A_n$ for variables $a_1, \ldots, a_n$ in $\mathtt{i}_1(A)$. By Lemma ?? the following holds: $(\mathtt{i}_1(A))(a_1/\mathtt{i}_1(A_1), \ldots, a_n/\mathtt{i}_1(A_n)) = \mathtt{i}_1(A(a_1/A_1, \ldots, a_n/A_n))$. Since $\mathbf{D}$ as a modal logic is closed on substitution, $(\Diamond \mathtt{i}_1(A))(a_1/A_1, \ldots, a_n/A_n) \in \mathbf{D}$, but by the last equation and the features of substitution, $(\Diamond \mathtt{i}_1(A))(a_1/\mathtt{i}_1(A_1), \ldots, a_n/\mathtt{i}_1(A_n)) = \Diamond((\mathtt{i}_1(A))(a_1/\mathtt{i}_1(A_1), \ldots, a_n/\mathtt{i}_1(A_n))) = \Diamond \mathtt{i}_1(A(a_1/A_1, \ldots, a_n/A_n))$. By definition of D_0, it means that $A(a_1/A_1, \ldots, a_n/A_n) \in D_0$. □

6. Conclusions

We gave a syntactic characterisation of the minimal discussive logic. This is as an initial step in our investigations on other variants of discussive logics obtained by other cases of relations that connect participants of a discussion.

Author Contributions: Conceptualization, K.M.-N. and M.N.; formal analysis, K.M.-N. and M.N.; investigation, K.M.-N. and M.N.; methodology, K.M.-N.; validation, K.M.-N.; writing–original draft preparation, K.M.-N. and M.N.; writing–review and editing, K.M.-N. and M.N.; supervision, M.N.; funding acquisition, M.N.

Funding: The authors of this work benefited from support provided by Polish National Science Centre (Narodowe Centrum Nauki), grant number 2016/23/B/HS1/00344.

Conflicts of Interest: The authors declare no conflict of interest.

References

- Mruczek-Nasieniewska, K.; Nasieniewski, M.; Pietruszczak, A. A modal extension of Jaśkowski's discussive logic D_2. *Log. J. IGPL* **2019**, *27*, 451–477. [CrossRef]
- Błaszczuk, J. J.; Dziobiak, W. An axiomatization of M^n-counterparts for some modal logics. *Reports Math. Log.* **1976**, *6*, 3–6.
- Furmanowski, T. Remarks on discussive propositional calculus. *Studia Log.* **1975**, *34*, 39–43. [CrossRef]
- Błaszczuk, J.J.; Dziobiak, W. Remarks on Perzanowski's modal system. *Bull. Sect. Log.* **1975**, *4*, 57–64.
- Perzanowski, J. On M-fragments and L-fragments of normal modal propositional logics. *Reports Math. Log.* **1975**, *5*, 63–72.
- Nasieniewski, M.; Pietruszczak, A. A method of generating modal logics defining Jaśkowski's discussive logic D2. *Studia Logica* **2011**, *97*, 161–182. [CrossRef]
- Nasieniewski, M.; Pietruszczak, A. On the weakest modal logics defining Jaśkowski's logic D2 and the D2-consequence. *Bull. Sect. Log.* **2012**, *41*, 215–232.
- Nasieniewski, M.; Pietruszczak, A. A method of generating modal logics defining Jaśkowski's discussive D2-consequence. In *Logic, Reasoning & Rationality*, Weber, E., Wouters, D., Meheus, J., Eds.; Springer: Dordrecht, The Netherlands, 2014; Volume 5, pp. 95–123. [CrossRef]
- Urchs, M. On the role of adjunction in para(in)consistent logic. In *Paraconsistency: The Logical Way to the Inconsistent, Proceedings of the Second World Congress on Paraconsistency, São Paulo, Brazil, May 12–19, 2000*; Carnielli, W. A., Coniglio, M. E., Loffredo D'Ottaviano, I. M., Eds.; Marcel Dekker: New York, NY, USA, 2002; pp. 487–499.
- Jaśkowski, S. Rachunek zdań dla systemów dedukcyjnych sprzecznych. *Studia Societatis Scientiarum Torunensis* **1948**, *1*, 57–77. (In English: Propositional calculus for contradictory deductive systems. *Studia*

Logica **1969**, *24*, 143–157, doi:10.1007/BF02134311. A propositional Calculus for inconsistent deductive systems. *Log. Log. Philos.* **1999**, *7*, 35–56, doi:10.12775/LLP.1999.003.) [CrossRef]

. Jaśkowski, S. O koniunkcji dyskusyjnej w rachunku zdań dla systemów dedukcyjnych sprzecznych. *Studia Societatis Scientiarum Torunensis* **1949**, *1*, 171–172. (In English: On the discussive conjunction in the propositional calculus for inconsistent deductive systems. *Log. Log. Philos.* **1999**, *7*, 57–59, doi:10.12775/LLP.1999.004.) [CrossRef]

. Bull, R. A.; Segerberg, K. Basic Modal Logic. In *Handbook of Philosophical Logic*; Gabbay, D. M., Guenthner, F., Eds.; D. Reidel Publishing Company: Dordrecht, Holland, 1984; Volume 3, pp. 1–88. [CrossRef]

. Segerberg, K. *An Essay in Classical Modal Logic*; Uppsala Universitet: Uppsala, Sweded, 1971; Volume 1–2.

. Kotas, J. The axiomatization of S. Jaśkowski's discussive system. *Studia Logica* **1974**, *33*, 195–200. [CrossRef]

. da Costa, N.C.A.; Dubikajtis, L. On Jaśkowski discussive logic. In *Non-Classical Logics, Model Theory and Computability*; Arruda, A. I., da Costa, N. C. A., Chuaqui, R., Eds.; North-Holland: New York, NY, USA, 1977; pp. 37–56.

. Ciuciura, J. A new real axiomatization of the discursive logic D2. In *Handbook of Paraconsistency*; Beziau, J. Y., Carnielli, W., Gabbay, D. M., Eds.; College Publications: London, UK, 2007; pp. 427–437.

. Omori, H.; Alama, J. Axiomatizing Jaśkowski's Discussive Logic $\mathbf{D_2}$. *Studia Logica* **2018**, *106*, 1163–1180. [CrossRef]

© 2019 by the authors. Licensee MDPI, Basel, Switzerland. This article is an open access article distributed under the terms and conditions of the Creative Commons Attribution (CC BY) license (http://creativecommons.org/licenses/by/4.0/).

Article

A Note on Fernández–Coniglio's Hierarchy of Paraconsistent Systems

Janusz Ciuciura

Department of Logic and Methodology of Science, Institute of Philosophy, Faculty of Philosophy and History, University of Łódź, Lindleya 3/5, 90–131 Łódź, Poland; janusz.ciuciura@uni.lodz.pl

Received: 8 October 2019; Accepted: 23 March 2020; Published: 30 March 2020

Abstract: A logic is called explosive if its consequence relation validates the so-called principle of ex contradictione sequitur quodlibet. A logic is called paraconsistent so long as it is not explosive. Sette's calculus P^1 is widely recognized as one of the most important paraconsistent calculi. It is not surprising then that the calculus was a starting point for many research studies on paraconsistency. Fernández–Coniglio's hierarchy of paraconsistent systems is a good example of such an approach. The hierarchy is presented in Newton da Costa's style. Therefore, the law of non-contradiction plays the main role in its negative axioms. The principle of ex contradictione sequitur quodlibet has been marginalized: it does not play any leading role in the hierarchy. The objective of this paper is to present an alternative axiomatization for the hierarchy. The main idea behind it is to focus explicitly on the (in)validity of the principle of ex contradictione sequitur quodlibet. This makes the hierarchy less complex and more transparent, especially from the viewpoint of paraconsistency.

Keywords: paraconsistent logic; paraconsistency; Sette's calculus; the law of explosion; the principle of ex contradictione sequitur quodlibet

---

1. Introduction

Let *var* denote a (non-empty) denumerable set of all propositional variables. The set of formulas $\mathcal{F}$ is inductively defined in the following way:

$$\varphi ::= p \mid \sim\alpha \mid \alpha \to \alpha$$

where $p \in var$, $\alpha \in \mathcal{F}$ and the symbols $\sim$, $\to$ denote negation and implication, respectively. A logic is a pair $\langle \mathcal{L}, \vdash \rangle$ consisting of a sentential language $\mathcal{L}$ and a consequence relation $\vdash$ defined on the (non-empty) set of formulas $\mathcal{F}$. A logic is called explosive if its consequence relation validates the principle of ex contradictione sequitur quodlibet, i.e., $\{\alpha, \sim\alpha\} \vdash \beta$, for any formulas α, β. "Paraconsistent logic is defined negatively: any logic is paraconsistent as long as it is not explosive" (cit.per [?]), or, to be more precise,

Definition 1. *A logic $\langle \mathcal{L}, \vdash \rangle$ is said to be paraconsistent if $\{\alpha, \sim \alpha\} \nvdash \beta$, for some formulas α, β.*

Already at first glance, it is striking that the definition is very broad as it includes some logics that have potentially nothing in common with paraconsistency (cf. [?], p. 19). Nonetheless, the definition reveals a tendency to view paraconsistent logic through the lens of negation understood as a connective symbol rather than a truth-function (cf. [?]. For a more extensive discussion on the paraconsistency, see, e.g., [? ? ?].).

In the early 1970s of the Twentieth Century, Sette published a paper devoted to one of the most remarkable paraconsistent calculi. The calculus, denoted as P^1, has some unusual properties: it behaves in a paraconsistent way only at the level of propositional variables, that is a pair of the formulas α and $\sim \alpha$ yields any β if, and only if the formula α is not a propositional variable.

The calculus P^1 is axiomatized by the following axiom schemas:

(A1) $\alpha \to (\beta \to \alpha)$
(A2) $(\alpha \to (\beta \to \gamma)) \to ((\alpha \to \beta) \to (\alpha \to \gamma))$
(A3) $(\sim\alpha \to \sim\beta) \to ((\sim\alpha \to \sim\sim\beta) \to \alpha)$
(A4) $\sim(\alpha \to \sim\sim\alpha) \to \alpha$
(A5) $(\alpha \to \beta) \to \sim\sim(\alpha \to \beta)$

and the rule of detachment (MP) $\alpha \to \beta, \alpha \,/\, \beta$.

The connectives of $\sim$ and $\to$ are taken here as primitives. As for the other connectives such as the conjunction, disjunction, and equivalence, they are introduced via the definitions ([?], pp. 178–179):

$$\alpha \wedge \beta =_{df} (((\alpha \to \alpha) \to \alpha) \to \sim((\beta \to \beta) \to \beta)) \to \sim(\alpha \to \sim\beta)$$
$$\alpha \vee \beta =_{df} (\alpha \to \sim\sim\alpha) \to (\sim\alpha \to \beta)$$
$$\alpha \leftrightarrow \beta =_{df} (\alpha \to \beta) \wedge (\beta \to \alpha).$$

The definitions are complex and often too awkward to handle. More user-friendly definitions are given in [?] (pp. 8–9 of the preprint) and [?] (p. 59):

$$\alpha \wedge \beta =_{df} \sim(\alpha \to \sim(\sim\beta \to \beta))$$
$$\alpha \vee \beta =_{df} \sim(\sim\alpha \to \alpha) \to \beta$$
$$\alpha \leftrightarrow \beta =_{df} (\alpha \to \beta) \wedge (\beta \to \alpha).$$

It is noteworthy that the disjunction connective can be also defined as in the three-valued Lukasiewicz logic, namely, $\alpha \vee \beta =_{df} (\alpha \to \beta) \to \beta$ (cf. [?], Section 2).

Many important theorems, which hold in the classical propositional calculus, can be proven for Sette's system, too. Below we recall some of them needed for our further discussion.

Theorem 1. *The deduction theorem holds for P^1.*

Proof. It is enough to observe that P^1 includes (A1), (A2), and the sole rule of inference in P^1 is (MP). □

Theorem 2. *For every $\Gamma, \Delta \subseteq \mathcal{F}$ and $\alpha, \beta, \gamma \in \mathcal{F}$, we have:*

1. *if $\alpha \in \Gamma$, then $\Gamma \vdash_{P^1} \alpha$,*
2. *if $\Gamma \subseteq \Delta$ and $\Gamma \vdash_{P^1} \alpha$, then $\Delta \vdash_{P^1} \alpha$,*
3. *if $\Delta \vdash_{P^1} \alpha$ and, for every $\beta \in \Delta$ it is true that $\Gamma \vdash_{P^1} \beta$, then $\Gamma \vdash_{P^1} \alpha$,*
4. *if $\Gamma \cup \{\alpha\} \vdash_{P^1} \gamma$ and $\Delta \vdash_{P^1} \alpha$, then $\Gamma \cup \Delta \vdash_{P^1} \gamma$*

   (in particular, if $\Gamma \cup \{\alpha\} \vdash_{P^1} \gamma$ and $\varnothing \vdash_{P^1} \alpha$, then $\Gamma \vdash_{P^1} \gamma$),
5. *$\Gamma \vdash_{P^1} \alpha$ iff for some finite $\Delta \subseteq \Gamma$, $\Delta \vdash_{P^1} \alpha$.*

Proof. The proof proceeds analogously to that of the classical propositional calculus. We refer the reader to [? ?] for details. □

Theorem 3. *Some (weaker) variants of the indirect deduction theorem hold for P^1, viz.:*

1. *if $\Gamma, \alpha \vdash_{P^1} \{\sim\beta, \sim\sim\beta\}$, then $\Gamma \vdash_{P^1} \sim\alpha$,*
2. *if $\Gamma, \sim\alpha \vdash_{P^1} \{\sim\beta, \sim\sim\beta\}$, then $\Gamma \vdash_{P^1} \alpha$,*

3. if $\Gamma, \alpha \to \beta \vdash_{P^1} \{\gamma \to \delta, \sim(\gamma \to \delta)\}$, then $\Gamma \vdash_{P^1} \sim(\alpha \to \beta)$,
4. if $\Gamma, \sim(\alpha \to \beta) \vdash_{P^1} \{\gamma \to \delta, \sim(\gamma \to \delta)\}$, then $\Gamma \vdash_{P^1} \alpha \to \beta$,

for every $\Gamma \subseteq \mathcal{F}$ and $\alpha, \beta, \gamma, \delta \in \mathcal{F}$. Note that the notation $\Gamma \vdash \{\phi, \psi\}$ is an abbreviation of '$\Gamma \vdash \{\phi\}$ and $\Gamma \vdash \{\psi\}$'.

Sette's calculus is sound and complete with respect to the matrix $\mathcal{M}_{P1} = \langle \{T_0, T_1, F\}, \{T_0, T_1\}, \sim, \to \rangle$, where $\{T_0, T_1, F\}$ and $\{T_0, T_1\}$ are the sets of logical and designated values, respectively. The connectives of $\to$ and $\sim$ are defined by the truth tables:

$\to$	T_0	T_1	F
T_0	T_0	T_0	F
T_1	T_0	T_0	F
F	T_0	T_0	T_0

$\sim$	
T_0	F
T_1	T_0
F	T_0

A P^1-valuation is any function v from the set of formulas to the set of logical values ($v : \mathcal{F} \longrightarrow \{T_0, T_1, F\}$, in symbols) compatible with the above truth-tables (see [?], pp. 176–178). A P^1-tautology is a formula that under every valuation v takes on the designated values $\{T_0, T_1\}$.

The logical meaning of the P^1-valuation is clear, but it was never stated in [?] how to interpret philosophically the three-valued semantics. This gave an impulse for further research, and several new semantics for the calculus were proposed (see, e.g., [? ? ? ? ?]). Notice that the principle of ex contradictione sequitur quodlibet does not play any significant part in P^1. Metaphorically speaking, paraconsistency is hidden somewhere between the lines of Sette's paper. Only at one point in his whole paper does Sette refer to paraconsistency: "(...) N.C.A da Costa presents a hierarchy C^n ($1 < n < \omega$) of propositional calculi which can be used as subjacent propositional logics for inconsistent (but not absolutely inconsistent) formal systems. The purpose of this note is to present a new propositional calculus P^1 which can be used as subjacent logic for inconsistent (but not absolutely inconsistent) formal systems (...)". ([?], p. 173.). In [?], we proposed an alternative axiomatization for P^1. The idea behind it was to focus explicitly on the (in)validity of ex contradictione sequitur quodlibet, or equivalently, the so-called law of explosion (DS) $\alpha \to (\sim\alpha \to \beta)$. This concept is directly reflected below in the axiomatization.

Remark 1. *The calculus P^1 can be axiomatized by the set of formulas:*
  (A1) $\alpha \to (\beta \to \alpha)$
  (A2) $(\alpha \to (\beta \to \gamma)) \to ((\alpha \to \beta) \to (\alpha \to \gamma))$
  (PL) $((\alpha \to \beta) \to \alpha) \to \alpha$
  (DS$^\sim$) $\sim\alpha \to (\sim\sim\alpha \to \beta)$
  (DS$^\to$) $(\alpha \to \beta) \to (\sim(\alpha \to \beta) \to \gamma))$
  (CM) $(\sim\alpha \to \alpha) \to \alpha$
with (MP) as the only primitive rule (see [?], for details).

In [?], an interesting hierarchy of the paraconsistent calculi starting from P^1 was proposed. It is based on a language more expressive than that which was given in Remark 1 and used by Sette. The hierarchy is obtained from the system C_ω of Newton da Costa, i.e.,
  (A1) $\alpha \to (\beta \to \alpha)$
  (A2) $(\alpha \to (\beta \to \gamma)) \to ((\alpha \to \beta) \to (\alpha \to \gamma))$
  (A3) $(\alpha \wedge \beta) \to \alpha$
  (A4) $(\alpha \wedge \beta) \to \beta$
  (A5) $\alpha \to (\beta \to (\alpha \wedge \beta))$
  (A6) $\alpha \to (\alpha \vee \beta)$
  (A7) $\beta \to (\alpha \vee \beta)$
  (A8) $(\alpha \to \gamma) \to ((\beta \to \gamma) \to (\alpha \vee \beta \to \gamma))$

(NN) $\sim\sim\alpha \to \alpha$
(ExM) $\alpha \vee \sim\alpha$
(MP) $\alpha \to \beta, \alpha \,/\, \beta$, (See [?], p. 501.)

by adding to it

(dC) $\sim(\beta \wedge \sim \beta) \to ((\alpha \to \beta) \to ((\alpha \to \sim\beta) \to \sim\alpha))$
(nC‡) $\sim((\alpha \ddagger \beta) \wedge \sim(\alpha \ddagger \beta))$, where $\ddagger \in \{\wedge, \vee, \to\}$
(nC$^{\sim n}$) $\sim(\sim^n \alpha \wedge \sim^{n+1} \alpha)$, where $n \in \mathbb{N}$ and $\sim^n \alpha$ denotes $\overbrace{\sim\sim ... \sim}^{n-times}\alpha$,

as new axiom schemas. Obviously, if $n = 0$, then P^0 is the classical propositional calculus; if $n = 1$, then P^1 is Sette's system (see [?], p. 9 of the preprint, for details). For a positive integer n, let P^n denote the calculus of the Fernández–Coniglio's hierarchy (P^n-hierarchy), henceforth.

Fernández and Coniglio proposed both a matrix and the so-called society semantics for the P^n-calculi. The former may be viewed as a generalization of $\mathcal{M}_{P^1}$ given by Sette in [?] (see p. 176), and da Costa in [?] (see, p. 499), that is,

$$\mathcal{M}_{P^n} = \langle X, D, \sim, \to \rangle,$$

where $X = \{T_0, T_1, T_2, \ldots T_n, F\}$ and $D = X - \{F\} = \{T_0, T_1, T_2, \ldots T_n\}$, $n \in \mathbb{N}$, are the sets of logical and designated values, respectively. The connectives of $\to$ and $\sim$ are defined in the following way $(i, k \in \mathbb{N}, i \leqslant n)$:

$\to$	T_0	T_i	F
T_0	T_0	T_0	F
T_k	T_0	T_0	F
F	T_0	T_0	T_0

$\sim$	
T_0	F
T_k	T_{k-1}
F	T_0

A P^n-valuation is any function $v : \mathcal{F} \longrightarrow X$ compatible with the above truth-tables. A P^n-tautology is a formula that under every valuation v takes on the designated values.

2. A New Axiomatization

The hierarchy discussed in this section is based on different criteria than those used to determine the P^n-hierarchy. Firstly, we assume that the connectives of conjunction, disjunction, and equivalence are treated as useful abbreviations, which formally do not appear in formulas; whereas $\sim$ and $\to$ will be taken as primitives. Secondly, the law of explosion is assumed to play a crucial role in defining the new hierarchy. The hierarchy will be obtained from that of Remark 1 by replacing (DS) with a more general schema, i.e.,

$$(DS^{\sim n}) \sim^n \alpha \to (\sim^{n+1} \alpha \to \beta),$$

where $n \in \mathbb{N}$ and $\sim^n \alpha$ is an abbreviation for $\overbrace{\sim\sim ... \sim}^{n}\alpha$; and adding to it the law of double negation, $\sim\sim\alpha \to \alpha$, as a new axiom schema. It is worth mentioning at this point that (NN) is provable in P^1 (see [?], pp. 174–175, and [?], p. 271), but it is not in any P^m, where $m > 1$. To put it more precisely, for each $n \in \mathbb{N}$, let S^n result from the implicational fragment of propositional intuitionistic logic by adding to it the following axiom schemas:

(PL) $((\alpha \to \beta) \to \alpha) \to \alpha$
(DS$^{\sim n}$) $\sim^n \alpha \to (\sim^{n+1}\alpha \to \beta)$
(DS$^\to$) $(\alpha \to \beta) \to (\sim(\alpha \to \beta) \to \gamma))$
(CM) $(\sim\alpha \to \alpha) \to \alpha$
(NN) $\sim\sim\alpha \to \alpha$

The other sentential connectives can be introduced by the definitions. Observe that if $n = 0$, then S^0 is the classical propositional calculus, and the axioms $(DS^\rightarrow)$, (NN) become redundant (cf. [?], p. 437); but if $n = 1$, then S^1 is equivalent to Sette's calculus, and (NN) is provable in S^1 (see [?], p. 268).

Definition 2. *Let $\alpha \in \mathcal{F}$ and $\Gamma \subseteq \mathcal{F}$. A formula α is provable from Γ within S^n ($\Gamma \vdash_{S^n} \alpha$, in symbols) iff there is a finite sequence of formulas, $\beta_1, \beta_2, ..., \beta_m$, such that $\beta_m = \alpha$, and for each $i \leq m$, at least one of the following is true:*

1. $\beta_i \in \Gamma$,
2. β_i is an axiom of S^n,
3. β_i is obtained from some of the previous β_j by application of the rule of detachment.

Definition 3. *A formula α is a thesis of S^n iff $\emptyset \vdash_{S^n} \alpha$.*

In what follows, we will need two lemmas to prove the key theorem:

Lemma 1. *Let $n \in \mathbb{N}$. Then:*

1. *The deduction theorem holds for S^n.*
2. *Some variants of the indirect deduction theorem hold for S^n, viz.:*

   a. *if $\Gamma, \alpha \vdash_{S^n} \{\sim^n\beta, \sim^{n+1}\beta\}$, then $\Gamma \vdash_{S^n} \sim\alpha$*

   b. *if $\Gamma, \sim\alpha \vdash_{S^n} \{\sim^n\beta, \sim^{n+1}\beta\}$, then $\Gamma \vdash_{S^n} \alpha$*

   c. *if $\Gamma, \alpha \rightarrow \beta \vdash_{S^n} \{\gamma \rightarrow \delta, \sim(\gamma \rightarrow \delta)\}$, then $\Gamma \vdash_{S^n} \sim(\alpha \rightarrow \beta)$*

   d. *if $\Gamma, \sim(\alpha \rightarrow \beta) \vdash_{S^n} \{\gamma \rightarrow \delta, \sim(\gamma \rightarrow \delta)\}$, then $\Gamma \vdash_{S^n} \alpha \rightarrow \beta$*

for every $\Gamma \subseteq \mathcal{F}$ and $\alpha, \beta, \gamma, \delta \in \mathcal{F}$.

Proof. 1. The proof is exactly the same as in Theorem 1.

2.a. Assume that $\Gamma, \alpha \vdash_{S^n} \{\sim^n\beta, \sim^{n+1}\beta\}$. Then, by the deduction theorem, we have $\Gamma \vdash_{S^n} \{\alpha \rightarrow \sim^n\beta, \alpha \rightarrow \sim^{n+1}\beta\}$. Since $\emptyset \vdash_{S^n} (\alpha \rightarrow \sim^n\beta) \rightarrow ((\alpha \rightarrow \sim^{n+1}\beta) \rightarrow \sim\alpha)$ (to prove this claim, apply the deduction theorem, $(DS^{\sim n})$, (HS), (C), $(CM2)$, and (MP)), then $\{\alpha \rightarrow \sim^n\beta, \alpha \rightarrow \sim^{n+1}\beta\} \vdash_{S^n} \sim\alpha$ by the deduction theorem. The relation $\vdash_{S^n}$ is transitive, so $\Gamma \vdash_{S^n} \sim\alpha$.

2.b. Suppose that $\Gamma, \sim\alpha \vdash_{S^n} \{\sim^n\beta, \sim^{n+1}\beta\}$, then $\Gamma \vdash_{S^n} \sim\sim\alpha$ (by 2.a). Since $\emptyset \vdash_{S^n} \sim\sim\alpha \rightarrow \alpha$, thus $\{\sim\sim\alpha\} \vdash_{S^n} \alpha$, and consequently, $\Gamma \vdash_{S^n} \alpha$.

2.c., 2.d. The proofs are similar to those of 2.a and 2.b. □

Lemma 2. *The (schemas of the) formulas:*

(IL) $\alpha \rightarrow \alpha$
(LoC) $(\alpha \rightarrow (\beta \rightarrow \gamma)) \rightarrow (\beta \rightarrow (\alpha \rightarrow \gamma))$
(HS) $(\alpha \rightarrow \beta) \rightarrow ((\beta \rightarrow \gamma) \rightarrow (\alpha \rightarrow \gamma))$
(C) $(\alpha \rightarrow (\alpha \rightarrow \beta)) \rightarrow (\alpha \rightarrow \beta)$
(LoE) $((\alpha \rightarrow \beta) \rightarrow \gamma) \rightarrow (\alpha \rightarrow (\beta \rightarrow \gamma))$
$(CM2)$ $(\alpha \rightarrow \sim\alpha) \rightarrow \sim\alpha$
$(DD^\rightarrow)$ $(\sim\phi \rightarrow \psi) \rightarrow ((\sim\phi \rightarrow \sim\psi) \rightarrow \phi)$, *where* $\phi := \alpha \rightarrow \beta$, $\psi := \gamma \rightarrow \delta$

are provable in S^n, $n \in \mathbb{N}$.

Proof. (IL), (LoC), (HS), and (C) immediately follow from the deduction theorem and (MP); (LoE) follows from the deduction theorem, $(A1)$ and (MP); $(CM2)$ from the deduction theorem, (NN) (HS), (CM), and (MP); and finally, $(DD^\rightarrow)$ can be easily obtained by the indirect deduction theorem and (MP). □

The lemmas will be particularly useful for proving the main result of this section.

Theorem 4. $S^n = P^n$, where $n \in \mathbb{N}$.

Proof. The proof is divided into two steps. The first is to demonstrate that each axiom schema of S^n is a P^n-tautology, and the rule (MP) preserves validity. This can be easily done with the help of the semantics for P^n. To illustrate the point, we show that (PL), $(DS^\rightarrow)$, and $(DS^{\sim n})$ are valid in $\mathcal{M}_{P^n}$.

(PL). Suppose that $((\alpha \to \beta) \to \alpha) \to \alpha$ is not a P^n-tautology. Thus, there is a P^n-valuation v such that $v(((\alpha \to \beta) \to \alpha) \to \alpha) = F$. There are two main cases to consider. Either $v((\alpha \to \beta) \to \alpha) = T_0$ and $v(\alpha) = F$, or $v((\alpha \to \beta) \to \alpha) = T_n$ and $v(\alpha) = F$, where $n \geqslant 1$. Case 1. If $v((\alpha \to \beta) \to \alpha) = T_0$ and $v(\alpha) = F$, then, no matter which value is assigned to β, $v(\alpha \to \beta) = T_0$, and consequently, $v((\alpha \to \beta) \to \alpha) = F$. But this gives a contradiction since $v((\alpha \to \beta) \to \alpha) = T_0$. Case 2. It follows from the truth-table for implication that $v((\alpha \to \beta) \to \alpha) = T_0$ or $v((\alpha \to \beta) \to \alpha) = F$, for any $\alpha, \beta \in \mathcal{F}$ and every P^n-valuation v. So it is not possible that $v((\alpha \to \beta) \to \alpha) = T_n$, where $n \neq 0$. Consequently, there is no P^n-valuation v such that $v(((\alpha \to \beta) \to \alpha) \to \alpha) = F$, which means that (PL) is a P^n-tautology.

$(DS^\rightarrow)$. Assume that $(\alpha \to \beta) \to (\sim(\alpha \to \beta) \to \gamma))$ is not a P^n-tautology. So there is a P^n-valuation v such that $v((\alpha \to \beta) \to (\sim(\alpha \to \beta) \to \gamma)) = F$. Hence, either $v(\alpha \to \beta) = T_0$ and $v(\sim(\alpha \to \beta) \to \gamma) = F$, or $v(\alpha \to \beta) = T_n$ and $v(\sim(\alpha \to \beta) \to \gamma) = F$. The latter is impossible due to the truth table for implication. Therefore, if $v(\alpha \to \beta) = T_0$, then $v(\sim(\alpha \to \beta)) = F$, and consequently, $v(\sim(\alpha \to \beta) \to \gamma) = T_0$. But this results in a contradiction because $v(\sim(\alpha \to \beta) \to \gamma) = F$. As a consequence, there is no P^n-valuation v such that $v((\alpha \to \beta) \to (\sim(\alpha \to \beta) \to \gamma)) = F$. The formula $(DS^\rightarrow)$ is a P^n-tautology.

$(DS^{\sim n})$. Suppose that $\sim^n \alpha \to (\sim^{n+1} \alpha \to \beta)$ is not a P^n-tautology, where $n \geqslant 1$. Then, there is a P^n-valuation v such that $v(\sim^n \alpha \to (\sim^{n+1} \alpha \to \beta)) = F$. As a result, either $v(\sim^n \alpha) = T_0$ and $v(\sim^{n+1} \alpha \to \beta) = F$, or $v(\sim^n \alpha) = T_{n-1}$ and $v(\sim^{n+1} \alpha \to \beta) = F$. Let $v(\sim^n \alpha) = T_0$ and $v(\sim^{n+1} \alpha \to \beta) = F$. Hence, $v(\sim^{n+1} \alpha) = F$ by the truth tables for negation. Since $v(\sim^{n+1} \alpha) = F$, then, no matter which value is assigned to β, $v(\sim^{n+1} \alpha \to \beta) = T_0$. But this entails a contradiction since $v(\sim^{n+1} \alpha \to \beta) = F$. Now, let $v(\sim^n \alpha) = T_{n-1}$ and $v(\sim^{n+1} \alpha \to \beta) = F$. Consequently, either $v(\sim^{n+1} \alpha) = T_0$ and $v(\beta) = F$, or $v(\sim^{n+1} \alpha) = T_n$ and $v(\beta) = F$. If $v(\sim^{n+1} \alpha) = T_0$, then, according to the truth table for negation, $v(\sim^n \alpha) = F$. But $v(\sim^n \alpha) = T_{n-1}$. On the other hand, if $v(\sim^{n+1} \alpha) = T_n$, then $v(\sim^n \alpha) = T_{n+1}$. But $v(\sim^n \alpha) = T_{n-1}$. Therefore, there is no P^n-valuation v such that $v(\sim^n \alpha \to (\sim^{n+1} \alpha \to \beta)) = F$. The formula $(DS^{\sim n})$ is a P^n-tautology.

For the second part of the proof, we have to demonstrate that each axiom schema of P^n is provable in S^n and (MP) is its admissible rule, where $n \in \mathbb{N}$. To begin with, notice that $(A1)$, $(A2)$, and (NN) are the axiom schemas of S^n, and (MP) is its sole rule of inference.

$(A3)$. We show that $(\alpha \wedge \beta) \to \alpha$ is a thesis of S^n, or, to be more precise, that $\sim(\alpha \to \sim(\sim\beta \to \beta)) \to \alpha$ is provable in S^n. To see that this claim is true, consider the following sequence of formulas:

1. $\sim(\alpha \to \sim(\sim\beta \to \beta))$ by the deduction theorem,
2. $(\alpha \to \sim (\sim\beta \to \beta)) \to (\sim (\alpha \to \sim (\sim\beta \to \beta)) \to \alpha)$ by $(DS^\rightarrow)$,
3. $\sim (\alpha \to \sim (\sim\beta \to \beta)) \to ((\alpha \to \sim (\sim\beta \to \beta)) \to \alpha)$ by (LoC), 2, (MP),
4. $(\alpha \to \sim (\sim\beta \to \beta)) \to \alpha$ by 1, 3, (MP),
5. α by (PL), 4, (MP),
6. $\sim (\alpha \to \sim (\sim\beta \to \beta)) \to \alpha$ by the deduction theorem,

and finally,

7. $(\alpha \wedge \beta) \to \alpha$ by the definition of $\wedge$.

$(A4)$. We prove that $(\alpha \wedge \beta) \to \beta$, i.e., $\sim(\alpha \to \sim(\sim\beta \to \beta)) \to \beta$, is a thesis of S^n. To see this, consider the sequence of formulas:

1. $\sim(\alpha \to \sim(\sim\beta \to \beta))$ by the deduction theorem,
2.–5. Proceed as in the preceding case,
6. $(\alpha \to \sim(\sim\beta \to \beta)) \to (\sim(\alpha \to \sim(\sim\beta \to \beta)) \to (\sim\beta \to \beta))$ by $(DS^\to)$,
7. $(\alpha \to \sim(\sim\beta \to \beta)) \to (\sim\beta \to \beta)$ by $(LoC), 6, 1, (MP)$,
8. $\alpha \to (\sim(\sim\beta \to \beta) \to (\sim\beta \to \beta))$ by $(LoE), 7, (MP)$,
9. $\sim(\sim\beta \to \beta) \to (\sim\beta \to \beta)$ by $5, 8, (MP)$,
10. $\sim\beta \to \beta$ by $(CM), 9, (MP)$,
11. β by $(CM), 10, (MP)$,
12. $\sim(\alpha \to \sim(\sim\beta \to \beta)) \to \beta$ by the deduction theorem,

and consequently,

13. $(\alpha \wedge \beta) \to \beta$ by the definition of $\wedge$.

(A5). We show that $\alpha \to (\beta \to (\alpha \wedge \beta))$, i.e., $\alpha \to (\beta \to \sim(\alpha \to \sim(\sim\beta \to \beta)))$, is provable in S^n. Consider the sequence of formulas:

1. α,
2. β,
3. $\sim\sim(\alpha \to \sim(\sim\beta \to \beta))$ by the indirect deduction theorem,
4. $\alpha \to \sim(\sim\beta \to \beta)$ by $(NN), 3, (MP)$,
5. $\sim(\sim\beta \to \beta)$ by $1, 4, (MP)$,
6. $\sim\beta \to \beta$ by $(A1), 2, (MP)$,

a contradiction (5, 6). This entails that:

7. $\sim(\alpha \to \sim(\sim\beta \to \beta))$,
8. $\alpha \to (\beta \to \sim(\alpha \to \sim(\sim\beta \to \beta)))$ by the deduction theorem 1, 2, 7, (MP),

and finally,

9. $\alpha \to (\beta \to (\alpha \wedge \beta))$ by the definition of $\wedge$.

(A6). We demonstrate that $\alpha \to (\alpha \vee \beta)$, i.e., $\alpha \to (\sim(\sim\alpha \to \alpha) \to \beta)$, is a thesis of S^n. To see that this claim holds, consider the sequence of formulas:

1. α,
2. $\sim(\sim\alpha \to \alpha)$ by the deduction theorem,
3. $\sim\alpha \to \alpha$ by $(A1), 1, (MP)$,
4. $(\sim\alpha \to \alpha) \to (\sim(\sim\alpha \to \alpha) \to \beta))$ by $(DS^\to)$,
5. β by $4, 3, 2, (MP)$,
6. $\alpha \to (\sim(\sim\alpha \to \alpha) \to \beta)$ by the deduction theorem,

and consequently,

7. $\alpha \to (\alpha \vee \beta)$ by the definition of $\vee$.

(A7). We show that $\beta \to (\alpha \vee \beta)$, i.e., $\beta \to (\sim(\sim\alpha \to \alpha) \to \beta)$, is provable in S^n. To see this, consider the sequence of formulas:

1. β,
2. $\sim(\sim\alpha \to \alpha)$ by the deduction theorem,
3. β by 1,
4. $\beta \to (\sim(\sim\alpha \to \alpha) \to \beta))$ by the deduction theorem,

and finally,

5. $\beta \to (\alpha \vee \beta)$ by the definition of $\vee$.

(A8). We prove that $(\alpha \to \gamma) \to ((\beta \to \gamma) \to (\alpha \vee \beta \to \gamma))$, i.e., $(\alpha \to \gamma) \to ((\beta \to \gamma) \to ((\sim(\sim\alpha \to \alpha) \to \beta) \to \gamma))$, is a thesis of S^n. To see that this claim is true, consider the following sequence of formulas:

1. $\alpha \to \gamma$,
2. $\beta \to \gamma$,
3. $\sim((\sim(\sim\alpha \to \alpha) \to \beta) \to \gamma)$ by the indirect deduction theorem,

Let $\phi := (\sim(\sim\alpha \to \alpha) \to \beta) \to \gamma$. Then,

4. $\phi \to (\sim\phi \to (\sim(\sim\alpha \to \alpha) \to \beta))$ by $(DS^{\to})$,
5. $\phi \to (\sim(\sim\alpha \to \alpha) \to \beta)$ by (LoC), 4, 3, (MP),
6. $(\phi \to (\sim(\sim\alpha \to \alpha) \to \beta)) \to (\sim(\sim\alpha \to \alpha) \to \beta)$ by (PL),
7. $\sim(\sim\alpha \to \alpha) \to \beta$ by 5, 6, (MP),
8. $\phi \to (\sim\phi \to \sim(\sim(\sim\alpha \to \alpha) \to \beta))$ by $(DS^{\to})$,
9. $\phi \to \sim(\sim(\sim\alpha \to \alpha) \to \beta))$ by (LoC), 8, 3, (MP).

If $\phi := (\sim(\sim\alpha \to \alpha) \to \beta) \to \gamma$, then,

10. $((\sim(\sim\alpha \to \alpha) \to \beta) \to \gamma) \to \sim(\sim(\sim\alpha \to \alpha) \to \beta))$,
11. $(\sim(\sim\alpha \to \alpha) \to \beta) \to (\gamma \to \sim(\sim(\sim\alpha \to \alpha) \to \beta))$ by (LoE), 10, (MP),
12. $\gamma \to \sim(\sim(\sim\alpha \to \alpha) \to \beta)$ by 11, 7, (MP),
13. $\beta \to \sim(\sim(\sim\alpha \to \alpha) \to \beta)$ by (HS), 2, 12, (MP),
14. $\sim(\sim\alpha \to \alpha) \to \sim(\sim(\sim\alpha \to \alpha) \to \beta)$ by (HS), 7, 13, (MP),
15. $\beta \to (\sim(\sim\alpha \to \alpha) \to \beta)$ by $(A1)$,
16. $\sim(\sim\alpha \to \alpha) \to (\sim(\sim\alpha \to \alpha) \to \beta)$ by (HS), 7, 15, (MP),

Let $\chi := \sim\alpha \to \alpha$ and $\psi := \sim(\sim\alpha \to \alpha) \to \beta$, then,

17. $(\sim\chi \to \psi) \to ((\sim\chi \to \sim\psi) \to \chi)$ by $(DD^{\to})$,
18. $(\sim\chi \to \sim\psi) \to \chi$ by 17, 16, (MP),
19. χ by 18, 14, (MP).

If $\chi := \sim\alpha \to \alpha$, then,

20. $\sim\alpha \to \alpha$,
21. α by (CM), 20, (MP),
22. γ by 21, 1, (MP),
23. $\gamma \to ((\sim(\sim\alpha \to \alpha) \to \beta) \to \gamma)$ by $(A1)$,
24. $(\sim(\sim\alpha \to \alpha) \to \beta) \to \gamma$ by 23, 22, (MP),

a contradiction (3, 24). This yields that:

25. $(\sim(\sim\alpha \to \alpha) \to \beta) \to \gamma$,
26. $(\alpha \to \gamma) \to ((\beta \to \gamma) \to ((\sim(\sim\alpha \to \alpha) \to \beta) \to \gamma))$ by the deduction theorem, and consequently,
27. $(\alpha \to \gamma) \to ((\beta \to \gamma) \to (\alpha \vee \beta \to \gamma))$ by the definition of $\vee$.

(ExM). We show that $\alpha \vee \sim\alpha$, i.e., $\sim(\sim\alpha \to \alpha) \to \sim\alpha$, is provable S^n.

1. $\sim(\sim\alpha \to \alpha)$ by the deduction theorem,
2. $(\sim\alpha \to \alpha) \to (\sim(\sim\alpha \to \alpha) \to \sim\alpha)$ by $(DS^{\to})$,
3. $(\sim\alpha \to \alpha) \to \sim\alpha$ by (LoC), 2, 1, (MP),
4. $((\sim\alpha \to \alpha) \to \sim\alpha) \to \sim\alpha$ by (PL),
5. $\sim\alpha$ by 4, 3, (MP),
6. $\sim(\sim\alpha \to \alpha) \to \sim\alpha$ by the deduction theorem,

and finally,

7. $\alpha \vee \sim\alpha$ by the definition of $\vee$.

(dC). We prove that $\sim(\beta \wedge \sim\beta) \to ((\alpha \to \beta) \to ((\alpha \to \sim\beta) \to \sim\alpha))$, i.e., $\sim\sim(\beta \to \sim(\sim\sim\beta \to \sim\beta)) \to ((\alpha \to \beta) \to ((\alpha \to \sim\beta) \to \sim\alpha))$, is a thesis of S^n.

1. $\sim\sim(\beta \to \sim(\sim\sim\beta \to \sim\beta))$,
2. $\alpha \to \beta$,
3. $\alpha \to \sim \beta$ by the deduction theorem,
4. $(\sim\sim \beta \to \sim \beta) \to (\sim (\sim\sim \beta \to \sim \beta) \to \sim \alpha)$ by $(DS^\to)$,
5. $\beta \to \sim (\sim\sim \beta \to \sim \beta)$ by (NN), 1, (MP),
6. $\beta \to (\sim (\sim\sim \beta \to \sim \beta) \to \sim \alpha)$ by (HS), 4, 5, (MP),
7. $\alpha \to (\sim (\sim\sim \beta \to \sim \beta) \to \sim \alpha)$ by (HS), 2, 6, (MP),
8. $\sim (\sim\sim \beta \to \sim \beta) \to (\alpha \to \sim \alpha)$ by (LoC), 7, (MP),
9. $\beta \to (\alpha \to \sim \alpha)$ by (HS), 5, 8, (MP),
10. $\alpha \to (\alpha \to \sim \alpha)$ by (HS), 2, 9, (MP),
11. $\alpha \to \sim \alpha$ by (C), 10, (MP),
12. $\sim \alpha$ by $(CM2)$, (11), (MP),
13. $\sim\sim (\beta \to \sim (\sim\sim \beta \to \sim \beta)) \to ((\alpha \to \beta) \to ((\alpha \to \sim \beta) \to \sim \alpha))$ by the deduction theorem, and consequently,
14. $\sim (\beta \wedge \sim \beta) \to ((\alpha \to \beta) \to ((\alpha \to \sim \beta) \to \sim \alpha))$ by the definition of $\wedge$.

$(nC^\ddagger)$. We demonstrate that $\sim((\alpha \ddagger \beta) \wedge \sim(\alpha \ddagger \beta))$, i.e., $\sim\sim((\alpha \ddagger \beta) \to \sim(\sim\sim(\alpha \ddagger \beta) \to \sim(\alpha \ddagger \beta)))$ and $\ddagger \in \{\wedge, \vee, \to\}$, is provable in S^n. Let $\phi := \alpha \to \beta$, if $\ddagger$ is $\to$; $\phi := \sim (\sim \alpha \to \alpha) \to \beta$, if $\ddagger$ is $\vee$; and $\phi := \sim (\alpha \to \sim (\sim \beta \to \beta))$, if $\ddagger$ is $\wedge$. As a result, we have:

1. $\sim\sim\sim (\phi \to \sim (\sim\sim \phi \to \sim \phi))$ by the indirect deduction theorem,
2. $\sim (\phi \to \sim (\sim\sim \phi \to \sim \phi))$ by (NN), 1, (MP),
3. $(\phi \to \sim (\sim\sim \phi \to \sim \phi)) \to (\sim (\phi \to \sim (\sim\sim \phi \to \sim \phi)) \to \phi))$ by $(DS^\to)$,
4. $(\phi \to \sim (\sim\sim \phi \to \sim \phi)) \to \phi$ by (LoC), 3, 2, (MP),
5. $((\phi \to \sim (\sim\sim \phi \to \sim \phi)) \to \phi) \to \phi$ by (PL),
6. ϕ by 5, 4, (MP),
7. $(\phi \to \sim (\sim\sim \phi \to \sim \phi)) \to (\sim (\phi \to \sim (\sim\sim \phi \to \sim \phi)) \to \sim\sim (\sim\sim \phi \to \sim \phi))$ by $(DS^\to)$,
8. $(\phi \to \sim (\sim\sim \phi \to \sim \phi)) \to \sim\sim (\sim\sim \phi \to \sim \phi)$ by (LoC), 7, 2, (MP),
9. $\phi \to (\sim (\sim\sim \phi \to \sim \phi) \to \sim\sim (\sim\sim \phi \to \sim \phi))$ by (LoE), 8, (MP),
10. $\sim (\sim\sim \phi \to \sim \phi) \to \sim\sim (\sim\sim \phi \to \sim \phi)$ by 6, 9, (MP),
11. $(\sim (\sim\sim \phi \to \sim \phi) \to \sim\sim (\sim\sim \phi \to \sim \phi)) \to \sim\sim (\sim\sim \phi \to \sim \phi)$ by $(CM2)$,
12. $\sim\sim (\sim\sim \phi \to \sim \phi)$ by 10, 11, (MP),
13. $\sim\sim \phi \to \sim \phi$ by (NN), 12, (MP),
14. $\sim \phi$ by (CM), 13, (MP),

a contradiction (6, 14). This entails that,

15. $\sim\sim (\phi \to \sim (\sim\sim \phi \to \sim \phi))$,

and finally,

16. $\sim (\phi \wedge \sim \phi)$.

However, if $\phi := \alpha \to \beta$, then $\sim ((\alpha \to \beta) \wedge \sim (\alpha \to \beta))$; if $\phi := \sim (\sim \alpha \to \alpha) \to \beta$, then $\sim ((\alpha \vee \beta) \wedge \sim (\alpha \vee \beta))$; and if $\phi := \sim (\alpha \to \sim (\sim \beta \to \beta))$, then $\sim ((\alpha \wedge \beta) \wedge \sim (\alpha \wedge \beta))$. Hence, $\sim ((\alpha \ddagger \beta) \wedge \sim (\alpha \ddagger \beta))$, where $\ddagger \in \{\wedge, \vee, \to\}$.

$(nC^{\sim n})$. We show that $\sim(\sim^n \alpha \wedge \sim^{n+1} \alpha)$, that is, $\sim\sim(\sim^n \alpha \to \sim(\sim^{n+2} \alpha \to \sim^{n+1} \alpha))$, where $n \in \mathbb{N}$, is provable in S^n.

1. $\sim\sim\sim (\sim^n \alpha \to \sim (\sim^{n+2} \alpha \to \sim^{n+1} \alpha))$ by the indirect deduction theorem,
2. $\sim (\sim^n \alpha \to \sim (\sim^{n+2} \alpha \to \sim^{n+1} \alpha))$ by (NN), 1, (MP),

Let $\phi := \sim^n \alpha \to \sim (\sim^{n+2} \alpha \to \sim^{n+1} \alpha)$. Then,

3. $\sim \phi$,
4. $\phi \rightarrow (\sim \phi \rightarrow \sim^n \alpha)$ by $(DS^\rightarrow)$,
5. $\phi \rightarrow \sim^n \alpha$ by (LoC), 4, 3, (MP),
6. $(\sim^n \alpha \rightarrow \sim (\sim^{n+2} \alpha \rightarrow \sim^{n+1} \alpha)) \rightarrow \sim^n \alpha$ by ϕ,
7. $\sim^n \alpha$ by (PL), 6, (MP),
8. $\phi \rightarrow (\sim \phi \rightarrow \sim\sim (\sim^{n+2} \alpha \rightarrow \sim^{n+1} \alpha))$ by $(DS^\rightarrow)$,
9. $\phi \rightarrow \sim\sim (\sim^{n+2} \alpha \rightarrow \sim^{n+1} \alpha)$ by (LoC), 8, 3, (MP),
10. $(\sim^n \alpha \rightarrow \sim (\sim^{n+2} \alpha \rightarrow \sim^{n+1} \alpha)) \rightarrow \sim\sim (\sim^{n+2} \alpha \rightarrow \sim^{n+1} \alpha)$ by ϕ,
11. $\sim^n \alpha \rightarrow (\sim (\sim^{n+2} \alpha \rightarrow \sim^{n+1} \alpha) \rightarrow \sim\sim (\sim^{n+2} \alpha \rightarrow \sim^{n+1} \alpha))$ by (LoE), 10, (MP),
12. $\sim (\sim^{n+2} \alpha \rightarrow \sim^{n+1} \alpha) \rightarrow \sim\sim (\sim^{n+2} \alpha \rightarrow \sim^{n+1} \alpha)$ by 11, 7, (MP),
13. $\sim\sim (\sim^{n+2} \alpha \rightarrow \sim^{n+1} \alpha)$ by $(CM2)$, 12, (MP),
14. $\sim^{n+2} \alpha \rightarrow \sim^{n+1} \alpha$ by (NN), 13, (MP),
15. $\sim^{n+1} \alpha$ by (CM), 14, (MP),
16. $\sim^n \alpha \rightarrow (\sim^{n+1} \alpha \rightarrow \phi)$ by $(DS^{\sim n})$,
17. ϕ by 16, 15, 7, (MP),

a contradiction (3, 17). This entails that,

18. $\sim\sim (\sim^n \alpha \rightarrow \sim (\sim^{n+2} \alpha \rightarrow \sim^{n+1} \alpha))$, and consequently,
19. $\sim (\sim^n \alpha \wedge \sim^{n+1} \alpha)$ by the definition of $\wedge$.

This finishes the proof of Theorem 4. □

3. Conclusions

In this paper, we proposed a new axiomatization for the P^n-hierarchy. The main idea behind it was to focus directly on the principle of ex contradictione sequitur quodlibet. This is a remarkable difference between Fernández–Coniglio's and our proposal, which makes the hierarchy less complex and more transparent from the viewpoint of paraconsistency. Additionally, we followed Sette's idea and the connectives of negation and implication were taken as primitives. In conclusion let us also mention that the several other hierarchies can be easily generated from P^n-hierarchy. For instance, by dropping $(DS^\rightarrow)$, we get the CB^n-hierarchy of the paraconsistent calculi (cf. [?]). The interested reader can also find a slightly different hierarchy in [?] (the so-called B^n-hierarchy).

Funding: This research received no external funding.

Acknowledgments: I am grateful to anonymous reviewers for their helpful comments on an earlier draft of this paper.

Conflicts of Interest: The author declare no conflict of interest.

References

1. Priest, G. Paraconsistent Logic. In *The Stanford Encyclopedia of Philosophy*; Winter 2012 ed.; Zalta, E.N., Ed.; 2012. Available online: https://plato.stanford.edu/entries/logic-paraconsistent/, (accessed on 10 November 2019).
2. Alves, E.H. The first axiomatization of a paraconsistent logic. *Bull. Sect. Log.* **1992**, *21*, 19–20.
3. Béziau, J.Y. Bivalence, excluded middle and non-contradiction. In *The Logica Yearbook 2003*; Běhounek, L., Ed.; Filosofia: Prague, Czech Republic, 2004; pp. 75–83.
4. Avron, A.; Arieli, O.; Zamansky, A. *Theory of Effective Propositional Paraconsistent Logics*; Studies in Logic, Mathematical Logic and Foundations, Volume 75; College Publications: London, UK, 2018.
5. Béziau, J.Y.; Carnielli, W.A.; Gabbay, D.M. (Eds.) *Handbook of Paraconsistency*; Studies in Logic, Book 9; College Publications: London, UK, 2007.
6. Carnielli, W.A.; Coniglio, M.E. *Paraconsistent Logic: Consistency, Contradiction and Negation*; Logic, Epistemology, and the Unity of Science, Volume 40; Springer International Publishing: Berlin/Heidelberg, Germany, 2016.

7. Sette, A.M. On the propositional calculus P1. *Math. Jpn.* **1973**, *18*, 89–128.
8. Fernández, V.L.; Coniglio, M.E. Combining Valuations with Society Semantics. *J. Appl. Non-Class. Log.* **2003**, *13*, 21–46. Available online: ftp://www.cle.unicamp.br/pub/professors/carnielli/articles/society.ps.gz (accessed on 1 October 2019). [CrossRef]
9. Marcos, J. On a problem of da Costa. In *Essays on the Foundations of Mathematics and Logic*; Sica, G., Ed.; Polimetrica: Monza, Italy, 2005; Volume 2, pp. 53–69.
10. Malinowski, G. *Many-Valued Logics*; Oxford Logic Guides, Book 25; Clarendon Press: Oxford, UK, 1994.
11. Pogorzelski, W.A.; Wojtylak, P. *Completeness Theory for Propositional Logics*; Studies in Universal Logic; Birkhäuser: Basel, Switzerland, 2008.
12. Wójcicki, R. *Theory of Logical Calculi. Basic Theory of Consequence Operations*; Springer: Dordrecht, The Netherlands, 1988; Synthese Library Volume 199.
13. Omori, H. Sette's Logics, Revisited. In *Logic, Rationality, and Interaction*; Baltag, A., Seligman, J., Yamada, T., Eds.; Springer: Berlin/Heidelberg, Germany, 2017; pp. 451–465.
14. Pynko, A.P. Algebraic Study of Sette's Maximal Paraconsistent Logic. *Studia Log.* **1995**, *54*, 173–180. [CrossRef]
15. Araujo, A.L.; Alves, E.H.; Guerzoni, J.A.D. Some Relations Between Modal and Paraconsistent Logic. *J. Non-Class. Log.* **1987**, *8*, 33–44.
16. Carnielli, W.A.; Lima-Marques, M. Society semantics and multiplevalued logics. In *Advances in Contemporary Logic and Computer Science: Proceedings of the Eleventh Brazilian Conference on Mathematical Logic, Salvador Da Bahia, Brazil, 6–10 May 1996*; Carnielli, W.A., D'Ottaviano, I.M., Eds.; American Mathematical Society: Providence, RI, USA, 1999; pp. 33–52.
17. Ciuciura, J. Paraconsistency and Sette's calculus P1. *Log. Log. Philos.* **2015**, *24*, 265–273. [CrossRef]
18. da Costa, N.C.A. On the theory of inconsistent formal systems. *Notre Dame J. Form. Log.* **1974**, *15*, 497–510. [CrossRef]
19. Imai, Y.; Iseki, K. On Axiom Systems of Propositional Calculi. I. *Proc. Jpn. Acad.* **1965**, *41*, 436–439. [CrossRef]
20. Ciuciura, J. On the system CB^1 and a lattice of the paraconsistent calculi. *Log. Log. Philos.* to appear.
21. Ciuciura, J. Paraconsistent heap. A Hierarchy of mbCn-systems. *Bull. Sect. Log.* **2014**, *43*, 173–182.

© 2020 by the authors. Licensee MDPI, Basel, Switzerland. This article is an open access article distributed under the terms and conditions of the Creative Commons Attribution (CC BY) license (http://creativecommons.org/licenses/by/4.0/).

Article

Deductive Systems with Multiple-Conclusion Rules and the Disjunction Property

Alex Citkin

Metropolitan Telecommunications, New York, NY 10041, USA; acitkin@gmail.com

Received: 9 July 2019; Accepted: 19 August 2019; Published: 30 August 2019

Abstract: Using the defined notion of the inference with multiply-conclusion rules, we show that in the logics enjoying the disjunction property, any derivable rule can be inferred from the single-conclusion rules and a single multiple-conclusion rule, which represents the disjunction property. Also, the conversion algorithm of single- and multiple-conclusion deductive systems into each other is studied.

Keywords: multiple conclusion rule; disjunction property; metadisjunction

1. Introduction

The question "What is inference rule?" is almost as profound as the question "What is truth?." Speaking very generally, inference rules are the acceptable means of reasoning. They give us a way to go from a set of accepted statements (propositions, judgments) to an acceptable statement (proposition, judgment). Rules can be given in different forms: Aristotle used rules in a form of moduses like [1] (§1 (6))

If	A is predicated of all B	premise
and	B is predicated of all C,	premise
then	A is predicated of all C.	conclusion

Nowadays, we would write such a rule as

$$\frac{A \text{ is predicated of all } B, \quad B \text{ is predicated of all } C}{A \text{ is predicated of all } C}.$$

In this paper, we consider only propositional logics. Thus, the premises and conclusions are propositions (propositional formulas). But even in this case, one may consider different forms of modus ponens [2]:

$$A, A \to B/B \quad \text{or} \quad A, A \text{ entails } B/B.$$

As we see, the former rule is about conditional, while the latter rule is about entailment. Clearly, the second form of modus ponens is a rule of meta-logic rather than a rule of logic. In this paper we confine ourselves to rules of the first form: a *finitary structural inference rule* or a *modus rule* is given by an ordered pair (Γ, A), where Γ is a finite (maybe empty) set of formulas and A is a formula, and we use Γ/A or $\frac{\Gamma}{A}$ to denote such a pair. Any pair Γ/A constitutes the modus rule that allows for any substitution σ, to infer $\sigma(A)$ from $\sigma(\Gamma)$.

Let us note that curiously enough, the rule of substitution "for any substitution, σ, from A infer $\sigma(A)$" is not structural: it allows to infer formula q from formula p, where p and q are propositional variables, but the structural rule p/q would allow to infer any formula from any formula.

By logic L we understand a (finitary structural) consequence relation $\vdash_L$ defined in the following way; for all sets of formulas Γ, Δ, and every formula A

$$
\begin{array}{lll}
(R): & \text{if } A \in \Gamma, \text{ then } \Gamma \vdash_L A & \text{reflexivity} \\
(M): & \text{if } \Gamma \subseteq \Delta \text{ and } \Gamma \vdash_L A, \text{ then } \Delta \vdash_L A & \text{monotonicity} \\
(T): & \text{if } \Gamma \vdash_L B \text{ and } \Gamma, B \vdash_L A, \text{ then } \Gamma \vdash_L A & \text{transitivity - cut} \\
(F): & \text{if } \Gamma \vdash_L A, \text{ then } \Delta \vdash_L A \text{ for some finite } \Delta \subseteq \Gamma & \text{finitarity} \\
(S): & \text{if } \Gamma \vdash_L A, \text{ then } \sigma(\Gamma) \vdash_L \sigma(A), \text{ for every } \sigma & \text{structurality}
\end{array}
\qquad (1)
$$

Given a logic L, we define a set of theorems $Th(L)$ of L to be

$$Th(L) := \{A : \varnothing \vdash_L A\}.$$

It is customary to define a logic L by a deductive system—a set of rules R and axioms. Because axioms can be viewed as rules without premises, we assume that a deductive system is a set of rules and a procedure for derivation such that $\Gamma \vdash_L A$ if and only if A can be derived from Γ by rules R.

We extend the notion of the deductive system by allowing to use multiple-conclusion rules. This requires to extend the notion of derivation, and we discuss this generalization in Section 3. Let us stress out that we do not consider multiple-conclusion logics in the sense of Shoesmith and Smiley [3], Carnap's logics of junctives [4], or hyperformulas [5]. Instead, we study regular logics defined by the deductive systems that admit multiple-conclusion rules. For instance, one can use the following rules to define the classical logic [3] (2.3) and further discussions in Chapter 18 of this book):

$$
\begin{array}{ccc}
\dfrac{A, A \to B}{B} & \dfrac{B}{A \to B} & \dfrac{\varnothing}{A, A \to B} \\[1em]
\dfrac{A \wedge B}{A} & \dfrac{A \wedge B}{B} & \dfrac{A, B}{A \wedge B} \\[1em]
\dfrac{A}{A \vee B} & \dfrac{B}{A \vee B} & \dfrac{A \vee B}{A, B} \\[1em]
\dfrac{A, \neg A}{\varnothing} & \dfrac{\varnothing}{A, \neg A} &
\end{array}
$$

The idea of using multiple-conclusion rules can be traced back at least to Carnap [6]. Much earlier, Peirce introduced the dialogisms, which are essentially the multiple-conclusion rules, but he preferred to replace them with the single-conclusion rules [7]. Gentzen's sequent also can be viewed as a multiple-conclusion construction. The following quotation from the authors of [3] explains why Carnap, and not Gentzen, perhaps, should be regarded as the one who introduced multiple-conclusion rules: "Its germ can be found in Gerhard Gentzen's celebrated Untersuchungen über das logische Schliessen (1934) if one is prepared to interpret his calculus of 'sequents' as a metatheory for a multiple-conclusion logic, but this is contrary to Gentzen's own interpretation, and it was Rudolf Carnap who first consciously broached the subject in his book Formalization of logic (1943)" [3] (Section 2.1, the historical note).

Carnap's motivation for introducing multiple-conclusion rules was as follows: if we want to syntactically characterize a two-valued classical semantics, this syntactical system should be valid (up to matrix isomorphisms) only in the two-element Boolean matrix. Let us consider matrices depicted in Figure 1 (the designated elements are marked by a bullet). It is not hard to see that any rule which is valid in matrix $\mathbf{B}_1$, is valid in all matrices $\mathbf{B}_n$ for all $n \geq 0$.

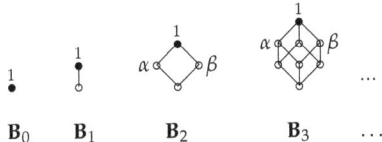

Figure 1. Boolean matrices.

Carnap's solution was to employ the rules of a different kind: a (structural) multiple-conclusion rule (or multiple-alternative rule, or m-rule for short) is an ordered pair Γ/Δ of finite sets of formulas. m-rule Γ/Δ is valid in a logical matrix $(\mathbf{A}, D)$ if for any valuation v such that $v(A) \in D$ for all $A \in \Gamma$, there is $B \in \Delta$ such that $v(B) \in D$, that is,

$$v(\Gamma) \subseteq D, \text{ entails } v(\Delta) \cap D \neq \emptyset. \qquad (2)$$

Let us consider the m-rules

$$t := p, \neg p/\emptyset \quad \text{and} \quad de := p \vee q/p, q.$$

Then, if we consider logical matrices $(\mathbf{2^0}, \{1\}), (\mathbf{2}, \{1\}), (\mathbf{2^n}, \{1\}), n \geq 2$ from Figure 1, we can see that

$(\mathbf{2^0}, \{1\}) \not\models t$	because $1 \wedge \neg 1 = 0 = 1$, while there is no conclusion taking value 1
$(\mathbf{2}, \{1\}) \models t$ and $(\mathbf{2}, \{1\}) \models de$	left for the reader to verify
$(\mathbf{2^n}, \{1\}) \not\models de$ for all $n > 1$	if $v(p) = \alpha$ and $v(q) = \beta$, then $\alpha \vee \beta = 1$, while $\alpha \neq 1$ and $\beta \neq 1$, that is, the premise takes a designated value, although both conclusions take the not-designated values cf. (2).

Thus, if we employ m-rules t and de as inference rules together with modus ponens and axioms of the Classical Logic Cl, there is only one logical matrix in which all these rules and axioms are valid, namely $(\mathbf{2}, \{1\})$.

In 1932 Gödel stated without proof (the proof is due to Gentzen) that intuitionistic propositional logic enjoys the disjunction property: for any formulas A, B, if formula $A \vee B$ is a theorem, then at least one of the formulas A, B must be a theorem. It is not true for the classical propositional logic: formula $p \vee \neg p$ is a theorem, while neither p, nor $\neg p$ are theorems. Let us point out that even though the classical logic does not enjoy the disjunction property, the rule de can be used as an inference rule without expanding the set of theorems.

In what follows, rule de plays a special role. In the setting of natural deduction, $\vee$-elimination rule is as depicted in Figure 2:

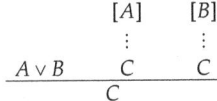

Figure 2. $\vee$-Elimination as a natural deduction rule.

That is, if we can derive $A \vee B$ and we can derive C separately from A and from B, then we can derive C.

In the multiple-conclusion setting (with the use of de) $\vee$-elimination can be expressed in a more natural way as depicted below in Figure 3:

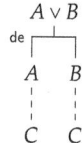

Figure 3. ∨-Elimination as a multiple-conclusion rule.

In this paper, we study how admissibility of de can be used to construct bases of admissible rules. Let us recall that a rule Γ/A is called admissible for a given logic L if for every substitution σ,

$$\sigma(\Gamma) \subseteq Th(\mathsf{L}) \text{ entails } \sigma(A) \in Th(\mathsf{L}).$$

It is not hard to see that a rule r is admissible for logic L defined by rules R if and only if rules $\mathsf{R} \cup \{\mathsf{r}\}$ define a logic L′ such that $Th(\mathsf{L}) = Th(\mathsf{L}')$.

We can extend the notion of admissibility to m-rules as follows; an m-rule Γ/Δ is admissible for a given logic L if every substitution σ,

$$\sigma(\Gamma) \subseteq Th(\mathsf{L}) \text{ entails } \sigma(\Delta) \cap Th(\mathsf{L}) \neq \varnothing.$$

The topic of the paper is the relations between the m-rules t and de and admissibility of rules. We divide m-rules into three categories: if $\mathsf{r} := \Gamma/\Delta$ is an m-rule, then

> r is conclusive if Δ consists of a single formula;
> r is inconclusive if Δ consists of more then one formula;
> r is terminating if $\Delta = \varnothing$.

A rule that has a nonempty set of alternatives is called proper. For instance,

> mp = $p, p \to q/q$ is a conclusive rule;
> de = $p \vee q/p, q$ is an inconclusive rule;
> t = $p \wedge \neg p/\varnothing$ is a terminating rule,

and mp and de are proper m-rules.

Successive application of rules leads to a notion of inference: from a given set of formulas Γ—assumptions—we infer a formula, A. Inferences from the empty set of assumptions (or from the axioms) are proofs.

In this paper, we focus on logics for which the ∨-elimination m-rule de is admissible (for instance, it is admissible for the intuitionistic logic and it is not admissible for the classical logic). Additionally, we will show that all m-rules except, perhaps, for de and t, can be eliminated from any base of admissible m-rules for such a logic.

2. Preliminaries

Let Fm be a set of all (propositional) formulas built in a usual way from a denumerable set of (propositional) variables Var and a finite set of connectives $\mathcal{C}$. The maps $\sigma : Var \longrightarrow \mathsf{Fm}$ are called substitutions. Given a substitution σ and a formula A, $\sigma(A)$ denotes the result of replacing each variable p occurring in A with formula $\sigma(p)$, and if Γ is a set of formulas, then $\sigma(\Gamma) := \{\sigma(A) : A \in \Gamma\}$.

Let $\mathcal{R}_{\mathsf{Fm}}$ be a class of all ordered pairs of finite (possibly empty) subsets of Fm. The members of $\mathcal{R}_{\mathsf{Fm}}$ are called multiple-conclusion rules, or multiple-alternative rules (m-rules for short). In the sequel, $\Gamma \Subset \mathsf{Fm}$ means that Γ is a finite subset of Fm, and if $\Gamma, \Delta \Subset \mathsf{Fm}$, the rule (Γ, Δ) is denoted as Γ/Δ. The members of Γ are called premises, while the members of Δ are called alternatives or conclusions. If $\{A_0,\ldots,A_{n-1}\}/\{B_0,\ldots,B_{m-1}\}$ is an m-rule, we drop curly brackets and write $A_0,\ldots,A_{n-1}/B_0,\ldots,B_{m-1}$. Also, if $\Gamma,\Gamma' \in \mathsf{Fm}$, we write Γ,Γ' to denote $\Gamma \cup \Gamma'$.

If $r := \Gamma/\Delta$ is an m-rule and σ is a substitution, then $\sigma(\Gamma)/\sigma(\Delta)$ is again an m-rule which is called an instance of r.

Let us note that m-rules allow empty sets of premises and alternatives/conclusions, and ∅ has a different meaning depending on whether it is a set of premises or a set of conclusions. To make it easier, we use ▾ for ∅ as a premise and we use ▴ for ∅ as a conclusion. We also assume that for every substitution σ, $\sigma(▾) = ▾$ and $\sigma(▴) = ▴$. Symbols ▾ are ▴ are merely notations and they are not elements of the language or metalanguage.

Formula A is valid in a given logic L (L-valid for short) or A is a theorem of L if $\vdash_L A$ (or $A \in Th(L)$), otherwise, A is called refuted in L.

We call logic L consistent if not every formula from Fm is a theorem of L. Let us note that because of structurality, i.e., because $\vdash_L$ obeys (S) from (1), a logic L is consistent if and only if $\nvdash_L q$, where q is a variable.

Let us observe that a rule $r := \Gamma/A$ is admissible for logic L (in symbols, $\Gamma \mid\!\sim_L A$) if any substitution that refutes A, refutes at least one member of Γ. If $\Gamma = ∅$, then rule Γ/A is admissible for L if and only if $A \in Th(L)$ [8]. For m-rules, an m-rule Γ/Δ is admissible for $\vdash_L$ (in symbols $\Gamma \mid\!\sim_L \Delta$) if every substitution σ that refutes all formulas from Δ, refutes at least one formula from Γ [8]. Thus, the rule ▾/▴ is not admissible in any logic.

If R is a set of all conclusive rules admissible for logic L, then R defines a logic $\widetilde{L}$ that has the same set of theorems as L. It is not hard to see that $\widetilde{L}$ is the biggest logic that has the same theorems as L, and we call $\widetilde{L}$ an admissible completion of L. In the book by Rybakov [9] (Definition 1.7.3), the term "admissible closure" is used; in our view, the latter term is a bit ambiguous, because "admissible closure" can refer to a consequence closure operator. Let us note that if R is a set of all conclusive rules admissible for L and R^m is a set of all m-rules admissible for L, then $\vdash_R$ and $\vdash_{R^m}$ define the same logic, namely, $\widetilde{L}$. Indeed, it is clear that a conclusive rule r is admissible for L if and only if r is admissible for L as m-rule.

If L is a logic defined by m-rules and r is a rule, by L + r we denote the smallest logic extending L and containing r. It is not hard to see that if r is admissible for L, then $Th(L+r) = Th(L)$.

Admissibility of a rule can be expressed in terms of L-unifiability. A set of formulas Γ is unifiable in L (or L-unifiable) if $\Gamma \not\mid\!\sim_L ▴$, otherwise Γ is nonunifiable in L. In other words, formulas Γ are L-unifiable if and only if there is such a substitution σ, that $\sigma(\Gamma) \subseteq Th(L)$, and in this case, σ is said to be an L-unifier of Γ. Thus, an m-rule Γ/Δ is admissible for L if and only if every L-unifier of Γ unifies at least one formula from Δ.

Logic L is strongly consistent if there is a finite set of formulas Γ such that $\Gamma \mid\!\sim_L ▴$, that is, Γ is nonunifiable in L. Note that if Γ is a nonunifiable in L finite set of formulas, then any rule Γ/Δ (including $\Gamma/▴$) is trivially admissible in L. The m-rules with nonunifiable set of premises are called passive (passive rules were introduced in [10]).

Example 1. *Throughout the paper we will use the examples from the following well-known logics.*

Int	*intuitionistic propositional logic;*
Pos	*positive fragment of* Int;
KP	*Kreisel-Putnam's logic;*
ML	*Medvedev's logic;*
Jhn	*Johansson's (or minimal) logic;*

and normal modal logics S4, D4, GL, K4, K4.1, S4, S4.1, Grz, Int, Dn *[11]. All these logics are consistent: formula p, where p is a propositional variable, is not a theorem. And all of them, except for* Pos, *are strongly consistent: formula* $p \wedge \neg p$ *is not unifiable, while in* Pos *every nonempty set of formulas is unifiable (substitute each variable with* $p \to p$).

3. Derivation

An m-inference (or m-derivation) of an m-rule from a given set of m-rules will be a generalization of a regular notion of Hilbert style inference (for instance, like in [12]). Since we allow to use the multiple-conclusion rules, inference cannot be a sequence anymore, and it is a tree: application of any inconclusive rule triggers branching. Each derived alternative shall be considered individually, like a separate case in a proof by cases. The introduced below notion of m-inference with use of m-rules reflects our everyday practice of making derivations from a set of assumptions: at each step we either refer to an assumption, or we apply a rule of inference and derive an (intermediate) conclusion, or we use a proof by cases and we lay down a set of alternatives that will be considered separately. The latter is captured by application of some inconclusive rule, when instead of a single conclusion we arrive at a set of alternatives to be considered (for different flavors of formalization of the proofs by cases can be fond in [13]). Our definition of m-inference slightly differs from the definition in [8,14]. An alternative approach to derivation in the multiple conclusion setting the reader can find in [15].

3.1. Basic Definitions: Derivation Trees

By (finite) tree we understand a partially ordered set $(\mathsf{Nd}, \leq)$ that has the biggest element (called a root) and for each $\mathfrak{n} \in \mathsf{Nd}$, the segment $\{\mathfrak{n}' \in \mathsf{Nd} : \mathfrak{n} \leq \mathfrak{n}'\}$ is a chain. Labeling of a tree $(\mathsf{Nd}, \leq)$ is a map $\lambda : \mathsf{Nd} \longrightarrow \{A : A \in \mathsf{Fm} \cup \{\blacktriangledown, \blacktriangle\}\}$, where $\blacktriangledown$ is allowed only in the root and $\blacktriangle$ is allowed only in a leaf. Moreover, the root is always labeled by $\blacktriangledown$. A tree together with labeling (that is the pair tree-labeling) is called a labeled tree. When we draw a labeled tree, to simplify notation, instead of node $\mathfrak{n}$ we will use its label $\lambda(\mathfrak{n})$. For instance, instead of left-hand side tree depicted in Figure 4 with labeling λ

node	$\mathfrak{n}_0$	$\mathfrak{n}_1$	$\mathfrak{n}_2$	$\mathfrak{n}_3$	$\mathfrak{n}_4$
λ	A	B	C	A	$\blacktriangle$

we use the right-hand side tree

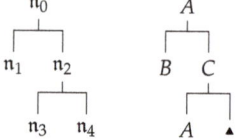

Figure 4. Labeling.

If $\mathfrak{n}$ is a node of a labeled tree, we let $\mathfrak{n}{\uparrow} := \{\mathfrak{n}' \in \mathsf{Nd} : \mathfrak{n} \leq \mathfrak{n}'\}$ and $\mathfrak{n}{\downarrow} := \{\mathfrak{n}' \in \mathsf{Nd} : \mathfrak{n}' < \mathfrak{n}\}$. Nodes from $\mathfrak{n}{\uparrow}$ are predecessors of $\mathfrak{n}$ and nodes from $\mathfrak{n}{\downarrow}$ are successors of $\mathfrak{n}$. A successor $\mathfrak{n}'$ of a node $\mathfrak{n}$ is called immediate if there are no nodes strongly between $\mathfrak{n}'$ and $\mathfrak{n}$. By $lv(\mathsf{Nd})$ we denote the set of all leaves of Nd, that is, $lv(\mathsf{Nd})$ is the set of all minimal elements of $(\mathsf{Nd}, \leq)$.

If $\mathsf{Nd}' \subseteq \mathsf{Nd}$, then $\lambda(\mathsf{Nd}') := \bigcup\{\lambda(\mathfrak{n}) : \mathfrak{n} \in \mathsf{Nd}'\}$. For instance, $\lambda(\mathfrak{n}{\uparrow})$ is a set of all formulas labeling all predecessors of $\mathfrak{n}$, and $\lambda(lv(\mathsf{Nd}))$ is a set of all formulas labeling all leaves of the tree $(\mathsf{Nd}, \leq)$.

A leaf labeled by $\blacktriangle$ is a terminal leaf, otherwise, the leaf is called extendable.

3.2. Definition of m-Inference

Now, we can introduce the notion of m-inference in the setting of m-rules. Our definition of m-inference is slightly different from the one introduced in works by the authors of [3,14], but as Theorem 2 shows, the classes of derivable m-rules coincide.

Definition 1. *Let* R *be a set of m-rules and* Γ *be a set of formulas (which may be empty). An m-inference from* Γ *by* R *(or* (Γ, R)*-inference for short) is a finite labeled tree, defined by induction:*

(a) *A tree containing only a root labeled by* $\blacktriangledown$ *is a* (Γ, R)*-inference;*

(b) If I is a (Γ, R)-inference, then a tree obtained from I by adjoining to an extendable leaf an immediate successor labeled by a formula from Γ is a (Γ, R)-inference;

(c) If I is a (Γ, R)-inference, and $\mathfrak{n}$ is an extendable leaf, then a tree, obtained from I by adjoining to $\mathfrak{n}$ immediate successors $\mathfrak{n}_0, \ldots, \mathfrak{n}_{m-1}$ labeled by formulas $B_0, \ldots, B_{m-1}$, is (Γ, R)-inference, provided there is an instance Δ/Π of a rule from R such that

$$\Delta \subseteq \lambda(\mathfrak{n}\uparrow) \text{ and } \Pi = \{B_0, \ldots, B_{m-1}\}.$$

(d) If I is a (Γ, R)-inference and $\mathfrak{n}$ is an expendable leaf, then a tree, obtained from I by adjoining to $\mathfrak{n}$ immediate successors $\mathfrak{n}_0$ labeled by ▲, is (Γ, R)-inference, provided there is an instance $\Delta/$▲ of a rule from R such that

$$\Delta \subseteq \lambda(\mathfrak{n}\uparrow).$$

For instance, suppose $\Gamma = \{C_0, \ldots, C_{k-1}\}$ and $A_0, \ldots, A_{n-1}/B_0, \ldots, B_{m-1}$ is an instance of a rule from R. Then, if a tree depicted in Figure 5 is a (Γ, R)-inference,

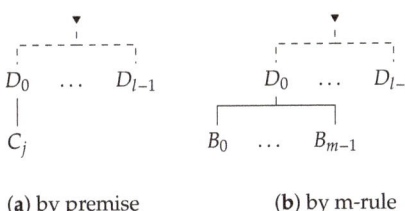

Figure 5. Initial inference.

then the trees depicted in Figure 6 are (Γ, R)-inferences, provided for (a) that $0 \leq j < k$, and for (b), that all premises $A_i, i < n$ can be found on the branch between leaf D_0 and the root.

Figure 6. Examples of Inferences.

Let us observe the following simple but important property of m-inferences.

Proposition 1. *Suppose that I and I' are (Γ, R)-inferences. Then the following assertions hold:*

(a) *a labeled tree obtained from I by omitting all successors of a given node is a (Γ, R)-inference;*

(b) *if we remove the root of I' and adjoin the remainder of I' to a leaf of I, the obtained labeled tree is a (Γ, R)-inference;*

(c) *for any substitution σ, the tree $\sigma(I)$ obtained from I by replacing in every node the labeling formula A by $\sigma(A)$, is a $\sigma(\Gamma)$, R-inference.*

The proof follows immediately from the definition of m-inference.

3.3. Derivations of m-Rules

Using the notion of m-inference from assumptions, we can define the notion of m-inference of m-rule.

Definition 2. *Let R be a set of m-rules and r := Γ/Δ be an m-rule. We say that r is derivable from rules R (in symbols R $\vdash$ r) if there is a (Γ, R)-inference I such that $\lambda(lv(I)) \subseteq \Delta \cup \{\blacktriangle\}$, i.e., every leaf is labeled by a formula from Δ or by ▲.*

The following proposition is an immediate consequence of the definition.

Proposition 2. *Let Γ be a finite set of formulas and R be a set of m-rules. Then any (Γ, R)-inference I is a derivation of m-rule Γ/Δ where $\Delta = \lambda(lv(I))$, i.e., Δ is a set of all formulas labeling all leaves of I.*

Corollary 1. *Suppose that I is an m-inference of an m-rule Γ/Δ from a set of m-rules R. Then if a formula $A \in \Delta$ labels a node n that is not a leaf, the tree I' obtained from I by omitting all nodes strongly below n, is also an m-inference of Γ/Δ from R.*

In other words, if R $\vdash$ Γ/Δ, then there is a (Γ, R)-inference, such that formulas from Δ label only leaves.

Theorem 1. *Suppose that R and R' are sets of m-rules, while r and r' are m-rules. Then the following hold;*

(i) *if $r \in R$, then R $\vdash$ r;*
(ii) *if R $\vdash$ r, then $R \cup R' \vdash r$;*
(iii) *if R $\vdash$ r and $R', r \vdash r'$, then R, $R' \vdash r'$.*

Proof. The proofs of (i) and (ii) are straightforward. Let us prove (iii).
Suppose that $r = \Gamma/\Delta$, $r' = \Gamma'/\Delta'$ and $\Delta = \{D_0, \ldots, D_{m-1}\}$. By assumption, there is a (Γ, R)-inference of Δ, which we denote by I, and there is a (Γ', R', r)-inference of Δ', which we denote by I'. To prove (iii) it suffices to observe that any application of rule r in I' can be replaced by an instance of the proof I (see Figure 7).

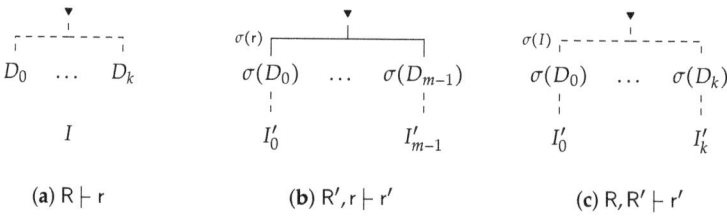

(a) R $\vdash$ r (b) $R', r \vdash r'$ (c) R, $R' \vdash r'$

Figure 7. Proof of (iii).

Let us observe that the set of leaves of the proof depicted in Figure 7c is a subset of the set of leaves of the proof depicted in Figure 7b. Thus, all the leaves of the proof depicted in Figure 7c contain formulas only from Δ'. □

Properties of m-Inference

Theorem 2. *Let R be a set of m-rules, Γ/Δ be an m-rule and A be a formula. Then*

$$R \vdash A/A; \qquad (R)$$
$$R \vdash \Gamma/\Delta \text{ entails } R \vdash \Gamma'/\Delta' \text{ for any } \Gamma' \supseteq \Gamma \text{ and } \Delta' \supset \Delta; \qquad (M)$$
$$\text{if } R \vdash \Gamma/\Delta, A \text{ and } R \vdash \Gamma, A/\Delta, \text{ then } R \vdash \Gamma/\Delta; \qquad (T)$$
$$\text{if } R \vdash \Gamma/\Delta, A, \text{ then } R \vdash \sigma(\Gamma)/\sigma(\Delta) \text{ for any substitution } \sigma \qquad (S).$$

Proof. Indeed, if A is a formula, then the tree that consists of a root, labeled by ▼, and its single immediate successor, labeled by A, is an m-inference of the rule A/A from R. Thus, (R) holds.

It follows immediately from the definition of inference that if I is an inference of a rule Γ/Δ from R, then I is at the same time an inference of the rule $\Gamma \cup \Gamma'/\Delta \cup \Delta'$ for any finite sets of formulas Γ', Δ'. That is, (M) holds.

Also straight from the definition it follows that if I is an inference of a rule Γ/Δ from R and σ is a substitution, then the tree $\sigma(I)$, obtained from I by replacing every label A with $\sigma(A)$, is an inference of $\sigma(\Gamma)/\sigma(\Delta)$ from R. Thus, (S) holds.

To demonstrate (T) we will show that, given an m-inferences of rules $\Gamma, A/\Delta$ and $\Gamma/A, \Delta$ from R, we can construct an m-inference of the rule Γ/Δ.

Suppose $\Delta = \{D_0, \ldots, D_{m-1}\}$ and we have inferences of $\Gamma, A/\Delta$ and $\Gamma/A, \Delta$ from R depicted respectively in Figure 8a,b:

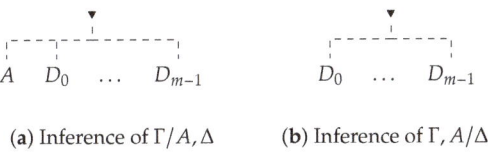

(a) Inference of $\Gamma/A, \Delta$ (b) Inference of $\Gamma, A/\Delta$

Figure 8. Proof of (T): the premises.

By Corollary 1, we can assume that A labels only leaves. Then we can construct an inference of Δ from Γ by adjoining inference (b) to every leaf labeled by A as depicted below in Figure 9:

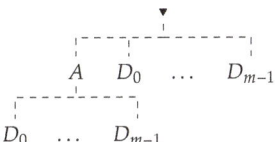

Figure 9. Proof of (T): the result.

and in such a way to obtain an m-inference of Γ/Δ. □

Let us also note the following property of passive rules which immediately follows from (M).

Corollary 2. *Let R be a set of m-rules containing rule $\Gamma/\blacktriangle$. Then, for any finite set of formulas Δ,*

$$R \vdash \Gamma/\Delta.$$

Corollary 3. *If R is a set of m-rules admissible for logic L and r is an m-rule such that $R \vdash r$, then r is also admissible for L.*

Proof. Suppose that $r = \Gamma/\Delta$ and σ is a substitution such that $\sigma(\Gamma) \in Th(L)$. By virtue of (S), there is an m-inference of $\sigma(\Gamma)/\sigma(\Delta)$ from R. From the definition of m-inference and admissibility of rules from R, any application of rule from R has at least one conclusion, which is a theorem of L. Hence, $\sigma(\Delta)$ contains a theorem of L, and hence rule Γ/Δ is admissible for L. Let us note that if some rule from R is terminating, its admissibility entails that L is not consistent and hence, every rule is admissible for it. □

3.4. m-Deductive Systems

Theorem 2 ensures that for any set of m-rules R, the restriction of $\vdash_R$ to the single-conclusion relation is a consequence relation. Thus, every set of m-rules R can be regarded as an m-deductive system $\vdash_R$. Recall that $Th(R)$ is the set of theorems: $Th(R) = \{A \in Fm : \vdash_R A\}$.

Let us point out that m-deductive systems lack some properties of regular deductive systems. For instance, a join of m-deductive systems with the same sets of theorems may be an m-deductive system with a strongly larger set of theorems. That is,

$$Th(R_0) = Th(R_1) \not\Rightarrow Th(R_0) = Th(R_1) = Th(R_0 \cup R_1).$$

We assume that the reader is familiar with Heyting algebras as models of intermediate logics.

Let us consider deductive system R defined by all formulas valid in algebra **C** depicted in Figure 10 and the rule mp (we regard formulas as rules of type $\blacktriangledown/A$); we constructed two m-deductive systems:

$$R_0 := R + de + dil \text{ and } R_1 := R + r + dir, \text{ where}$$

$$r := (\neg\neg p \to p) \to (p \vee \neg p)/\neg p \vee \neg\neg p, \quad dil := p/q \vee p \text{ and } dir := p/p \vee q.$$

Our goal is to verify that $Th((R)) = Th((R_0)) = Th((R_1))$ and formula $A := (\neg p \vee \neg\neg p) \vee (\neg\neg p \to p)$ is not a theorem of R_0 and R_1, while A is a theorem of $R_0 \cup R_1$, that is, $Th(R_0) = Th(R_1) \subsetneq Th(R_0 \cup R_1)$.

We start with an observation that algebra **C**' (see Figure 10) is (isomorphic to) a Lindenbaum algebra of R on one variable. Hence, both algebras **C** and **C**' are models for R.

Next, we observe (and we left for the reader to perform this routine check) that (a) rules de and dil are valid in **C**, hence, **C** is a model of R_0; (b) rules r and dir are valid in **C**', hence, **C**' is a model of R_1; and (c) all three m-logics have the same sets of theorems, that is,

$$Th(L) = Th(L_0) = Th(L_1).$$

Also, as we can see in Figure 10, formula A is refuted in **C** and hence, $A \notin Th(R_0)$ and $A \notin Th(R_1)$. Thus, we only need to show that $A \in Th(R_0 \cup R_1)$.

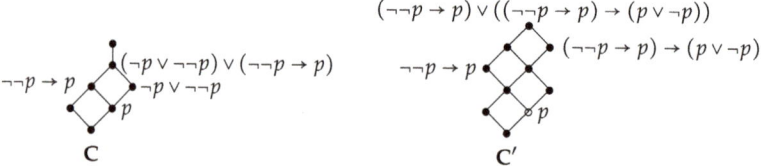

Figure 10. Example.

Indeed, because algebra **C**' is a Lindebaum algebra of R on one variable and any formula in one variable is a theorem of R precisely when it is valid on the generator of **C**' depicted as ○. Thus, because formula $((\neg\neg p \to p) \to (p \vee \neg p)) \vee (\neg\neg p \to p)$ is valid on this generator, it is a theorem of R, as well as it is a theorem of R_0 and R_1. Thus, we can construct an m-inference of A from $((\neg\neg p \to p) \to (p \vee \neg p)) \vee (\neg\neg p \to p)$, which is presented in Figure 11.

$$\blacktriangledown$$
$$|$$
$$((\neg\neg p \to p) \to (p \vee \neg p)) \vee (\neg\neg p \to p)$$
$$\text{de}$$
$$((\neg\neg p \to p) \to (p \vee \neg p)) \quad (\neg\neg p \to p)$$
$$r| \qquad\qquad dil|$$
$$(\neg p \vee \neg\neg p) \quad (\neg p \vee \neg\neg p) \vee (\neg\neg p \to p)$$
$$\text{dir}|$$
$$(\neg p \vee \neg\neg p) \vee (\neg\neg p \to p)$$

Figure 11. m-Inference of formula A.

Thus, formula A is a theorem of L'.

3.5. Admissible Bases

Suppose that L is a logic, R is a set of m-rules and r is an m-rule. We say that r is derivable from m-rules R relative to m-logic L (in symbols $R \vdash_L r$) if $R \cup L \vdash r$.

Let us observe that, by Corollary 3, if R is a set of admissible for L m-rules and $R \vdash_L r$, then r is admissible for L m-rule. Thus, we can use m-inferences to axiomatize admissible completion of logics.

Definition 3. *Suppose L is a logic and R is a set of admissible for L conclusive rules. Then R is an admissible relative to L base if every admissible for L conclusive rule r is derivable from R relative to L, that is,*

$R \vdash_L r$ *for every conclusive admissible for L rule r.*

Example 2. *For Int the Visser rules*

$$v_n := \frac{A^{(n)} \to (p_n \vee q_n)}{\bigvee_{i=0}^{n}(A^{(n)} \to p_i)},$$

where $A^{(n)} := \bigwedge_{i=0}^{n-1}(p_i \to q_i), n \geq 1$, *form a relative admissible base (cf. the work by the authors of [16]).*

In a natural way, the notion of admissible base can be extended to m-rules.

Definition 4. *Suppose L is a logic and R is a set of admissible for L m-rules. Then R is an admissible relative to L m-base if every admissible for L m-rule r is derivable from R relative to L, that is,*

$R \vdash_L r$ *for every admissible for L m-rule r.*

And an admissible m-base R for L is independent if neither proper subset of R is an admissible m-base for L.

Definition 5. *Suppose L is a logic and R is a set of admissible for L m-rules. Then R is an admissible relative to L extended base if every admissible for L conclusive rule r is derivable from R relative to L, that is,*

$R \vdash_L r$ *for every admissible for L m-rule r.*

And an admissible extended base R for L is independent if neither proper subset of R is an admissible m-base for L.

In Sections 5 and 6 we see that in the logics with the disjunction property, the admissible bases and m-bases are closely related.

Remark 1. *There is a difference in the properties of admissible conclusive and inconclusive rules. Namely, in contrast to conclusive rules, an inconclusive rule can be derivable and not admissible. Indeed, if Cl is the logic defined by axiom schemes of the classical logic and mp as a single inference rule and* $Cl^\vee := Cl + de$, *then,* $Th(Cl) = Th(Cl^\vee)$: *it is clear that* $Th(Cl) \subseteq Th(Cl^\vee)$, *on the other hand, every theorem of* $Cl^\vee$ *is valid in the two-element Boolean algebra, because all axioms and rules of* $Cl^\vee$ *are valid in it. Thus, rule de is trivially derivable in* $Cl^\vee$ *but it is not admissible:* $\vdash_{Cl^\vee} p \vee \neg p$, *while* $\nvdash_{Cl^\vee} p$ *and* $\nvdash_{Cl^\vee} \neg p$.

4. Introducing Meta-Disjunction

If our language contains disjunction with regular properties, m-Rule

$$de := \frac{A \vee B}{A, B}$$

plays a very special role in constructing deductive systems. Indeed, if R is a set of proper m-rules such that $R \vdash de$, we can replace the set of m-rules R by de and the set of conclusive rules

$$R^{(1)} := \{\Gamma/\bigvee \Delta : \Gamma/\Delta \in R\},$$

where $\bigvee \Delta := \bigvee_{A \in \Delta} A$. Indeed, if in an inference we apply m-rule Γ/Δ, instead, we can apply rule $\Gamma/\bigvee \Delta$, and then apply $n - 1$ times m-rule de (where $n = |\Delta|$). In this section, we discuss the sufficient conditions for logics to have an analog of ∨-elimination.

4.1. m-Protodisjunction

Let $\nabla(p,q)$ be a formula (a nonempty finite set of formulas) in two variables (in the sequel we write $p \nabla q$ to make the meaning more transparent). Then we let

$$\begin{aligned}
\text{de} &:= p\nabla q/p, q & \text{(Disjunction Elimination)} \\
\text{dir} &:= p/p\nabla q & \text{(Disjunction Introduction-Right)} \\
\text{dil} &:= q/p\nabla q & \text{(Disjunction Introduction-Left)}
\end{aligned}$$

In the case when ∇ contains more then one formula,

$$\begin{aligned}
\text{de} &:= p\nabla q/p, q & \text{(Disjunction Elimination)} \\
\text{dir} &:= p/D(p,q) \text{ for each } D \in \nabla & \text{(Disjunction Introduction-Right)} \\
\text{dil} &:= q//D(p,q) \text{ for each } D \in \nabla & \text{(Disjunction Introduction-Left)}
\end{aligned}$$

We will use the following rules capturing the properties of ∇.

$$\begin{aligned}
\text{dc} &:= (p\nabla q)/(q\nabla p) & \text{(Commutativity)} \\
\text{dra} &:= ((p\nabla q)\nabla r)/(p\nabla(q\nabla r)) & \text{(Right associativity)} \\
\text{dla} &:= ((p\nabla q)\nabla r)/(p\nabla(q\nabla r)) & \text{(Left associativity)} \\
\text{dd} &:= ((p\nabla r)\nabla(q\nabla r))/(p\nabla(q\nabla r)) & \text{(Self-distributivity)} \\
\text{di} &:= (p\nabla p)/p & \text{(Idempotency)}
\end{aligned}$$

Let

$$D := \{\text{dc}, \text{dra}, \text{dla}, \text{dd}, \text{di}\} \qquad (3)$$

be the set of rules representing the properties of ∇.

Proposition 3. *The following hold;*

$$\begin{aligned}
DC &: \text{de}, \text{dir}, \text{dil} \vdash \text{dc} \\
DRA &: \text{de}, \text{dir}, \text{dil} \vdash \text{dra} \\
DLA &: \text{de}, \text{dir}, \text{dil} \vdash \text{dla} \\
DD &: \text{de}, \text{dir}, \text{dil} \vdash \text{dd} \\
DI &: \text{de}, \text{dir}, \text{dil} \vdash \text{di}
\end{aligned}$$

Proof. The proofs of (DC), (DRA), and (DD) are depicted in Figure 12. Proof of (DLA) is similar to the proof of (DRA), and (DI) is an immediate consequence of de and the definition of m-inference.

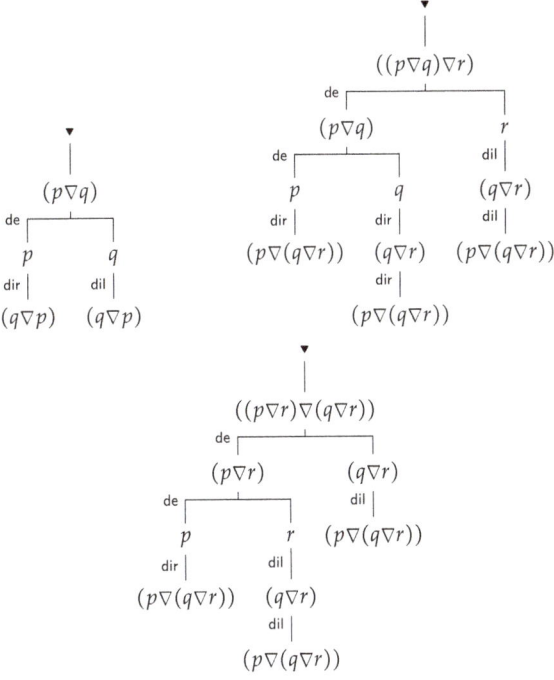

Figure 12. Proof of $(DC), (DRA), (DD)$.

Definition 6. *A set of m-rules R is m-disjunctive if for some formula $\nabla(p,q)$ (some nonempty finite set of formulas) in two variables, rules dir, dil, and de are derivable from R, and we call formula(s) ∇ an m-protodisjunction for R (comp. with the notion of protodisjunction in works by the authors of [13,17]).*

Definition 7. *An m-logic L has the disjunction property (DP for short) if rules dir, dil, and de are admissible in L, that is, if $\widetilde{L}$ is m-disjunctive.*

For instance, modal logic S4 has the Disjunction property relative to m-protodisjunction $\nabla := (\Box p \vee \Box q)$.

Let us observe that if R is an m-disjunctive set (or L enjoys the DP), then rules dc, dra, dla, dd, di are derivable from R (or respectively, these rules are admissible for L).

It is important that m-protodisjunction is defined uniquely up to R-equivalence in the following sense.

Let R be a set of m-rules. Sets of formulas Γ and Δ are said to be R-equivalent if

$$\Gamma \vdash_R D, \text{ for each } D \in \Delta \text{ and } \Delta \vdash_R A, \text{ for each } A \in \Gamma.$$

Proposition 4. *Let R be an m-disjunctive set of rules, and ∇_0 and ∇_1 be m-protodisjunctions for R. Then ∇_0 and ∇_1 are R-equivalent.*

Proof. Suppose that $D \in \nabla_1$ and $\nabla_0 = \{D_i, i < n\}$. An m-inference of D from ∇_0 is depicted in Figure 13. □

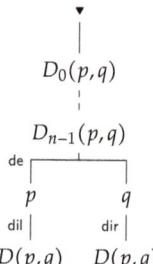

Figure 13. Proof of Proposition 4.

Let us note, that if a logic has conjunction with regular properties, ∇ can be reduced to a single formula.

Note 1. *In order not to add an extra layer of complexity, in this paper we consider only the case when ∇ consists of a single formula, even though the main results hold in a general case.*

If R is an m-disjunctive set of m-rules or if $\vdash_L$ is an m-logic with the DP, we always assume that meta-disjunction is expressible by a formula ∇.

Example 3. *Let us consider intuitionistic propositional logic* Int *and normal modal logic* S4. *For* Int *we can take $p \nabla q = p \vee q$. For* S4 *we can take $p \nabla q = \Box p \vee \Box q$. It is clear that rules* dir *and* dil *are derivable in these logics, that is, $p \vdash_L p \nabla q$ and $q \vdash_L p \nabla q$, where $L \in \{$Int, S4$\}$, and all three rules (*dir*, *dil*, and* de*) are admissible for* Int *and* S4 *[11]. For logic* BCK *one can take $p \nabla q = (p \to q) \to q$. Let us point out that rule* dc $:= p \nabla q / q \nabla p$ *is admissible for* BCK*, but it is not derivable [18] (Theorem 4.2).*

4.2. Properties of m-Protodisjunction

In this section, we prove that with respect to $\vdash$, m-protodisjunction has the properties which disjunction is expected to have.

Proposition 5. *Suppose ∇ is an m-protodisjunction for a set of m-rules* R. *Then for any formulas $A, B \in$ Fm, and any $\Gamma, \Delta \subseteq$ Fm,*

$$\Gamma, A \vdash_R \Delta \text{ and } \Gamma, B \vdash_R \Delta \text{ entails } \Gamma, A \nabla B \vdash_R \Delta. \tag{4}$$

Proof. One can apply de to $\Gamma, A \nabla B$ and obtain two cases to consider: Γ, A and Γ, B. In each of these cases one can derive Δ. □

Immediately from Proposition 5 it follows that if $\vdash_L$ is an m-logic enjoying the DP, then for any formulas $A, B \in$ Fm and any $\Gamma, \Delta \subseteq$ Fm,

$$\Gamma, A \vdash_L \Delta \text{ and } \Gamma, B \vdash_L \Delta \text{ entails } \Gamma, A \nabla B \vdash_L \Delta \tag{5}$$

and consequently,

$$\Gamma, A \vdash_L C \text{ and } \Gamma, B \vdash_L C \text{ entails } \Gamma, A \nabla B \vdash_L C,, \tag{6}$$

and

$$A \vdash_L C \text{ and } B \vdash_L C \text{ entails } A \nabla B \vdash_L C. \tag{7}$$

If $\Delta := \{B_0, \ldots, B_{m-1}\}$ is a finite set of formulas, we let

$$\begin{aligned} \nabla(\varnothing) &:= \varnothing & &, \text{when } m = 0 \\ \nabla(\{B_0\}) &:= B_0 & &, \text{when } m = 1 \\ \nabla(\Delta) &:= B_0 \nabla \ldots \nabla B_{m-1} & &, \text{when } m > 1. \end{aligned} \quad (8)$$

Thus, ∇ converts any finite nonempty set of formulas into a single formula. m-Protodisjunction has the following property.

Corollary 4. *Suppose Δ is a nonempty finite set of formulas and R is a set of m-rules with m-protodisjunction ∇ (or $\vdash_L$ is an m-logic with the DP). Then*

$$\vdash_R \nabla(\Delta) \text{ if and only if } \vdash_R B \text{ for some } B \in \Delta$$

(accordingly, $\vdash_L \nabla(\Delta)$ if and only if $\vdash_L B$ for some $B \in \Delta$).

Proof. The corollary can be proven by a simple induction on cardinality of Δ. □

Let us introduce the following notations: suppose that A is a formula, Γ, Δ are sets of formulas, $r := \Gamma/\Delta$, and q is a variable not occurring in A and formulas from Γ and Δ. Then

$$\begin{aligned} &(a) & A^q &:= A \nabla q; \\ &(b) & &\text{if } \Gamma \neq \varnothing, \text{ then } \Gamma^q := \{A^q : A \in \Gamma\}; \\ &(c) & \blacktriangledown^q &:= \blacktriangledown \nabla q = \blacktriangledown \text{ (because } \blacktriangledown \text{ represents meta-truth)}; \\ &(d) & \blacktriangle^q &:= \blacktriangle \nabla q = q \text{ (because } \blacktriangle \text{ represents meta-falshood)}; \\ &(e) & r^q &:= \Gamma^q/\Delta^q; \\ &(f) & \nabla(r) &:= \Gamma/\nabla(\Delta); \\ &(g) & \check{r} &:= \Gamma^q/\nabla(\Delta^q), \text{ i.e., } \check{r} = \nabla(r^q). \end{aligned} \quad (9)$$

If R is a set of m-rules and q is a variable not occurring in any m-rule from R,

$$R^q := \{r^q : r \in R\} \quad \text{and} \quad \check{R} := \{\check{r} : r \in R\}$$

Remark 2. Note that if $r := \Gamma/\blacktriangle$ is a terminating rule and $q \notin Var(r)$, then, $r^q = \Gamma^q/q$ and the latter is a conclusive rule. Thus, for any m-rule $r := \Gamma/\Delta$ and any $q \notin Var(r)$, the rule $\check{r} = \Gamma^q/\nabla(\Delta^q)$ is always a conclusive rule: even if $\Delta = \varnothing$, we have $\blacktriangle^q = q$ and $\nabla(q) = q$ (cf. Equation (8)). If r is a conclusive rule, the rules r^q and $\check{r}$ coincide. Hence, if R is a set of conclusive rules, $R^q = \check{R}$.

Example 4. If $r := (\neg p_0 \to (p_1 \vee p_2))/((\neg p_0 \to p_1) \vee (\neg p_0 \to p_2))$, then

$$r^q := \frac{(\neg p_0 \to (p_1 \vee p_2)) \vee q}{((\neg p_0 \to p_1) \vee (\neg p_0 \to p_2)) \vee q}.$$

Rule r is admissible for any extension of Int [19], while rule r^q is admissible for any extension of Int enjoying the disjunction property.

If $r := \Diamond p \wedge \Diamond \neg p / \blacktriangle$, then

$$r^q := \frac{(\Diamond p \wedge \Diamond \neg p) \nabla q}{q}, \text{ where } \nabla \text{ is the strong disjunction: } A \nabla B := \Box A \vee \Box B.$$

Rule r is admissible for any normal modal logic extending D4 [10], while rule r^q is admissible for any normal modal logic extending D4 enjoying the disjunction property.

Proposition 6. *Let R be a set of conclusive rules, Γ be a finite set of formulas and A, B, C be formulas. Then,*

$$R \vdash \frac{\Gamma, A}{C} \quad \text{entails} \quad D, R^q \vdash \frac{\Gamma, A \nabla B}{C \nabla B}.$$

Proof. Let I be a single-conclusion inference of C from Γ, A. Then C is the last formula in I. There are three cases to consider: (a) $C = A$; (b) $C \in \Gamma$; (c) C is obtained from some preceding formulas by a conclusive rule from R.

Case (a). Is trivial.

Case (b). If $C \in \Gamma$, then by the definition of m-inference, $R \vdash \Gamma/C$. Thus, $R \vdash \Gamma, A \nabla B/C$ and, by dir, we get $\vdash \Gamma, A \nabla B/C \nabla B$.

Case (c). Suppose C is obtained by an instance $C_0, \ldots, C_{m-1}/C$ of some rule $r \in R$. Then,

$$C_0 \nabla B, \ldots, C_{m-1} \nabla B / C \nabla B$$

is an instance of r^q and we can easily convert inference I into the inference of $C \nabla B$ (see Figure 14).

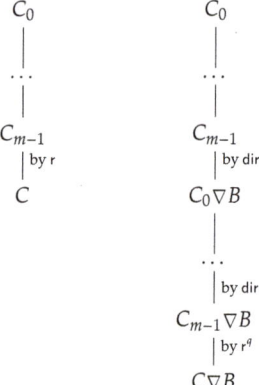

Figure 14. Proof of Prop. 6.

□

Corollary 5. *Let R be a set of conclusive rules, Γ be a finite set of formulas and $A_0, \ldots, A_{m-1}, C$ be formulas. Then, for any $i < m$,*

$$R \vdash \frac{\Gamma, A_i}{C} \quad \text{entails} \quad D, R^q \vdash \frac{\Gamma, A_0 \nabla \ldots \nabla A_{m-1}}{A_0 \nabla \ldots \nabla A_{i-1} \nabla C \nabla A_{i+1} \nabla \ldots \nabla A_{m-1}}.$$

Proof. Indeed, for $i = 0$ we can take $B = A_1 \nabla \ldots \nabla A_{m-1}$ and use Proposition 6. In the case when $i > 0$, we can use commutativity and associativity of ∇ and reduce this case to the case $i = 0$. □

5. From Admissible m-Base to Admissible Base

Our goal is to demonstrate that for any consistent m-logic with the DP, the problem of admissibility of any given proper m-rule can be reduced to the problem of admissibility of some conclusive rule.

Theorem 3. *Suppose $\vdash_L$ is a consistent m-logic with the DP. Let $r := \Gamma/\Delta$ be a proper m-rule and q be a variable not occurring in r. Then the following are equivalent.*

(a) *m-Rule r is admissible for $\vdash_L$;*

(b) m-Rule r^q is admissible for $\vdash_L$;
(c) Rule $\breve{r} = \Gamma^q/\nabla(\Delta^q)$ is admissible for $\vdash_L$;
(d) Rule $\nabla(r) = \Gamma/\nabla(\Delta)$ is admissible for $\vdash_L$.

Proof. (a) $\Longrightarrow$ **(b).** Assume that proper m-rule $r := \Gamma/\Delta$ is admissible for $\vdash_L$. Because r is proper, $\Delta \neq \varnothing$. Let us consider the three following cases:

$\Gamma = \varnothing$, that is, $r = \mathbf{v}/\Delta$ and $r^q = \mathbf{v}/\Delta^q$;
$\Gamma \neq \varnothing$, that is $r^q = \Gamma^q/\Delta^q$.

Case $\Gamma = \varnothing$. Admissibility of $\mathbf{v}/\Delta$ entails that there is $B \in \Delta$ such that $\vdash_L B$ and consequently, for every substitution σ, $\vdash_L \sigma(B)$. Hence, because dir is admissible for $\vdash_L$, for every substitution σ, we have $\vdash_L \sigma(B) \nabla \sigma(q)$ and therefore, rule $\mathbf{v}/\Delta^q$ is admissible for $\vdash_L$.

Case $\Gamma \neq \varnothing$. We need to prove that for any substitution σ,

if $\vdash_L \sigma(A) \nabla \sigma(q)$ for all $A \in \Gamma$, then there is $B \in \Delta$ such that $\vdash_L \sigma(B) \nabla \sigma(q)$.

Let σ be a unifier of Γ^q, that is $\vdash_L \sigma(A) \nabla \sigma(q)$ for all $A \in \Gamma$. Let us consider two subcases:

(i) σ unifies Γ;
(ii) σ does not unify Γ.

Proof of (i). Recall that r is admissible for $\vdash_L$. Therefore, if σ unifies Γ, that is, $\vdash_L \sigma(A)$ holds for all $A \in \Gamma$, by admissibility, there is $B \in \Delta$ such that $\vdash_L \sigma(B)$. Hence, because rule dir is admissible for $\vdash_L$, we can apply it and obtain $\vdash_L \sigma(B) \nabla \sigma(q)$. □

Proof of (ii). Suppose that σ does not unify Γ. Then, there is $A \in \Gamma$ such that $\not\vdash_L \sigma(A)$. On the other hand, σ unifies Γ^q and hence, $\vdash_L \sigma(A) \nabla \sigma(q)$. Then, by the Disjunction Property, $\vdash_L \sigma(q)$. Rule dil is admissible for $\vdash_L$, hence, $\vdash_L \sigma(q)$ entails $\vdash_L \sigma(B) \nabla \sigma(q)$ for every $B \in \Delta$. □

(b) $\Longrightarrow$ **(c).** Suppose m-rule $r^q := \Gamma^q/\Delta^q$ is admissible for $\vdash_L$. We need to prove that rule $\Gamma^q/\nabla(\Delta^q)$ is admissible for $\vdash_L$. Let σ be a substitution which unifies Γ^q, that is, σ makes all premises Γ^q derivable. Then, our assumption entails that $\vdash_L \sigma(B^q)$ holds for some $B \in \Delta$. Hence, by virtue of Corollary 4, $\vdash_L \nabla(\sigma(\Delta^q))$ which means that $\vdash_L \sigma(\nabla(\Delta^q))$ holds.

(c) $\Longrightarrow$ **(d).** Suppose that rule $\Gamma^q/\nabla(\Delta^q)$ is admissible for $\vdash_L$. We need to prove that rule $\Gamma/\nabla(\Delta)$ is admissible for $\vdash_L$. Let us consider two cases: $\Gamma = \varnothing$ and $\Gamma \neq \varnothing$.

Case $\Gamma = \varnothing$, that is, rule $\mathbf{v}/\nabla(\Delta^q)$ is admissible for $\vdash_L$. Hence, if Δ consists of a single formula B, we have $\vdash_L B^q$. If Δ contains more than one formula, by the disjunction property, there is a formula $B \in \Delta$ such that $\vdash_L B^q$. If we take into account that $\not\vdash_L q$ and apply the disjunction property to $\vdash_L B \nabla q$, we can conclude that $\vdash_L B$ holds and hence, $\vdash_L \sigma(B)$ holds for every substitution σ. Thus, rule $\mathbf{v}/\nabla(\Delta)$ is admissible for $\vdash_L$.

Case $\Gamma \neq \varnothing$ and we need to show that rule $\Gamma/\nabla(\Delta)$ is admissible for $\vdash_L$. Let σ be a substitution that unifies Γ, that is $\vdash_L \sigma(A)$ for all $A \in \Gamma$.

Recall that $\vdash_L$ is consistent, therefore there is a formula C such that $\not\vdash_L C$. Take the substitution

$$\sigma'(p) = \begin{cases} \sigma(p), & \text{if } p \neq q \\ C & \text{otherwise.} \end{cases}$$

Because q does not occur in formulas from Γ, for all $A \in \Gamma$, $\sigma'(A) = \sigma(A)$, and, consequently, $\sigma'(A^q) = \sigma(A) \nabla C$. By the assumption, $\vdash_L \sigma(A)$, and hence by dir, $\vdash_L \sigma(A) \nabla C$, that is, $\vdash_L \sigma'(A^q)$. Thus, σ' unifies Γ^q. Now, we can apply the assumption that rule Γ^q/Δ^q is admissible, and we can conclude that for some formula $B \in \Delta$, formula $\sigma'(B^q)$ is derivable. Thus, because $\sigma'(B^q) = \sigma'(B \triangledown q) = \sigma'(B) \triangledown C$, formula $\sigma'(B) \triangledown C$ is a theorem. Recall that we selected C not to be a theorem, hence, by the Disjunction Property, formula $\sigma'(B)$ is a theorem. And because variable q does not occur in B, we have $\sigma'(B) = \sigma(B)$, and it means that $\sigma(B)$ is a theorem and rule Γ/Δ is admissible.

(d) $\implies$ (a). Suppose that the rule $\Gamma/\triangledown(\Delta)$ is admissible for $\vdash_L$. Let σ be a unifier for Γ. Then, by the admissibility of $\Gamma/\triangledown(\Delta)$, we have $\vdash \sigma(\triangledown(\Delta))$. Now, we can apply Corollary 4 and conclude that for some $B \in \Delta$, $\vdash_L \sigma(B)$. Thus, the rule Γ/Δ is admissible in $\vdash_L$. □

Remark 3. *For any consistent m-logic $\vdash_L$ with the disjunction property, the rules r and r^q are either both admissible, or both not admissible. Moreover, r can be derived from r^q, while the converse needs not to be true: the restricted Visser rule V_n^- is derivable from the Visser rule V_n [20], while the converse does not hold [21] (Corollary 2). If the m-rules dir, dil, and de are not just admissible (which is required by the disjunction property), but they are derivable in $\vdash_L$, then r^q is derivable from r.*

Logic L is a-decidable if the problem of admissibility of conclusive rules in L is decidable, and logic L is am-decidable if the problem of admissibility of m-rules in L is decidable.

Corollary 6. *For every consistent m-logic with the DP, the problems of a- and am-admissibility are equivalent.*

Example 5. *It is well known from [11] that logics Int, S4, K4, Grz, GL enjoy the DP. Therefore, because the problem of a-admissibility for them is decidable [22–24], the problem of m-admissibility for these logics is decidable too. In algebraic terms, for each of these logics, the universal theory of the Lindenbaum algebra is decidable [24] (Theorem 10).*

5.1. A Note on Terminating Rules

The goal of this section is to show that terminating rules can be eliminated from any m-inference of any proper rule.

If R is a set of m-rules, by $\overline{R}$ we denote a set of proper m-rules obtained from R by replacing every terminating rule $t := A_1, \ldots, A_n/\blacktriangle \in R$ with conclusive rule $\bar{t} := A_1, \ldots, A_n/q$, where q does not occur in any of formulas $A_i, i = 1, \ldots, n$.

Proposition 7. *Suppose that there is an m-inference I of a proper rule Γ/Δ from R. Then, there is an inference $\bar{I}$ of Γ/Δ from $\overline{R}$ that does not contain terminal leaves.*

Proof. Let I be an m-inference of Γ/Δ from R. By assumption, rule Γ/Δ is proper, hence, Δ is not empty. Suppose that $B \in \Delta$.

Let us observe that, by the definition of m-inference, application of any terminating rule gives a leaf of this m-inference. Thus, if $t = A_1, \ldots, A_n/\blacktriangle \in R$ and $\sigma(A_1), \ldots, \sigma(A_n)/\blacktriangle$ is an instance of t that we have been used in I, instead, we can use the instance $\sigma(A_1), \ldots, \sigma(A_n)/B$ of a proper rule $\bar{t}$ that belongs to $\overline{R}$. And in such a way, we can eliminate all applications of the terminating rules from I. □

From Proposition 7 it immediately follows that for strongly consistent logics any m-base (finite m-base) can be converted into an m-base (finite m-base) containing at most one terminating m-rule.

Proposition 8. *Let R be an m-base (a relative m-base) of strongly consistent m-logic in which formula A is not unifiable and let $R' := \overline{R} \cup \{A/\blacktriangle\}$. Then R' is an m-base (relative m-base).*

Proof. First, from Proposition 7, it follows that every conclusive admissible rule is derivable from $\overline{\mathsf{R}}$. Next, if m-rule $\Gamma/\blacktriangle$ is admissible, then rule Γ/A is admissible too, and hence it is derivable from $\overline{\mathsf{R}}$. Thus, m-rule $\Gamma/\blacktriangle$ is derivable from $\overline{\mathsf{R}}$. □

5.2. Converting Admissible m-Base into Admissible Base

To convert a given relative admissible m-base R of a given logic L with the DP into a relative base, one can do the following: convert every rule $\mathsf{r} = \Gamma/\Delta$ from R into rule $\Gamma^q/\nabla(\Delta^q)$; the obtained set of rules we denote $\check{\mathsf{R}}$. The set $\check{\mathsf{R}}$ consists of conclusive rules (cf. Remark 2). From Theorem 3 we know that each rule from $\check{\mathsf{R}}$ remains admissible. Below, we show that every admissible for L conclusive rule can be derived from $\check{\mathsf{R}}$. Let us note that as a result of such a conversion some m-rules become trivial, for instance, $\widehat{\mathsf{de}}$:

$$p_1 \nabla p_2/p_1, p_2 \text{ becomes } p_1 \nabla p_2 \nabla q/p_1 \nabla p_2 \nabla q.$$

Theorem 4. *Suppose that L is a logic with the DP and R is an admissible m-base (a relative m-base) of L, then, $\check{\mathsf{R}}$ is an admissible base (respectively, a relative base) of L.*

To prove Theorem 4, we prove a bit more general Theorem 5, which holds not only for the logics enjoying the DP. Recall from Proposition 3 that all rules from D (cf. Equation (3)) representing the properties of ∇ are derivable in any logic with the DP. Hence, we can take Δ in Theorem 5 to be a singleton, and we will obtain Theorem 4 as a corollary.

Theorem 5. *Suppose R is a set of proper m-rules, and $\mathsf{r} := \Gamma/\Delta$ is a proper rule. If*

$$\mathsf{R} \vdash \Gamma/\Delta,$$

then

$$\mathsf{D}, \check{\mathsf{R}} \vdash \Gamma/\nabla(\Delta).$$

In other words, any m-inference of Γ/Δ from R can be converted into inference of $\Gamma/\nabla(\Delta)$ from $\check{\mathsf{R}}$ which is a single-conclusion inference, because all rules from D and $\check{\mathsf{R}}$ are conclusive.

Proof. Let I be a (Γ, R)-inference and Δ is a set of formulas which appear in the leaves of I. Without loss of generality we can assume that each formula from Δ appears in a leaf of the inference: if $\lambda(lv(I)) = \Delta' \subsetneq \Delta$, after we have derived $\nabla(\Delta')$, we can apply (multiple times if necessary) rule dir and derive $\nabla(\Delta)$.

Let k be the number of nodes of I having more than one immediate successor (in other words, k represents the number of applications of inconclusive rules). By induction on k we prove that for any (Γ, R)-inference, such that $\lambda(lv(I)) = \Delta$, there is a $(\Gamma, \mathsf{D} \cup \check{\mathsf{R}})$-inference, I', such that $\lambda(lv(I')) = \nabla(\Delta)$. Let us note that because all members of $\mathsf{D} \cup \check{\mathsf{R}}$ are conclusive rules, I' is a linear inference and has a single leaf labeled by $\nabla(\Delta)$.

Basis. If $k = 0$, then I is a linear inference. By the assumption, R is a set of proper m-rules and therefore, it does not contain any terminating rules. Hence, the leaf of I contains a formula D from Δ, and using the rules from D one can easily extend I to derive $\nabla(\Delta)$.

Assumption. Assume that for any (Γ, R)-inference I having less than k branching nodes, there is an $(\mathsf{D} \cup \check{\mathsf{R}}, \Gamma)$- inference I' the leaf of which contains $\nabla(\Delta)$.

Step. Let I be a (Γ, R)-inference having k branching nodes. Let $\mathfrak{n}$ be a branching node having no branching successors. Suppose that $B_0, \ldots, B_{m-1}$ is a list of formulas in the leaves below $\mathfrak{n}$, and $B_m, \ldots, B_{n-1}$ is a list of all formulas in the leaves that are not successors of $\mathfrak{n}$. Assume also that $\lambda(\mathfrak{n}) = B$ (see Figure 15) and alternatives $C_0, \ldots, C_{m-1}$ are obtained by application of inconclusive rule $\mathsf{r} := A_0, \ldots, A_{s-1}/C_0, \ldots, C_{m-1}$ from R.

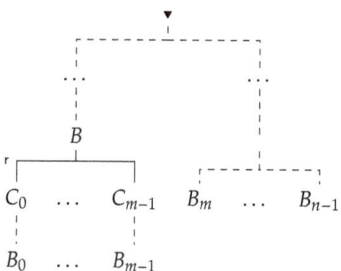

Figure 15. Prop. 5: Initial Inference.

Let us remove all successors of $\mathfrak{n}$ from I. By Proposition 1(a), the resulting tree is also a (Γ, R)-inference with $B, B_m, \ldots, B_{n-1}$ in its leaves (see the left-hand side of Figure 16).

Observe that $\check{r} = A_0 \nabla q, \ldots, A_{n-1} \nabla q / C_0 \nabla \ldots \nabla C_{m-1} \nabla q$, hence

$$A_0 \nabla \overline{C}, \ldots, A_{s-1} \nabla \overline{C} / C_0 \nabla \ldots \nabla C_{m-1} \nabla \overline{C}, \tag{10}$$

where $\overline{C} := C_0 \nabla \ldots \nabla C_{m-1}$, is an instance of $\check{r}$. As all formulas $A_0, \ldots, A_{s-1}$ are in the nodes preceding $\mathfrak{n}$, we can extend the reduced inference (see Figure 16).

By Corollary 5, from $C_0 \nabla C_1 \nabla \ldots \nabla C_{m-1}$ we can derive

$$B_0 \nabla C_1 \nabla \ldots \nabla C_{m-1}.$$

Then we can apply Corollary 5 again and get

$$B_0 \nabla B_1 \nabla \ldots \nabla C_{m-1}.$$

And so on, until we got

$$B_0 \nabla B_1 \nabla \ldots \nabla B_{m-1}$$

In such a way we obtain the inference depicted in Figure 17.

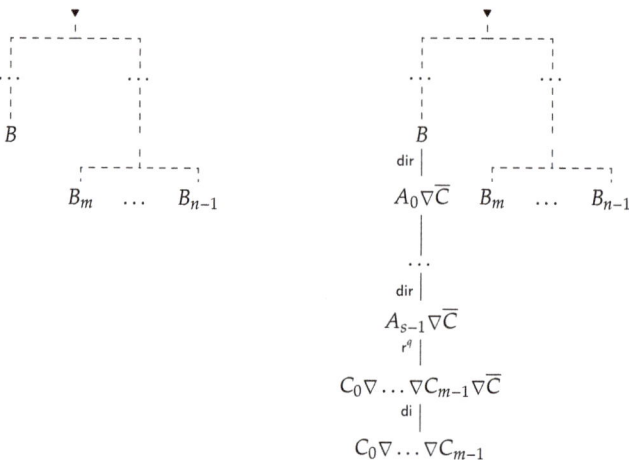

Figure 16. Theorem 5: Step.

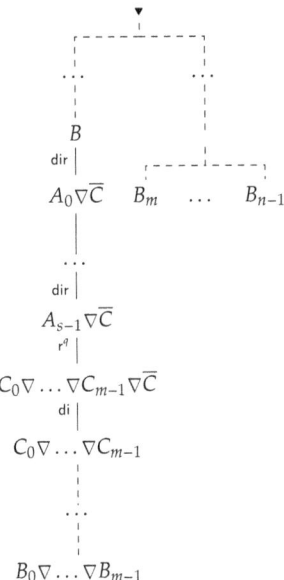

Figure 17. Theorem 5: Finishing Step.

Let us observe that the obtained inference contains $k-1$ branching nodes, therefore we can apply the induction assumption and convert the obtained inference into an inference of $B_0 \nabla \ldots \nabla B_{n-1}$, using, if necessary, rules from D. □

Corollary 7. *Suppose that* L *is a logic enjoying the* DP. *Then, if* L *has a finite admissible m-base (relative m-base), then,* L *has a finite admissible base (relative base).*

Example 6. *It was proven in [22] by Rybakov that logics* Int *and* S4 *have no finite relative admissible bases. Hence, these logics have no finite relative admissible m-bases either.*

6. From Admissible Base to Admissible m-Base

In the previous section we saw how to convert a given (relative) admissible m-base R of a logic with the DP into a (relative) admissible base Ř. In this section, we show how to convert a given (relative) admissible base into a (relative) admissible m-base.

Theorem 6. *Suppose that* L *is a logic and* R *is an admissible base (relative base). Then, if* L *is strongly consistent and there is a nonunifiable in* L *finite set of formulas* Ξ, *the set* R ∪ {DP, Ξ/▲} *is an admissible m-base (relative m-base); otherwise,* R ∪ {DP} *is an admissible m-base (relative m-base).*

Proof. First, let us prove that every proper admissible rule is derivable from R ∪ {DP} for L. Indeed, suppose that $r := \Gamma/\Delta$ is a proper m-rule admissible in L. If Δ consists of a single formula, m-rule r is a conclusive rule, and hence by the definition of the base, r is derivable from R. Suppose $|\Delta| > 1$. Let us consider rule $\Gamma/\nabla(\Delta)$. By Theorem 3(d), $\Gamma/\nabla(\Delta)$ is admissible for L, and hence, $\Gamma/\nabla(\Delta)$ is derivable from R. Now, we can $|\Delta| - 1$ times apply de and obtain Γ/Δ. Thus, Γ/Δ is derivable from R ∪ {DP}.

For relative bases one can take a set of conclusive rules L ∪ R and repeat the preceding argument.

Next, we prove that if Ξ is a finite nonunifiable in L set of formulas, then every admissible for L terminating m-rule can be derived from R ∪ {Ξ/▲}.

Indeed, suppose that m-rule $r = \Gamma/\blacktriangle$ is admissible for L. Then, Γ is a nonunifiable in L finite set of formulas and hence, conclusive rule Γ/A is admissible in L for every formula A. In particular, m-rules Γ/B are admissible for each $B \in \Xi$. Therefore, because R is a base, rule Γ/B is derivable from R for each $B \in \Xi$. Thus, each premise of rule $\Xi/\blacktriangle$ is derivable from Γ and by transitivity of consequence relation, $\Gamma/\blacktriangle$ is derivable from R, $\Xi/\blacktriangle$.

For relative bases one can take a set of conclusive rules $L \cup R$ and repeat the preceding argument. □

Corollary 8. *Suppose that L is a logic enjoying the DP. Then, if L has a finite admissible base (relative base), then L has a finite admissible m-base (relative m-base).*

Combining together Corollaries 7 and 8, we obtain the following.

Corollary 9. *A logic with the DP has a finite admissible base (relative base) if and only if it has a finite admissible m-base (relative m-base).*

7. Some Applications

(1) Logics K4, K4.1, S4, S4.1, Grz, Int, Dn, $n \geq 1$ are a-decidable [9]. Hence, by Corollary 6, all these logics are am-decidable.

(2) Logics Pos and Jhn are a-decidable [25]; hence, by Corollary 6, these logics are am-decidable.

(3) A relative admissible base for conclusive rules for Int was established in [16]. Adding m-rules $p \vee q/p, q$ and $p, \neg p/\blacktriangle$ to this relative base (or any relative base for this matter) gives us a relative m-base described in [26].

(4) It is known from [19] that Medvedev's logic ML (which enjoys the DP) is structurally complete. Hence, m-rules $p \vee q/p, q$ and $p, \neg p/\blacktriangle$ form a relative to ML admissible m-base.

(5) Gabbay-de Jongh logics $BB_n, n > 1$ enjoy the DP [27]. The relative m-bases for these logics have been constructed in [28] (Theorem 5.36):

$$r_n := (\vee_{i=1}^n (p_i \to p) \to \vee_{j=1}^n p_j)/\vee_{j=1}^n ((\vee_{i=1}^n p_i \to p) \to p_j), \text{de, t},$$

where $t := p, \neg p/\blacktriangle$. By Theorem 4, $\check{r}_n$ (or r_n^q, because rules r_n are conclusive) gives a relative admissible base for BB_n:

$$\check{r} := (\vee_{i=1}^n (p_i \to p) \to \vee_{j=1}^n p_j) \vee q/\vee_{j=1}^n ((\vee_{i=1}^n p_i \to p) \to p_j) \vee q.$$

Note that we did not include de and ť because these rules are trivial:

$$\check{de} = p_1 \vee p_2 \vee q/p_1 \vee p_2 \vee q \text{ and } \check{t} = p \vee q, \neg p \vee q/q.$$

Funding: This research received no external funding.

Conflicts of Interest: The author declares no conflict of interest.

References

1. Łukasiewicz, J. *Aristotle's Syllogistic from the Standpoint of Modern Formal Logic*; Clarendon Press: Oxford, UK, 1951; p. xi+141.
2. Scott, D.S. Rules and derived rules. In *Logical Theory and Semantic Analysis, Essays dedicated to Stig Kanger*; Stenlund, S., Ed.; D.Reidel Publishing Company: Dordrecht, The Netherlands, 1974; pp. 147–161.
3. Shoesmith, D.J.; Smiley, T.J. *Multiple-Conclusion Logic*; Reprint of the 1978 original [MR0500331]; Cambridge University Press: Cambridge, UK, 2008; p. xiv+396.
4. Carnap, R. *Formalization of Logic*; Harvard University Press: Cambridge, MA, USA, 1943; p. xviii+159.
5. Bezhanishvili, N.; Ghilardi, S. Multiple-conclusion rules, hypersequents syntax and step frames. In *Advances in Modal Logic*; College Publications: London, UK, 2014; Volume 10, pp. 54–73.

6. Carnap, R. *Introduction to Semantics*; Harvard University Press: Cambridge, MA, USA, 1942; p. xii+263.
7. Chris, B. A Note on Peirce and Multiple Conclusion Logic. *Trans. Charles S. Peirce Soc.* **1982**, *18*, 349–351.
8. Iemhoff, R. On rules. *J. Philos. Log.* **2015**, *44*, 697–711. [CrossRef]
9. Rybakov, V.V. *Admissibility of Logical Inference Rules*; Studies in Logic and the Foundations of Mathematics; North-Holland Publishing Co.: Amsterdam, The Netherlands, 1997; Volume 136, p. ii+617.
10. Rybakov, V.V.; Terziler, M.; Gencer, C. Unification and passive inference rules for modal logics. *J. Appl. Non-Class. Logics* **2000**, *10*, 369–377. [CrossRef]
11. Chagrov, A.; Zakharyaschev, M. *Modal Logic*; Oxford Logic Guides; Oxford Science Publications; The Clarendon Press: Oxford, UK; Oxford University Press: New York, NY, USA, 1997; Volume 35, p. xvi+605.
12. Wójcicki, R. *Theory of Logical Calculi*; Synthese Library; Kluwer Academic Publishers Group: Dordrecht, The Netherlands, 1988; Volume 199, p. xviii+473.
13. Cintula, P.; Noguera, C. The proof by cases property and its variants in structural consequence relations. *Stud. Log.* **2013**, *101*, 713–747. [CrossRef]
14. Iemhoff, R. Consequence relations and admissible rules. *J. Philos. Log.* **2015**, 1–22. [CrossRef]
15. Skura, T.; Wiśniewski, A. A system for proper multiple-conclusion entailment. *Log. Log. Philos.* **2015**, *24*, 241–253. [CrossRef]
16. Iemhoff, R. On the admissible rules of intuitionistic propositional logic. *J. Symb. Log.* **2001**, *66*, 281–294. [CrossRef]
17. Tommaso, M. A study of truth predicates in matrix semantics. *arXiv* **2019**, arXiv:1908.01661.
18. Kowalski, T. BCK is not structurally complete. *Notre Dame J. Form. Log.* **2014**, *55*, 197–204. [CrossRef]
19. Prucnal, T. On two problems of Harvey Friedman. *Stud. Log.* **1979**, *38*, 247–262. [CrossRef]
20. Iemhoff, R. Intermediate logics and Visser's rules. *Notre Dame J. Form. Log.* **2005**, *46*, 65–81. [CrossRef]
21. Citkin, A. A note on admissible rules and the disjunction property in intermediate logics. *Arch. Math. Log.* **2012**, *51*, 1–14. [CrossRef]
22. Rybakov, V.V. Bases of admissible rules of the logics S4 and Int. *Algebra Log.* **1985**, *24*, 87–107. [CrossRef]
23. Rybakov, V.V. Logical equations and admissible rules of inference with parameters in modal provability logics. *Stud. Log.* **1990**, *49*, 215–239. [CrossRef]
24. Rybakov, V.V. Decidability of logical equations in the modal system Grz and in intuitionistic logic. *Sib. Mat. Zh.* **1991**, *32*, 140–153, 213. [CrossRef]
25. Odintsov, S.; Rybakov, V. Unification and admissible rules for paraconsistent minimal Johanssons' logic **J** and positive intuitionistic logic **IPC**$^+$. *Ann. Pure Appl. Log.* **2013**, *164*, 771–784. [CrossRef]
26. Jeřábek, E. Admissible rules of modal logics. *J. Log. Comput.* **2005**, *15*, 411–431. [CrossRef]
27. Gabbay, D.M.; De Jongh, D.H.J. A sequence of decidable finitely axiomatizable intermediate logics with the disjunction property. *J. Symb. Log.* **1974**, *39*, 67–78. [CrossRef]
28. Goudsmit, J. Intuitionistic Rules Admissible Rules of Intermediate Logics. Ph.D. Thesis, Utrech University, Utrecht, The Netherlands, 2015.

© 2019 by the author. Licensee MDPI, Basel, Switzerland. This article is an open access article distributed under the terms and conditions of the Creative Commons Attribution (CC BY) license (http://creativecommons.org/licenses/by/4.0/).

Article

Minimal Systems of Temporal Logic

Dariusz Surowik

University of Bialystok, Swierkowa 20 B, 15-328 Białystok, Poland; surowik@uwb.edu.pl

Received: 6 April 2020; Accepted: 26 May 2020; Published: 16 June 2020

Abstract: The article discusses minimal temporal logic systems built on the basis of classical logic as well as intuitionistic logic. The constructions of these systems are discussed as well as their basic properties. The **K**$_t$ system was discussed as the minimal temporal logic system built based on classical logic, while the **IK**$_t$ system and its modification were discussed as the minimal temporal logic system built based on intuitionistic logic.

Keywords: temporal logic; intuitionistic logic; minimal system; knowledge

1. Temporal Logic

Temporal logic is the logic in which they appear, as logical constants, expressions whose meaning is determined by a reference to time. In its wide sense, temporal logic includes all logical problems of temporal representation of information. The task of temporal logic is to define and systematize inference rules for reasoning carried out in a language in which the same expression in terms of shape is used to pronounce sentences whose logical value may not be the same in different temporal contexts of their use.

The precursor of temporal logics was A. N. Prior. One of Prior's basic concepts was the temporal interpretation of modal operators. The enriched language of temporal logic was to enable formalization of reasoning regarding situations changing in time. Originally, temporal logic was to be a tool for formalizing philosophical, linguistic and semiotic considerations. Currently, apart from these applications, temporal logic is also widely used in computer science.

Among temporal logics, tense logic stands out, i.e., logic in a language whose only specific time operators are grammatical operators.

2. K$_t$—Minimal Tense Logic

The basic deductive system of logic of time is the **K**$_t$ system (**K**$_t$ is a temporal analogue of the **K** system (minimal deductive system for modal logic).). **K**$_t$ is a tense logic system built over classical propositional calculus by enriching this logic with specific axioms and rules. This is the minimal system. Therefore, the theses of this system are all and only those sentences that are true regardless of what properties time has (In fact, one assumption is made about the structure of time, namely it is assumed that a semantic time in the **K**$_t$ has a point structure.).

The **K**$_t$ system, as a minimal system, can be expanded by adding additional rules and specific axioms. In this sense, the minimality of **K**$_t$ means that any other temporal logic system built above classical propositional logic is richer than the **K**$_t$. In the tense logics we have the tense operators: F, G, P, H understood as follows:

$F\varphi$ - *it will be that φ,*

$P\varphi$ - *it was that φ,*

$G\varphi$ - *it always will be that φ,*

$H\varphi$ - *it always was that φ.*

However, usually the operators F and G and operators P and H are mutually definable (The mutual definability of operators F and G as well as P and H occur in temporal logic systems based on classical logic. In temporal logic systems based on intuitionistic or multi-valued logic, the mutual definability of these operators usually does not take a place.).

Definition 1 (The alphabet of the language $\mathfrak{L}_{K_t}$).

- *countable set of propositional letters $\mathcal{AP}$,*
- *connectives: $\neg, \rightarrow$,*
- *temporal operators: G, H (In some tense logic systems, as a primary operators are assumed F and P.),*
- *brackets: $), ($.*

A set of sentences is defined as follows:

Definition 2 (A set of sentences). *The set of sentences is the smallest set $FOR(\mathfrak{L}_{K_t})$ such that:*

- $\mathcal{AP} \subseteq FOR(\mathfrak{L}_{K_t})$,
- *if $\varphi, \psi \in FOR(\mathfrak{L}_{K_t})$, then $\neg\varphi, \varphi \rightarrow \psi, G\varphi, H\varphi \in FOR(\mathfrak{L}_{K_t})$.*

In the language $\mathfrak{L}_{K_t}$, all boolean symbols retain their meaning. However, there are additional specific operators in this language. Therefore, when we speak about the validity of propositions due to the meaning of classical propositional connectives, then we mean the sentences in which new operators occur.

We accept the following abbreviations:

(a) $\varphi \vee \psi \quad \equiv \quad \neg\varphi \rightarrow \psi$,

(b) $\varphi \wedge \psi \quad \equiv \quad \neg(\varphi \rightarrow \neg\psi)$,

(c) $\varphi \leftrightarrow \psi \quad \equiv \quad \neg[(\varphi \rightarrow \psi) \rightarrow \neg(\psi \rightarrow \varphi)]$,

(d) $F\varphi \quad \equiv \quad \neg G \neg \varphi$,

(e) $P\varphi \quad \equiv \quad \neg H \neg \varphi$.

Axioms

The **K$_t$** system is axiomatizable (The axiomatic system is one of many possible forms of a deductive system. This approach to construction of a deductive system has many advantages when it comes to methodological research. However, in case of axiomatic systems, we have some problems when it comes to practical command. This is due to the unstructured axiomatic systems. The structure of the sentence does not indicate the method of proving this sentence. In the case of other approaches to construction of a deductive system, e.g., sequent calculus, natural deduction or semantic tables, it is different.). Various sets of axioms and rules of this system were proposed. These differences are primarily due to the decision on a set of specific primitive symbols. Usually, the set of these symbols consists of the symbols G and H, while F and P are defined. When building a set of axioms for invariant systems, i.e., systems without the rule of substitution for sentence letters, apart from specific axiom schemes, either all tautologies of classical propositional logic or only selected tautology schemes are taken, but they are selected in such a way that all tautologies of classical propositional logic can be obtained. In this work, we used the second option and for the purposes of our considerations regarding $\mathbf{K_t}$ we will adopt the following set of axioms:

Axioms:

For any sentences $\varphi, \psi \in \mathfrak{L}_{K_t}$ ($\mathbf{K_t}$ can be axiomatizable in many ways. The completeness of the K_t with respect of these set of axioms was demonstrated by J. F. A. K. van Benthem [?].).

1. All tautologies of the classicall propositional calculus of the language $\mathfrak{L}_{K_t}$,
2. $G(\varphi \to \psi) \to (G\varphi \to G\psi)$,
3. $H(\varphi \to \psi) \to (H\varphi \to H\psi)$,
4. $\varphi \to GP\varphi$,
5. $\varphi \to HF\varphi$.

Rules

$$MP: \frac{\varphi \to \psi, \varphi}{\psi}. \qquad RG: \frac{\varphi}{G\varphi} \qquad RH: \frac{\varphi}{H\varphi}.$$

The specific $\mathbf{K_t}$ axioms are the 2–5 axioms. Axioms 2–3 are temporal equivalents of the $\mathbf{K}$ axiom for modal logics. These axioms apply only to the properties of G and H, respectively. Axioms 4–5 bind the operators G and P as well as H and F respectively.

The proof in $\mathbf{K_t}$ is understood in the usual way.

Definition 3 (Proof in $\mathbf{K_t}$). *Let Σ be any set of sentences of the language $\mathfrak{L}_{K_t}$. The sentence string $\varphi_0, \varphi_1, ..., \varphi_n$ is a proof of the sentence φ from the set Σ, (we write $\Sigma \vdash_{K_t} \varphi$) if and only if $\varphi = \varphi_n$ and for any i such that $0 \leq i \leq n$ at least one of the following conditions holds:*

1. *φ_i is an element of the set Σ,*
2. *φ_i is an axiom,*
3. *φ_i is obtained from their predecessors by MP, RG or RH, respectively.*

The sentence φ, which is derived from the empty set Σ, or $\emptyset \vdash_{K_t} \varphi$, is the thesis of the system $\mathbf{K_t}$. Instead of writing $\emptyset \vdash_{K_t} \varphi$, we will write $\vdash_{K_t} \varphi$.

In the $\mathbf{K_t}$ system, if a subsentences φ of the sentence ϕ is equivalent to the sentence ψ, entering ϕ in the place of the sentence φ as the inscription of the sentence ψ, $\phi(\psi/\varphi)$, gives the sentence equivalent to ϕ.

Theorem 1. *If $\Sigma \vdash_{K_t} \varphi \leftrightarrow \psi$, then $\Sigma \vdash_{K_t} \phi \leftrightarrow \phi(\psi/\varphi)$. (This theorem is not just the $\mathbf{K_t}$ theorem. It is the theorem of tense priorist logic.)*

Proof. We will prove by induction due to the length of the sentence ϕ. Let $\Sigma \vdash_{K_t} \varphi \leftrightarrow \psi$. Let ϕ be a propositional letter p. The only subsentence of a sentence ϕ is the propositipnal letter p. Then φ is equal p. Result of replacement φ in the ϕ by ψ will be the sentence ψ. Because by assumption we have $\Sigma \vdash_{K_t} \varphi \leftrightarrow \psi$, then:

$$\Sigma \vdash_{K_t} \phi \leftrightarrow \phi(\varphi/\psi).$$

As an induction assumption, we assume that for any sentence ϕ_i witch length is not greater than k the thesis is true, i.e.,

$$\Sigma \vdash_{K_t} \phi_i \leftrightarrow \phi_i(\varphi/\psi).$$

We will show that this thesis is also true for sentences of length $k+1$.

Let the string $\varphi_1, \varphi_2, ..., \varphi_n(= \phi_i \leftrightarrow \phi_i(\psi/\varphi))$ be a proof of the sentence: $\phi_i \leftrightarrow \phi_i(\psi/\varphi)$. We add the following sentences to this proof:

n+1. $\neg\phi_i(\varphi/\psi) \leftrightarrow \neg\phi_i$ TRANS, n
n+2. $(\neg\phi_i(\varphi/\psi) \leftrightarrow \neg\phi_i) \to (\phi_i \leftrightarrow \phi_i(\varphi/\psi))$ axiom 1
n+3. $\phi_i \leftrightarrow \phi_i(\varphi/\psi)$ MP,n+1,n+2

The sentence $\neg\phi_i(\varphi/\psi)$ is $(\neg\phi_i)(\varphi/\psi)$, then:

$$\Sigma \vdash_{K_t} \neg\phi_i \leftrightarrow (\neg\phi_i)(\varphi/\psi).$$

Let it now ϕ will be according to the character $\phi_i \to \phi_j$, with sentences ϕ_i and ϕ_j meet the induction assumption, i.e.,

$$\Sigma \vdash_{K_t} \phi_i \leftrightarrow \phi_i(\varphi/\psi)$$

and

$$\Sigma \vdash_{K_t} \phi_j \leftrightarrow \phi_j(\varphi/\psi).$$

Let the string $\varphi_1, \varphi_2, ..., \varphi_k(= \phi_i \leftrightarrow \phi_i(\psi/\varphi))$ be a proof of the sentence $\phi_i \leftrightarrow \phi_i(\psi/\varphi)$, while the string $\varphi_{k+1}, \varphi_{k+2}, ..., \varphi_n(= \phi_j \leftrightarrow \phi_j(\psi/\varphi))$ be a proof of the sentence: $\phi_j \leftrightarrow \phi_j(\psi/\varphi)$. To the sequence of the sentences $\varphi_1, \varphi_2, ..., \varphi_k, \varphi_{k+1}, \varphi_{k+2}, ..., \varphi_n$ we add sentences:

n+1. $(\phi_i \leftrightarrow \phi_i(\psi/\varphi)) \to \{(\phi_j \leftrightarrow \phi_j(\psi/\varphi)) \to [(\phi_i \to \phi_j) \leftrightarrow (\phi_i(\psi/\varphi) \to \phi_j(\psi/\varphi))]\}$ axiom 1
n+2. $(\phi_j \leftrightarrow \phi_j(\psi/\varphi)) \to [(\phi_i \to \phi_j) \leftrightarrow (\phi_i(\psi/\varphi) \to \phi_j(\psi/\varphi))]$ MP,k,n+1
n+3. $[(\phi_i \to \phi_j) \leftrightarrow (\phi_i(\psi/\varphi) \to \phi_j(\psi/\varphi))]$ MP,n,n+2

$(\phi_i(\psi/\varphi) \to \phi_j(\psi/\varphi))]$ is the sentence $(\phi_i(\psi \to \phi_j(\psi/\varphi))]$, so we received proof that

$$\Sigma \vdash_{K_t} (\phi_i \to \phi_j) \leftrightarrow (\phi_i \to \phi_j(\psi/\varphi)).$$

Now let us consider the case when the sentence ϕ is the sentence of the form $G\phi_i$, with the sentence ϕ_i is a sentence satisfying the induction assumption, i.e., $\Sigma \vdash_{K_t} \phi_i \leftrightarrow \phi_i(\varphi/\psi)$. Let the string $\varphi_1, \varphi_2, ..., \varphi_n$ be a proof of the sentence $\phi_i \leftrightarrow \phi_i(\varphi/\psi)$ from the sentence Σ. To the proof we add:

n+1. $G\phi_i \leftrightarrow G\phi_i(\varphi/\psi)$.

$G\phi_i(\varphi/\psi)$ is the sentence $(G\phi_i)(\varphi/\psi)$. So we received proof that

$$\Sigma \vdash_{K_t} G\phi_i \leftrightarrow G\phi_i(\varphi/\psi).$$

The case where the sentence ϕ is according to the form $H\phi_i$ is similar to the case when ϕ is the sentence $G\phi_i$. □

The Theorem **??** will be used in the proof of the next Theorem, which says that one of the K_t inference rules is the *REQ* replacement rule. This rule is a very useful rule in proving the theses of the K_t system.

Theorem 2 (Rule REQ). *If* $\Sigma \vdash_{K_t} \varphi \leftrightarrow \psi$, *then* $\dfrac{\phi}{\phi(\psi/\varphi)}$.

Proof. Let $\Sigma \vdash_{K_t} \varphi \leftrightarrow \psi$ and $\Sigma \vdash_{K_t} \phi$. According to the Theorem **??** there is a proof of the sentence $\phi \leftrightarrow \phi(\varphi/\psi)$ from the set Σ. To this proof we add the proof of the sentence ϕ. We add to the proof sequence the sentence $\phi(\varphi/\psi)$, which is a result from applying the Modus Ponens rule to sentences: ϕ and $\phi \leftrightarrow \phi(\varphi/\psi)$. □

In addition to the three inference rules proposed in this version of the axiomatics of the $\mathbf{K_t}$ system can be used to derive in this system the rules corresponding to the regularity rule for modal logics.

Theorem 3. *The RRG rule:* $\quad \dfrac{\varphi \to \psi}{G\varphi \to G\psi} \;$ *is a rule of* $\mathbf{K_t}$.

Proof. To demonstrate that RRG is a secondary rule $\mathbf{K_t}$, it must be demonstrated that

$$\text{if } \Sigma \vdash_{K_t} \varphi \to \psi, \text{ then } \Sigma \vdash_{K_t} G\varphi \to G\psi.$$

Let $\Sigma \vdash_{K_t} \varphi \to \psi$. Let the sequence $\varphi_1, ..., \varphi_n$ will prove the sentence $\varphi \to \psi$ from the set Σ. To this we add the following sentences:

n+1.	$G(\varphi \to \psi)$	RG,n
n+2.	$G(\varphi \to \psi) \to (G\varphi \to G\psi)$	axiom 2
n+3.	$G\varphi \to G\psi$	MP,n+1,n+2

The resulting sequence is a proof of the sentence $G\varphi \to G\psi$ from the set Σ. □

Theorem 4. *The RRH rule:* $\quad \dfrac{\varphi \to \psi}{H\varphi \to H\psi} \;$ *is a secondary rule of* $\mathbf{K_t}$.

Proof. Analogical to the proof of the previous theorem (using the axiom 3 and the rule RH). □

Based on Theorems **??** and **??** two further inference rules can be derived in $\mathbf{K_t}$.

Theorem 5. *The RF rule:* $\quad \dfrac{\varphi \to \psi}{F\varphi \to F\psi} \;$ *is a secondary rule of* $\mathbf{K_t}$.

Proof. Let $\Sigma \vdash_{K_t} \varphi \to \psi$. Let the sequence: $\varphi_1, ..., \varphi_n$ will prove the sentence $\varphi \to \psi$ from the set Σ. To this we add the following sentences:

n+1.	$\neg\psi \to \neg\varphi$	TRANS,n
n+2.	$G\neg\psi \to G\neg\varphi$	RRG,n+1
n+3.	$\neg G\neg\varphi \to \neg G\neg\psi$	TRANS,n+2
n+4.	$F\varphi \to F\psi$	REQ($\neg G\neg\varphi/F\varphi$), REQ($\neg G\neg\psi/F\psi$).

The resulting sequence is proof of the sentence $F\varphi \to F\psi$ from the set Σ. □

Theorem 6. *The RP rule:* $\quad \dfrac{\varphi \to \psi}{P\varphi \to P\psi} \;$ *is a secondary rule of* $\mathbf{K_t}$.

Proof. Analogical to the proof of the Theorem **??**. □

Operators H, P and G, F have the *Mirror Image Property*.

Definition 4 (Mirror Image Property). *The mirror image of the φ formula is created by simultaneously replacing each instance of the H operator with the G operator and the G operator with the H operator in the φ formula, and simultaneously replacing each instance of the P operator with the F operator and the F operator with the P operator.*

The Mirror Image of the φ we will mean by $MI(\varphi)$. E.g: $MI(\varphi \to GP\varphi) = \varphi \to HF\varphi$. The mirror image of the set of Σ is the mirror image set of the Σ elements. We mean the mirror image of Σ by $MI(\Sigma)$ and define as follows:

Definition 5 (A mirror image of a set of formulas). $M(\Sigma) = \{MI(\varphi) : \varphi \in \Sigma\}$.

If φ is derivable from Σ, then mirror image of φ is derivable from mirror image of the Σ.

Theorem 7. *For any* $\Sigma(\subset FOR(\mathfrak{L}_{K_t}))$: *if* $\Sigma \vdash_{K_t} \varphi$, *then* $MI(\Sigma) \vdash_{K_t} MI(\varphi)$.

Proof. Let $\Sigma \vdash_{K_t} \varphi$. Let the sequence $\varphi_1, \varphi_2, ..., \varphi_n$ will be a proof of φ from the Σ. We will show that the sequence $MI(\varphi_1), MI(\varphi_2), ..., MI(\varphi_n)$ is a prooof of the sentence $MI(\varphi)$ from the $MI(\Sigma)$, $MI(\Sigma) \vdash_{K_t} MI(\varphi)$. We will carry out the proof by induction due to the length of the proof of the sentence φ.

If φ_1 is an axiom, then $MI(\varphi_1)$ is also an axiom. If φ_1 is an element of Σ, then $MI(\varphi_1)$ is also an element of $MI(\Sigma)$. Then if $\Sigma \vdash_{K_t} \varphi_1$, then $MI(\Sigma) \vdash_{K_t} MI(\varphi_1)$.
Let us assume that for $i, i \leq k$:

$$\text{if } \Sigma \vdash_{K_t} \varphi_i, \text{ then } MI(\Sigma) \vdash_{K_t} MI(\varphi_i).$$

We will show that if $\Sigma \vdash_{K_t} \varphi_{k+1}$, then $MI(\Sigma) \vdash_{K_t} MI(\varphi_{k+1})$. Let $\Sigma \vdash_{K_t} \varphi_{k+1}$. The sentence φ_{k+1} can be an axiom or an element of a set Σ. There are cases discussed for the sentence φ_1. Now let us consider the cases where the sentence φ_{k+1} was obtained using one of the inference rules. Let them φ_{k+1} will be a sentence derived from sentences φ_m and $\varphi_m \to \varphi_{k+1}$ by applying the rule MP. By induction, we have that

$$MI(\Sigma) \vdash_{K_t} MI(\varphi_m)$$

and

$$MI(\Sigma) \vdash_{K_t} MI(\varphi_m \to \varphi_{k+1}).$$

Because $MI(\varphi_m \to \varphi_{k+1})$ has the form $MI(\varphi_m) \to MI(\varphi_{k+1})$, so applying the rule MP to the sentences $MI(\varphi_m) \to MI(\varphi_{k+1})$ and $MI(\varphi_m)$, we obtain $MI(\varphi_{k+1})$. Let it now φ_{k+1} will be the sentence derived from the sentence φ_m by applying the rule RG. By induction, we have that $MI(\Sigma) \vdash_{K_t} MI(\varphi_m)$. After applying the rule RH to the sentence $MI(\varphi_m)$ we obtain $HMI(\varphi_m)$. However, this sentence is equal to the sentence $MI(G\varphi_m)$. Then $MI(\Sigma) \vdash_{K_t} MI(G\varphi_m)$. The case when the sentence φ_{k+1} was obtained by applying the RH rule to the sentence φ_k is similar to the previous case. □

Corollary 1. *Let* $MI(\Sigma) \subseteq \Sigma$.

$$\text{If } \Sigma \vdash_{K_t} \varphi, \text{ then } \Sigma \vdash_{K_t} MI(\varphi)$$

or

$$\frac{\varphi}{MI(\varphi)}$$

is a secondary rule.

Corollary 2. *Let* $MI(\Sigma) \subseteq \{\varphi : \Sigma \vdash_{K_t} \varphi\}$.

$$\text{If } \Sigma \vdash_{K_t} \varphi, \text{ then } \Sigma \vdash_{K_t} MI(\varphi)$$

or

$$\frac{\varphi}{MI(\varphi)}$$

is a secondary rule.

3. IK$_t$—Minimal Intuitionistic Temporal Logic

Now we will discuss a system of temporal logic over intuitionistic propositional logic. It is a system of minimal intuitionistic temporal logic **IK$_t$** (**IK$_t$** is the intuitionistic analogue of the system **K$_t$** - minimal temporal logic built over classical propositional logic.).

This system can be used to formally describe knowledge that changes over time, although there are no explicit epistemic operators in the language of this system. Knowledge representation is not implemented at the syntactic level, but because of the properties of intuitionistic logic, knowledge is represented at the semantic level. This is the result of semantics proposed for intuitionistic logic, using terms such as *proof* (It was proposed by Kolmogorov.), *information*, or *knowledge* (Kripke-style semantics.).

Kripke-style semantics are proposed for intuitionistic temporal logic. Thus, in Kripke models we have a set of worlds W and the relationship R. In the case of intuitionistic logic, we do not speak about elements of the W set as possible worlds, but rather as information states, states of knowledge, etc. The reachability relationship between the elements w and v (i.e., wRv) is interpreted as *w has access to v*, which means that the v information state is available from the w information state. The key difference between Kripke models for intuitionistic logic and Kripke models for modal logic built over classical logic lies in the fact that in the case of modal logic built over classical logic, the R relation is only used to interpret modal operators, and in the case of intuitionistic logic, this relation is used to interpret the intuitionistic negation and implication.

The formula $\neg\varphi$ is true (In intuitionistic logic the term *forced* is also used.) in some information state w if and only if there is no information state available from w in which φ is true. In other words, the formula $\neg\varphi$ is true in the state w if there is no possibility that φ is true in any information state accessible from the state w.

The same is true with the intuitionistic implication. The formula $\varphi \to \psi$ is true in the information state w, if and only if, in any information state available from the state w, the truth of φ implies the truth of ψ. In addition, Kripke models assume monotonicity for intuitionistic logic. The formula fulfilled in a given information state remains fulfilled in any extension of this state.

Modality in intuitionistic logic can be seen on the example of the syntactic definition of intuitionistic negation. The $\neg\varphi$ formula is equivalent to the $\varphi \to \bot$ formula. Intuitionistic negation can therefore be seen as a kind of impossibility operator.

Kripke's intuitionistic model is a triangle $\mathfrak{M} = \langle W, R, V \rangle$, where $V : \mathcal{AP} \to 2^W$. The formula φ is satysfied in the model $\mathfrak{M}$, in the state w, when:

$\mathfrak{M}, w \models \varphi \quad \equiv \quad w \in V(\varphi)$, when $\varphi \in \mathcal{AP}$,

$\mathfrak{M}, w \models \neg\varphi \quad \equiv \quad$ for any $wRw' : \mathfrak{M}, w' \nvDash \varphi$,

$\mathfrak{M}, w \models \varphi \land \psi \quad \equiv \quad \mathfrak{M}, w \models \varphi$ and $\mathfrak{M}, w \models \psi$,

$\mathfrak{M}, w \models \varphi \lor \psi \quad \equiv \quad \mathfrak{M}, w \models \varphi$ or $\mathfrak{M}, w \models \psi$,

$\mathfrak{M}, w \models \varphi \to \psi \quad \equiv \quad$ for any w' such that wRw', if $\mathfrak{M}, w' \models \varphi$, then $\mathfrak{M}, w' \models \psi$.

In intuititionistic logic from the truth of the $\neg\varphi$ formula in the current information state, we do not only know that φ is not true in the current information state (such information is obtained in the case of classical logic), but we also know that the formula φ will never be true, and our *never* applies to all available extensions of the current information state. In addition to the information provided explicitly, we therefore have an additional *information* in intuitionistic logic. This feature of intuitionistic logic van Benthem calls *knowledge implicite* [?]. No additional specific operators are needed to express it in intuitionistic logic. Despite similar semantics, this feature definitely distinguishes intuitionistic logic from epistemic logic built on classical logic. The language of epistemic logic is used to represent *knowledge explicitly*, and to represent it, in addition to classical sentence connectives, the epistemic operator K is used. The language of intuitionistic logic allows expressing certain concepts without explicitly referring to epistemic operators. For example, based on the truth of the formula $\neg\neg\varphi$, we say that for each information state there is such an extension in which φ is true. Apart from details, it is very close to that *we know that φ must be true*.

In Kripke semantics for epistemic logic built over classical propositional calculus, the formula $K\varphi$ in the $\mathfrak{M}$ model, in the w information state, was defined as follows:

$$\mathfrak{M}, w \models K\varphi \quad \equiv \quad \text{for any } wRw' : \mathfrak{M}, w' \models \varphi.$$

Let us consider the truth of the formula $K\neg\varphi$ in the model $\mathfrak{M}$, in the state w. In accordance with the condition of satisfy with the operator K we have:

$$\mathfrak{M}, w \models K\neg\varphi \quad \equiv \quad \text{for any } wRw' : \mathfrak{M}, w' \models \neg\varphi.$$

Taking into account the condition of fulfilling of the negation in epistemic logic built over classical logic, we have:

$$\mathfrak{M}, w \models K\neg\varphi \quad \equiv \quad \text{for any } wRw' : \mathfrak{M}, w' \not\models \varphi.$$

The condition of fulfilling of the intuitionistic negation, i.e.,

$$\mathfrak{M}, w \models \neg\varphi \quad \equiv \quad \text{for any } wRw' : \mathfrak{M}, w' \not\models \varphi$$

Indicates that intuitionistic negation ($\neg$) can be seen as a combination of the K operator and classical negation ($K\neg$). Similarly, it can be shown that the intuitionistic formula $\varphi \Rightarrow \psi$ can be seen, aside from the details, as *modalized implication $K(\varphi \to \psi)$*, i.e., a combination of the K epistemic operator and the classic implication.

IK$_t$ (The construction of the **IK$_t$** system and proof of the system's completeness with respect to the proposed semantics was provided by W.B. Ewald [?].) is a system of temporal logic built over intuitionistic propositional calculus. The language $\mathfrak{L}_{\mathbf{IK}_t}$ is the language of intuitionistic propositional logic enriched with temporal operators: G, H, F, P.

Definition 6. *The set of sentences $FOR(\mathfrak{L}_{\mathbf{IK}_t})$ is the smallest set of finite sequences of elements of the language alphabet $\mathfrak{L}_{\mathbf{IK}_t}$ such that:*

1. *if $\varphi \in \mathcal{AP}$, then $\varphi \in FOR(\mathfrak{L}_{\mathbf{IK}_t})$,*
2. *if $\varphi, \psi \in FOR(\mathfrak{L}_{\mathbf{IK}_t})$, then $\neg\varphi, G\varphi, F\varphi, H\varphi, P\varphi, (\varphi \wedge \psi), (\varphi \vee \psi), (\varphi \to \psi), (\varphi \leftrightarrow \psi) \in FOR(\mathfrak{L}_{\mathbf{IK}_t})$.*

In the **IK$_t$** system, the operators G and F as well as H and P, unlike systems built over classical logic, are not mutually definable.

4. Semantics for IK$_t$ Proposed by Ewald

The construction of semantics for **IK$_t$** is based on a partially ordered set of states of knowledge, which is considered by the cognitive subject. Each state of knowledge is assigned a set of time moments and temporal order. When the cognitive subject reaches a greater state of knowledge (According to Ewald [?], the cognitive subject moves to a greater states of knowledge.), retains all the information that he had in lower states of knowledge. To define semantics for this system, Ewald constructs an intuitionistic temporal structure.

Definition 7 (intuitionistic temporal structure [?]). *An intuitionistic temporal structure $\mathfrak{M}$ is an ordered quintuple*

$$\langle S, \leq, \{T_s\}_{s \in S}, \{\mu_s\}_{s \in S}, \{R_t^s\}_{s \in S, t \in T_s} \rangle$$

where:

- $(S, \leq)$ *is a partially ordered set,*
- T_s *is a non-empty set,*
- μ_s *is a binary relation to T_s,*
- R_t^s *is a formula relation that satisfies the conditions:*

1. $R_t^s(\varphi) \equiv R_t^{s'}(\varphi)$, when $\varphi \in \mathcal{AP}$ and $s \leq s'$,
2. $R_t^s(\varphi \wedge \psi) \equiv R_t^s(\varphi)$ and $R_t^s(\psi)$,
3. $R_t^s(\varphi \vee \psi) \equiv R_t^s(\varphi)$ or $R_t^s(\psi)$,
4. $R_t^s(\neg \varphi) \equiv$ for any $s \leq s'$ it is not true that $R_t^{s'}(\varphi)$,
5. $R_t^s(\varphi \rightarrow \psi) \equiv$ for any $s \leq s'$ (if $R_t^{s'}(\varphi)$, then $R_t^{s'}(\psi)$),
6. $R_t^s(F\varphi) \equiv$ there is t', $t\mu_s t' : R_{t'}^s(\varphi)$,
7. $R_t^s(P\varphi) \equiv$ there is t', $t'\mu_s t : R_{t'}^s(\varphi)$,
8. $R_t^s(G\varphi) \equiv$ for any s', t' such that: $s \leq s'$, $t' \in T_{s'}$, $t\mu_{s'} t' : R_{t'}^{s'}(\varphi)$,
9. $R_t^s(H\varphi) \equiv$ for any s', t' such that: $s \leq s'$, $t' \in T_{s'}$, $t'\mu_{s'} t : R_{t'}^{s'}(\varphi)$,

We will now give intuitions related to individual elements of the above structure. The $(S, \leq)$ pair is a partially ordered set of states of knowledge. T_s is a set of time moments in the state s. μ_s is a binary relation on the set T_s. In addition, to fulfill the postulate that the cognitive entity, achieving a greater state of knowledge, retains all information from smaller states, it is required that for $s \leq s'$ the following conditions holds: $T_s \subseteq T_{s'}$ and $\mu_s \subseteq \mu_{s'}$. In other words, a cognitive subject achieving a higher state of knowledge maintains a set of time moments and temporal order from smaller states of knowledge.

The truth of a formula in an intuitionistic temporal structure and the truth of the formula are defined as follows:

Definition 8 (the truth in an intuitionistic temporal structure). $\mathfrak{M} \models \varphi$, *the formula φ is true in the intuitionistic temporal structure $\mathfrak{M}$, if and only if for any $s \in S$ and any $t \in T_s : R_t^s(\varphi)$.*

Definition 9 (the truth of the formula). $\models \varphi$, *formula φ is true if and only if, for any $\mathfrak{M} : \mathfrak{M} \models \varphi$.*

5. Axioms IK_t

(1) φ, if φ is a tautology of the intuitionistic logic of the language $\mathfrak{L}_{IK_t}$.

(2) $G(\varphi \to \psi) \to (G\varphi \to G\psi)$
(2') $H(\varphi \to \psi) \to (H\varphi \to H\psi)$

(3) $G(\varphi \wedge \psi) \leftrightarrow (G\varphi \wedge G\psi)$
(3') $H(\varphi \wedge \psi) \leftrightarrow (H\varphi \wedge H\psi)$

(4) $F(\varphi \vee \psi) \leftrightarrow (F\varphi \vee F\psi)$
(4') $P(\varphi \vee \psi) \leftrightarrow (P\varphi \vee P\psi)$

(5) $G(\varphi \to \psi) \to (F\varphi \to F\psi)$
(5') $H(\varphi \to \psi) \to (P\varphi \to P\psi)$

(6) $(G\varphi \wedge F\psi) \to F(\varphi \wedge \psi)$
(6') $(H\varphi \wedge P\psi) \to P(\varphi \wedge \psi)$

(7) $G\neg\varphi \to \neg F\varphi$
(7') $H\neg\varphi \to \neg P\varphi$

(8) $FH\varphi \to \varphi$
(8') $PG\varphi \to \varphi$

(9) $\varphi \to GP\varphi$
(9') $\varphi \to HF\varphi$

(10) $(F\varphi \to G\psi) \to G(\varphi \to \psi)$
(10') $(P\varphi \to H\psi) \to H(\varphi \to \psi)$

(11) $F(\varphi \to \psi) \to (G\varphi \to F\psi)$
(11') $P(\varphi \to \psi) \to (H\varphi \to P\psi)$

Rules: MP, RH, RG.

Ewald [?] proves the adequacy of the **IK$_t$** system with respect to the class of intuitionistic temporal structures. For the purposes of proof of adequacy, the concept of consistent pair of sets is introduced.

Definition 10 (consistent pair of sets). *The (X,Y) pair of set of sentences is consistent if and only if such finite subsets do not exist $X_0 (= \{\varphi_1, \varphi_2, ..., \varphi_m\}) \subseteq X$ and $Y_0 (= \{\psi_1, \psi_2, ..., \psi_n\}) \subseteq Y$ such that $\vdash (\varphi_1 \wedge \varphi_2 \wedge ... \wedge \varphi_m) \to (\psi_1 \vee \psi_2 \vee ... \vee \psi_n)$*

In the **IK$_t$** we can to prove the intuitionistic equivalent of the Lindenbaum lemma, namely:

Theorem 8. *If the pair (X,Y) is consistent, then there is the consistent pair of (X', Y') such that:*

1. $X \subseteq X'$ and $Y \subseteq Y'$,
2. $X' \cap Y' = \emptyset$,
3. for any formula $\varphi : \varphi \in X'$ or $\varphi \in Y'$.

The pair that fulfills these conditions is *maximum consistent pair*. Each (X,Y) maximum consistent pair can be represented by a valuation $v : v : FOR(IK_t) \to \{0,1\}$, such that $v(\varphi) = 1$ iff $\varphi \in X$. Ewald proves for the **IK$_t$** system the strong completeness Theorem in the following version:

Theorem 9 (Adequacy **IK$_t$** [?]). *For any **IK$_t$**− valuation v there is an intuititionistic structure $\mathfrak{M} = \langle S, \leq, \{T_s\}_{s \in S}, \{u_s\}_{s \in S}, \{R^s_t\}_{s \in S, t \in T_s}\rangle$, state on knowledge $s \in S$ and moment $t \in T_s$ such that for any formula $\varphi \in FOR(\mathfrak{L}_{IK_t})$ holds $R^s_t(\varphi)$ iff $v(\varphi) = 1$.*

In the semantic of the **IK$_t$** system, we did not impose any conditions on the temporal order in intuitionistic temporal structures. The **IK$_t$** system is therefore an analogue of the **K$_t$** system, i.e., it is a minimal system of intuitionistic temporal logic.

6. Modified Semantics for IK$_t$

We will consider the modified semantics for **IK$_t$** and examine its basic properties. **IK$_t$** is used to describe states of knowledge that change as knowledge gains. Acquiring knowledge in **IK$_t$** is understood as moving to states of knowledge; however, as in the **IK$_t$** system, it is assumed that all knowledge from a given state of knowledge is available in any state of knowledge not lesser than contemplated. Therefore, the monotonicity of the knowledge acquisition process is assumed. We achieve knowledge by enriching our knowledge with new facts. This can occur in several cases.

We can enrich our knowledge when by research we describe events from the past that took place at times that were not known in a given state of knowledge. We did not have any information about these events in this state of knowledge. In this case, the temporal structure in not lesser state of knowledge expands into the past and is a superset of the temporal structure of a given state of knowledge. For the same reasons, the time structure of the state of knowledge may expand into the future.

The expansion of the temporal structure (regardless of whether it takes place in the past or in the future) causes a change in the domain of the relationship. Therefore, in the new state of knowledge, the changed relation between moments of time should be considered.

Another possible option to achieve knowledge is the situation when the set of moments of time does not change, but the powers of sets of formulas increase, which we can determine if they are fulfilled in given time moments. Therefore, in this case there is no expansion of the time structure, neither into the past nor into the future, but by getting to know the present, past or future better within the known temporal structure, we attribute to moments more numerous sets of formulas fulfilled in these moments.

In the proposed semantics, the state of knowledge consists of a set of facts, which are semantic correlates of formulas, a set of moments of time, and the relationship at the set of moments of time. A subset of the set of facts assigned to a specific moment is understood as the set of facts known at that moment.

Achievable states of knowledge are different in their level of knowledge. The level of knowledge is determined by its constituent elements, namely: a set of moments of time , the temporal order relation and sets of formulas fulfilled at individual time moments. We will say that the state of knowledge of m'' has not lesser level of knowledge than the state of knowledge of m', if and only if the following conditions are satisfied:

1. The set of moments of time in the state m' is included in the set of moments of time in the state m''. (Changing the number of moments of time causes a change in the level of knowledge.)
2. In the m'' , there are - occurring between moments of time - *earlier-later* relationships that existed in the m' state of knowledge. Also, in the m'' , such relationships can occur that did not take place in the state m'.
3. All events that are known in the state of knowledge m' are also known in the state of knowledge m''. (What is known does not cease to be known also when new known events occur.) In addition at the moments of time of the state of knowledge m'', may be known some events that are not known in the equivalents of these moments in the state of knowledge m'.

There are specific relationships between conditions 1, 2 and 3. Fulfillment of condition 1 implies fulfillment of condition 2, because we skip situations in which *new* moments of time are not in any relationship *earlier-later* with other moments. A change in the set of moments of time therefore entails a change in the relationship between the moments of time. It is not the other way round. Changing the relationship between the moments of time does not have to involve changing the set of time moments. In the state of knowledge with no less level of knowledge, new relationships *earlier-later* can occur between time moments in the state of knowledge with a lower level of knowledge. Therefore, fulfillment of condition 2 does not entail fulfillment of condition 1. Similarly, fulfillment of condition 3 does not entail fulfillment of condition 1 or 2, because new facts may be known without new time moments or new relationships *earlier-later*.

Each moment is assigned a non-empty set of known events. If there are new moments, there are also new facts known. The fulfillment of condition 1 implies the fulfillment of condition 3.

The existence of new relationships *earlier-later*, on the other hand, entails the existence of new facts known at the times in which new relationships *earlier-later* take place. Thus, as in the case of condition 1, the fulfillment of condition 2 implies the fulfillment of condition 3.

We have two types of time. The first is the time that is assigned to the state of knowledge. It is a structure consisting of a set of moments of time and relationship *earlier-later* of a given state of knowledge. The other is time that is not relativized to any state of knowledge. This time is the sum of the times assigned to all possible states of knowledge.

We write theese intuitions in a formal way.

- I is a non-empty set (indexes of state of knowledge).
- T_i ($i \in I$) is a non-empty set (of moments in the state of knowledge indexed by i).
- $R_i (\subseteq T_i \times T_i)$ is a binary relation defined on a set of moments of time in the state of knowledge indexed by i. Relation R_i is understood as the relation *earlier-later* on the set of moments of time of state of knowledge indexed by i.
- $\mathfrak{T}_i = \langle T_i, R_i \rangle$. It is a time in the state of knowledge indexed by i.
- $T = \bigcup_{i \in I} T_i$ is a set of all time moments existing in any state of knowledge.
- $R = \bigcup_{i \in I} R_i$ is a binary relation on the set T. This relation is understood as the *earlier-later* relation for a time not relativized to any state of knowledge. We note that $R \subseteq T \times T$.
- $\mathfrak{T} = \langle T, R \rangle$ it is a time not relativized to any state of knowledge.
- $V_i \subseteq T_i \times 2^{AP}$, where $i \in I$. V_i is a function that assigns $t \in T_i$ subsets $V_i(t)$ to a set of sentence letters.
- $\mathcal{F} = \{V_i : i \in I\}$ is a set of valuations.
- $m_i = \langle T_i, R_i, V_i \rangle$ where $i \in I$. (m_i is the state of knowledge indexed by i.)

- $\mathfrak{M} = \{\langle T_i, R_i, V_i \rangle : V_i \in \mathcal{F}, i \in I\}$, or $\mathfrak{M} = \{m_i : i \in I\}$. $\mathfrak{M}$ is a model based on the $\mathfrak{T}$ and class $\mathcal{F}$ function.

We define the relationship $\leq (\subseteq \mathfrak{M} \times \mathfrak{M})$

Definition 11. *For any $i, j \in I$:*
$m_i \leq m_j$ *iff* $(T_i \subseteq T_j$ *and* $R_i \subseteq R_j$ *and for any* $t \in T_i : V_i(t) \subseteq V_j(t))$.

That for the states of knowledge m_i, m_j the relation $\leq$ ($m_i \leq m_j$) is understood as follows: state of knowledge m_j has no lower level of knowledge than the state of knowledge m_i.

The relationship $\leq$ is determined by the inclusions of a set of moments of time, the relationship between the moments of time and sets of events known at particular moments of time. The $\leq$ relation is therefore reflexive and transitive.

Theorem 10 ([?]). *For any $m_i (\in \mathfrak{M}) : m_i \leq m_i$.*

Theorem 11 ([?]). *For any $m_i, m_j, m_k (\in \mathfrak{M})$:*

$$if\ (m_i \leq m_j\ and\ m_j \leq m_k),\ then\ m_i \leq m_k.$$

The relationship $\leq$ partially organizes the set of states of knowledge. In the states of knowledge, various relationships may occur between sets of time moments, earlier-later relations and valuations. Let us consider some of them.

The first possible situation is:

$$T_i = T_j \text{ and } R_i = R_j \text{ and } \underset{t \in T_i}{\forall} \left(V_i(t) \subseteq V_j(t) \right).$$

This situation occurs when sets of time moments of states of knowledge m_i and m_j are the same ($T_i = T_j$). The relations ($R_i = R_j$) are the same in both states of knowledge. The state of knowledge m_j, as a state of knowledge with no lower level of knowledge than the state of knowledge m_i, is created by changing the value of the function V_i that assigns moments to subsets of the set $\mathcal{AP}$. In other words, in this case, the state of knowledge about a not lower level of knowledge is created by increasing the amount of facts known at particular times.

The second possible situation may be as follows:

$$T_i \subseteq T_j, R_i \subseteq R_j \text{ and } \underset{t \in T_i}{\forall} \left(V_i(t) = V_j(t) \right).$$

In this case, the m_j, as a state of knowledge with not lesser level of knowledge than the m_i, is created by adding to the structure of the state of knowledge m_i new moments of time. For any time $t (\in T_i)$ does not change the set $V_i(t)$. The change in the level of knowledge is that in the state of knowledge m_j new time moments appear (in the future or in the past). Due to the new time moments, in the state of knowledge m_j all the components change. The set of time moments changes. The relation *earlier-later* is changing, because certain time moments of the state of knowledge m_i will be in relation *earlier-later* with new time moments. The evaluating function is also changing, assigning subsets of the sentence letter set to moments of time because its domain is changing (subsets of the set sentence letters will be assigned new time moments).

Yet another option is:

$$T_i = T_j, R_i \subseteq R_j \text{ and } \underset{t \in T_i}{\forall} \left(V_i(t) \subseteq V_j(t) \right).$$

It may also be that the change in the level of knowledge of the state of knowledge does not consist of changing the set of time moments known in the state of knowledge m_i but on the change of the property of time in the state of knowledge m_i. In other words, the change of ownership of the relationship in this state of knowledge. Such a change, however, entails a change in the number of facts known at these times.

Further states of knowledge - with an increasingly higher level of knowledge—can arise by increasing the level of knowledge regarding the various components of the state of knowledge.

To shorten the entries we will introduce the designation:
Mark
$m_i^* (= \langle T_i^*, R_i^*, V_i^* \rangle)$ (where $i \in I$) is any $m_j (\in \mathfrak{M})$ such that $m_i \leq m_j$.

Definition 12 (the truth of a formula in the state of knowledge at some moment of time). *The truth of the formula $\varphi (\in FOR(\mathfrak{L}_{IK_t}))$ in the model $\mathfrak{M}$, state of knowledge $m_i (= \langle T_i, R_i, V_i \rangle)$, at the moment $t (\in T_i)$ we define as follows:*

1. $\mathfrak{M}, m_i, t \models \varphi$ $\equiv$ $\varphi \in V_i(t), \text{if } \varphi \in \mathcal{AP}$,
2. $\mathfrak{M}, m_i, t \models \neg\varphi$ $\equiv$ for any $m_i^* \in \mathfrak{M} : \mathfrak{M}, m_i^*, t \not\models \varphi$
3. $\mathfrak{M}, m_i, t \models \varphi \vee \psi$ $\equiv$ $\mathfrak{M}, m_i, t \models \varphi$ or $\mathfrak{M}, m_i, t \models \psi$,
4. $\mathfrak{M}, m_i, t \models \varphi \wedge \psi$ $\equiv$ $\mathfrak{M}, m_i, t \models \varphi$ and $\mathfrak{M}, m_i, t \models \psi$,
5. $\mathfrak{M}, m_i, t \models \varphi \rightarrow \psi$ $\equiv$ for any $m_i^* \in \mathfrak{M} : (\mathfrak{M}, m_i^*, t \not\models \varphi$ or $\mathfrak{M}, m_i^*, t \models \psi)$,
6. $\mathfrak{M}, m_i, t \models F\varphi$ $\equiv$ there exists $t' \in T_i, tR_i t' : \mathfrak{M}, m_i, t' \models \varphi$,
7. $\mathfrak{M}, m_i, t \models G\varphi$ $\equiv$ for any $m_i^*(\in \mathfrak{M})$, for any $t'(\in T_i^*)$ such that $tR_i^* t': \mathfrak{M} \models m_i^*, t'\varphi$,
8. $\mathfrak{M}, m_i, t \models P\varphi$ $\equiv$ there exists $t' \in T_i, t'R_i t : \mathfrak{M}, m_i, t' \models \varphi$,
9. $\mathfrak{M}, m_i, t \models H\varphi$ $\equiv$ for any $m_i^*(\in \mathfrak{M})$, for any $t'(\in T_i^*)$ such that $t'R_i^* t: \mathfrak{M} \models m_i^*, t'\varphi$,

The necessary condition for the sentence $F\varphi$ to be true in the state of knowledge m_i, at the time of t ($\in T_i$) is the existence in the time structute of the state of knowledge m_i the moment t' ($\in T_i$), later than t ($tR_i t'$), in which the sentence φ is true. If such a moment exists in the structure of time of m_i, then from the definition of the relationship $\leq$ and the theory of multiplicative properties of inclusions it follows that such a moment also exists in the structure of time of each state of knowledge with a level of knowledge not less than the level of state of knowledge m_i. Hence verification of the truth of the sentence $F\varphi$ in the state of knowledge m_i can be limited to the state of knowledge m_i. Please note that if the sentence $F\varphi$ is not true at the time t it does not mean that in t the sentence $F\neg\varphi$ is true.

For the G operator the situation is different. According to understanding the G operator, the sentence $G\varphi$ reads: *it will always be in the future that φ*. For the sentence $G\varphi$ to be true in the state of knowledge m_i at $t (\in T_i)$, it is necessary that the sentence φ is true in any state of knowledge m_i^* at any time t' ($\in T_i^*$) later than t ($tR_i^* t'$). The truth of the sentence $G\varphi$ cannot be considered only within the temporal limits of a given state of knowledge. Just because the sentence φ is always true in the future means that φ is true at any point in the future. Since the state of knowledge m_i is assigned only a certain fragment of the time structure, when defining the concept of the truth for a sentence built using the operator G, all states of knowledge with a level of knowledge not lower than the level of knowledge of state m_i.

If the definition of the truth of the sentence $G\varphi$ were in the form that was adopted in the system, e.g., in the system T_m [?] (intuitionistic temporal logic of unchanging time (By *unchanging time* (in accepted terminology) is understood a time such that for any $i, j \in I: (T_i = T_j$ and $R_i = R_j)$.)), i.e.,

$$\mathfrak{M}, m_i, t \models G\varphi \text{ iff for any } t' \in T_i, \text{ such that } tR_i t' : \mathfrak{M}, m_i, t' \models \varphi$$

this would lead to contradictions. It would be possible that in some state of knowledge m_i would occur at the moment t

$$\mathfrak{M}, m_i, t \models G\varphi. \tag{1}$$

and at some level of knowledge m_j, with a level of knowledge not lesser than the level of knowledge of the state of knowledge m_i, i.e., $m_i \leq m_j$, there would be a moment $t_1 (\in T_j)$ such that: $t_1 \notin T_i$, $tR_j t_1$ and $\mathfrak{M}, m_j, t_1 \not\models \varphi$. Therefore, we have:

$$\mathfrak{M}, m_j, t \not\models G\varphi. \tag{2}$$

What is known does not cease to be known when the level of knowledge increases. Since the state of knowledge of m_j is a state of knowledge with a level of knowledge of not less than the level of knowledge of the state of m_i, so that $\mathfrak{M}, m_i, t \models G\varphi$ we conclude that $\mathfrak{M}, m_j, t \models G\varphi$. This is contrary to (2).

The understanding of the truth of the formula $G\varphi$, in the state of knowledge m_i, at the moment t excludes the situation described above.

We will now give some basic definitions.

Definition 13. $\mathfrak{M} \models \varphi$, φ is true in the model $\mathfrak{M}$, iff for any state of knowledge $m_i(\in \mathfrak{M})$ and for any $t(\in T_i) : \mathfrak{M}, m_i, t \models \varphi$.

Definition 14. $\mathfrak{T} \models \varphi$, φ is true in time $\mathfrak{T}$, iff φ is true in the model $\mathfrak{M}$ for any non-empty class $\mathcal{F}(= \{V_i : i \in I\})$ of function.

Definition 15. $\models \varphi$, φ is true iff for any $\mathfrak{T} : \mathfrak{T} \models \varphi$.

In some sciences (e.g., empirical sciences) it happens that sentences considered to be true at some time, with the development of scientific theories, turn out to be false. It happens that certain laws of empirical sciences in force in a given period are subject to verification and are changed, and sometimes even rejected, as laws that inaccurately or even misrepresent the state of the world. Such verification is possible due to the increase in the level of knowledge. In our terminology, we would write this fact as follows: the sentence true in some state of knowledge m_i, in some state of knowledge which level of knowledge is not lesser than the level of knowledge of m_i may not be true. In the $\mathbf{IK_t}$ system, this is not possible. What is true in the state of knowledge m_i is also true in any state of knowledge, with a level of knowledge not lesser than the level of knowledge of m_i.

There are many differences between temporal logic systems based on classical logic and temporal logic systems based on intuitionistic logic. One of them is that failing to the truth of φ does not entail the truth of $\neg\varphi$.

Let us consider the following situation. The sentence φ is not known in the state of knowledge m_i at the moment $t(\in T_i)$, while is known at this moment in a state of knowledge m_j, whose level knowledge is not lesser than the level of knowledge in the state m_i. If the sentence φ is not known at the time t in the state m_i, it would be considered that at the time t the sentence $\neg\varphi$ is known, then—according to the accepted condition of fulfilling $\neg\varphi$ - the sentence φ could not be known at the time of t in any state of knowledge with a level of knowledge not lesser than the level of knowledge of m_i. In particular, the sentence φ could not be known at the time t, in the state of knowledge m_j. This leads to a contradiction, since we get that φ is known at the time of t, in the state m_j, and we conclude that it is known and unknown at the same time. When the sentence φ is known at some moment of time, in some state of knowledge m_i, then in any state of knowledge with the level of knowledge not lesser than the level of knowledge of state m_i at this moment the sentence φ is known. However, when $\neg\varphi$ is not known at some moment of time, it does not mean that at this moment, in any state of knowledge with a level of knowledge no lesser than the level of knowledge of m_i, is known φ. It only means that it is not true that in every state of knowledge in which the level of knowledge is not lesser than the level of knowledge of m_i, φ is currently unknown.

We will prove a lemma that expresses the monotonicity of knowledge in the IK_t system. What is known in the state of knowledge m_i is also known in every state of knowledge whose level of knowledge is not lesser than the level of knowledge of the state m_i.

Lemma 1. *For any formula* $\varphi(\in FOR(\mathfrak{L}_{\mathbf{IK_t}}))$, *for any* $m_i, m_j(\in \mathfrak{M}) :$

$$\text{if } (m_i \leq m_j \text{ and } \mathfrak{M}, m_i, t \models \varphi), \text{ then } \mathfrak{M}, m_j, t \models \varphi.$$

Proof. We will prove by induction, due to the length of the formula φ. Suppose that $m_i \leq m_j$.

($\varphi \in \mathcal{AP}$) Let us first consider the case when φ is a sentence letter.
By Definition ?? if $m_i \leq m_j$, then for any $t \in T_i$ holds

$$V_i(t) \subseteq V_j(t). \tag{3}$$

If $\mathfrak{M}, m_i, t \models \varphi$, then from the Definition ??

$$\varphi \in V_i(t). \tag{4}$$

From (3) and (4) we receive

$$\varphi \in V_j(t). \tag{5}$$

Because φ is a sentence letter, so from (5) and the definition of ?? we have $\mathfrak{M}, m_j, t \models \varphi$.

Induction assumption: Let φ, ψ be such that :

(a) if $\mathfrak{M}, m_i, t \models \varphi$, then $\mathfrak{M}, m_j, t \models \varphi$,
and
(b) if $\mathfrak{M}, m_i, t \models \psi$, then $\mathfrak{M}, m_j, t \models \psi$.

We will consider complex formulas built from the formulas φ, ψ using sentence connectives and temporal operators.

($\neg \varphi$) Let us assume that $\mathfrak{M}, m_i, t \models \neg \varphi$.
From the definition of the condition for negation (Definition ??) we have:

$$\text{for any } m_k, \text{ such that } m_i \leq m_k : \mathfrak{M}, m_k, t \not\models \varphi. \tag{6}$$

Let us consider any state of knowledge m_l with a level of knowledge not lesser than the level of m_j, i.e.,

$$m_j \leq m_l. \tag{7}$$

From (7), the assumption that $m_i \leq m_j$ and the transitivity of the $\leq$, we have that $m_i \leq m_l$. Therefore, from (6) we have: $\mathfrak{M}, m_l, t \not\models \varphi$. Because m_l is any state of knowledge whose level of knowledge is not lesser than the level of knowledge of m_j, we get:

$$\text{for any } m_l \text{ such that } m_j \leq m_l \text{ we have: } \mathfrak{M}, m_l, t \not\models \varphi. \tag{8}$$

From (8) and the condition for negation (Definition ??) we have: $\mathfrak{M}, m_j, t \models \neg \varphi$.
($\varphi \wedge \psi$) Let us assume that $\mathfrak{M}, m_i, t \models \varphi \wedge \psi$.
So from the condition for the conjunction (Definition ??) we have:

$$\mathfrak{M}, m_i, t \models \varphi, \tag{9}$$

and

$$\mathfrak{M}, m_i, t \models \psi. \tag{10}$$

From (9) and point a) of the induction assumption we get:

$$\mathfrak{M}, m_j, t \models \varphi. \tag{11}$$

Similarly, from (10) and point b) of the induction assumption we get:

$$\mathfrak{M}, m_j, t \models \psi. \tag{12}$$

From (11), (12) and the condition for the conjunction (Definition ??) we get $\mathfrak{M}, m_j, t \models \varphi \wedge \psi$.
($\varphi \vee \psi$) Reasoning analogous to conjunction.
($\varphi \to \psi$) Let us assume that $\mathfrak{M}, m_i, t \models \varphi \to \psi$.
From the condition for the implication (Definition ??) we have:

$$\text{for any } m_i^* (\in \mathfrak{M}) : (\mathfrak{M}, m_i^*, t \not\models \varphi \text{ or } \mathfrak{M}, m_i^*, t \models \psi), \tag{13}$$

Let us consider the state of knowledge m_l with a level of knowledge not lesser than the level of knowledge of m_j, i.e.,

$$m_j \leq m_l. \tag{14}$$

From (14), the assumption that $m_i \leq m_j$ and the transitivity of the relationship $\leq$ we get that $m_i \leq m_l$. From (13) we have: $\mathfrak{M}, m_l, t \not\models \varphi$ or $\mathfrak{M}, m_l, t \models \psi$. Because m_l is any state of knowledge in which the level of knowledge is not lesser than the level of knowledge in the state m_j, we get:

$$\text{for any } m_l \text{ such that } m_j \leq m_l : \mathfrak{M}, m_l, t \not\models \varphi \text{ or } \mathfrak{M}, m_l, t \models \psi. \tag{15}$$

From (15) and the condition for the implications (Definition ??) we get $\mathfrak{M}, m_j, t \models \varphi \to \psi$.

($G\varphi$) Suppose $\mathfrak{M}, m_i, t \models G\varphi$. From the condition for the G operator (Definition ??) we have:

$$\text{for any } m_i^* (\in \mathfrak{M}), \text{ for any } t_1 (\in T_i^*) \text{ such that } tR_i^* t_1 : \mathfrak{M}, m_i^*, t_1 \models \varphi, \tag{16}$$

Let us consider any state of knowledge m_l witch a level of knowledge is not lesser than the level of knowledge of the state m_j, i.e.,

$$m_j \leq m_l. \tag{17}$$

From (17), the assumption that $m_i \leq m_j$ and the transitivity of the relationship $\leq$, we get that $m_i \leq m_l$. Som from (16) we get :

$$\text{for any } t_1 (\in T_l) \text{ such that } tR_l t_1 \text{ holds: } \mathfrak{M}, m_l, t \models \varphi. \tag{18}$$

Because the state of knowledge m_l is a state of knowledge with a level of knowledge not lower than the level of knowledge in the state m_j we have:

$$\text{for any } m_l, \text{ for any } t_1 (\in T_l) \text{ if } (m_j \leq m_l \text{ and } tR_l t_1), \text{ then } \mathfrak{M}, m_l, t_1 \models \varphi. \tag{19}$$

From (19) and the condition for the G operator (Definition ??) we obtain: $\mathfrak{M}, m_j, t \models G\varphi$
($H\varphi$) Reasoning similar to the G operator.
($F\varphi$) Let us assume that $\mathfrak{M}, m_i, t \models F\varphi$. From the condition for the operator F (Definition ??) there is the moment $t_1 (\in T_i), tR_i t_1$, such that:

$$\mathfrak{M}, m_i, t_1 \models \varphi. \tag{20}$$

From (2) and point a) of the induction assumption we have:

$$\mathfrak{M}, m_j, t_1 \models \varphi. \tag{21}$$

Assuming that $m_i \leq m_j$ and the definition of ?? we get that:

$$t \in T_j, t_1 \in T_j, tR_j t_1. \tag{22}$$

From (21), (22) and the condition for the Γ operator (Definition ??) we obtain $\mathfrak{M}, m_j, t \models F\varphi$.
($P\varphi$) Reasoning similar to the F operator.
□

We have therefore shown that what is true in a given state of knowledge m_i it is also true in any state of knowledge in which the level of knowledge is not lesser than the level of knowledge in the state m_i.

7. Simplified Axiomatics IK_t

The axioms proposed by Ewald $\mathbf{IK_t}$ are dependent axioms. Some axioms can be derived from other axioms. Proofs of dependencies of selected axioms were provided by Surowik [?]. We offer a simplified set of axioms for $\mathbf{IK_t}$:

- A1) φ, if φ is a tautology of the intuitionistic logic of the language $\mathcal{L}_{IK_t}$.
- (A2) $G(\varphi \to \psi) \to (G\varphi \to G\psi)$ (A2') $H(\varphi \to \psi) \to (H\varphi \to H\psi)$
- (A3) $F(\varphi \vee \psi) \to (F\varphi \vee F\psi)$ (A3') $P(\varphi \vee \psi) \to (P\varphi \vee P\psi)$
- (A4) $G(\varphi \to \psi) \to (F\varphi \to F\psi)$ (A4') $H(\varphi \to \psi) \to (P\varphi \to P\psi)$
- (A5) $F\varphi \to \neg G \neg \varphi$ (A5') $P\varphi \to \neg H \neg \varphi$
- (A6) $FH\varphi \to \varphi$ (A6') $PG\varphi \to \varphi$
- 9A7) $\varphi \to GP\varphi$ (A7') $\varphi \to HF\varphi$
- (A8) $(F\varphi \to G\psi) \to G(\varphi \to \psi)$ (A8') $(P\varphi \to H\psi) \to H(\varphi \to \psi)$

Rules: MP, RH, RG.

We will prove that this axiomatics is equivalent to the axiomatics proposed by Ewald. To demonstrate the derivability of some $\mathbf{IK_t}$ axioms with the other axioms of this system, the following Theorems will be useful.

Theorem 12.

(a)   The RRG rule: $\dfrac{\varphi \to \psi}{G\varphi \to G\psi}$ is a rule of $\mathbf{IK_t}$.

(b)   The RRH rule: $\dfrac{\varphi \to \psi}{H\varphi \to H\psi}$ is a rule of $\mathbf{IK_t}$.

Proof. We will prove only (a). Proof (b) is analogous.

(a)

1. $\vdash_{IK_t} \varphi \to \psi$ assumption
2. $\vdash_{IK_t} G(\varphi \to \psi)$ 1,RG
3. $\vdash_{IK_t} G(\varphi \to \psi) \to (G\varphi \to G\psi)$ A2
4. $\vdash_{IK_t} G\varphi \to G\psi$ 2,3,MP

□

Theorem 13.

(a) The RF rule: $\dfrac{\varphi \to \psi}{F\varphi \to F\psi}$ is a rule of $\mathbf{IK_t}$.

(b) The RP rule: $\dfrac{\varphi \to \psi}{P\varphi \to P\psi}$ is a rule of $\mathbf{IK_t}$.

The proof of this theorem is obtained in a manner analogous to the proof of the theorem of the previous one, with the difference that instead of the axiom $A2$ ($A2\,'$) we use the $A4$ ($A4'$) axiom.

We will show that in $\mathbf{IK_t}'$ "old" axioms $3, 3', 6, 6', 7, 7', 11, 11'$ are inferable. The implications of the "old" 4 and $4'$ axioms are also inferable.

Lemma 2. $\vdash_{\mathbf{IK_t}} G(\varphi \wedge \psi) \leftrightarrow (G\varphi \wedge G\psi)$

Proof.

(A) $\vdash_{\mathbf{IK_t}} G(\varphi \wedge \psi) \to (G\varphi \wedge G\psi)$

1.	$\vdash_{\mathbf{IK_t}} (\varphi \wedge \psi) \to \varphi$	A1
2.	$\vdash_{\mathbf{IK_t}} (\varphi \wedge \psi) \to \psi$	A1
3.	$\vdash_{\mathbf{IK_t}} G(\varphi \wedge \psi) \to G\varphi$	1, RRG
4.	$\vdash_{\mathbf{IK_t}} G(\varphi \wedge \psi) \to G\psi$	2, RRG
5.	$\vdash_{\mathbf{IK_t}} (G(\varphi \wedge \psi) \to G\varphi) \to \left((G(\varphi \wedge \psi) \to G\psi) \to (G(\varphi \wedge \psi) \to (G\varphi \wedge G\psi)) \right)$	A1
6.	$\vdash_{\mathbf{IK_t}} (G(\varphi \wedge \psi) \to G\psi) \to (G(\varphi \wedge \psi) \to (G\varphi \wedge G\psi))$	3,5,MP
7.	$\vdash_{\mathbf{IK_t}} G(\varphi \wedge \psi) \to (G\varphi \wedge G\psi)$	4,6,MP

(B) $\vdash_{\mathbf{IK_t}} (G\varphi \wedge G\psi) \to G(\varphi \wedge \psi)$

1.	$\vdash_{\mathbf{IK_t}} \varphi \to (\psi \to (\varphi \wedge \psi))$	A1
2.	$\vdash_{\mathbf{IK_t}} G\varphi \to G(\psi \to (\varphi \wedge \psi))$	1,RRG
3.	$\vdash_{\mathbf{IK_t}} G(\psi \to (\varphi \wedge \psi)) \to (G\psi \to G(\varphi \wedge \psi))$	A2
4.	$\vdash_{\mathbf{IK_t}} G\varphi \to (G\psi \to G(\varphi \wedge \psi))$	2,3,SYLL
5.	$\vdash_{\mathbf{IK_t}} \left(G\varphi \to (G\psi \to G(\varphi \wedge \psi))\right) \to ((G\varphi \wedge G\psi) \to G(\varphi \wedge \psi))$	A1
6.	$\vdash_{\mathbf{IK_t}} (G\varphi \wedge G\psi) \to G(\varphi \wedge \psi)$	4,5,MP

With (**A**) and (**B**) we get a thesis. □

The next lemma is proved similarly.

Lemma 3. $\vdash_{\mathbf{IK_t}} H(\varphi \wedge \psi) \leftrightarrow (H\varphi \wedge H\psi)$

Lemma 4. $\vdash_{\mathbf{IK_t}} (F\varphi \vee F\psi) \to F(\varphi \vee \psi)$

Proof.

1.	$\vdash_{\mathbf{IK_t}} \varphi \to (\varphi \vee \psi)$	A1
2.	$\vdash_{\mathbf{IK_t}} \psi \to (\varphi \vee \psi)$	A1
3.	$\vdash_{\mathbf{IK_t}} F\varphi \to F(\varphi \vee \psi)$	1,RF

4. $\vdash_{\mathbf{IK_t}} F\psi \to F(\varphi \vee \psi)$ 2,RF
5. $\vdash_{\mathbf{IK_t}} (F\varphi \to F(\varphi \vee \psi)) \to ((F\psi \to F(\varphi \vee \psi)) \to ((F\varphi \vee F\psi) \to F(\varphi \vee \psi)))$ A1
6. $\vdash_{\mathbf{IK_t}} (F\psi \to F(\varphi \vee \psi)) \to ((F\varphi \vee F\psi) \to F(\varphi \vee \psi))$ 3,5,MP
7. $\vdash_{\mathbf{IK_t}} (F\varphi \vee F\psi) \to F(\varphi \vee \psi)$ 4,6,MP

□

Lemma 5. $\vdash_{\mathbf{IK_t}} (P\varphi \vee P\psi) \to P(\varphi \vee \psi)$

Proof analogous to the proof of the previous lemma.

Lemma 6. $\vdash_{\mathbf{IK_t}} (G\varphi \wedge F\psi) \to F(\varphi \wedge \psi)$

Proof.

1. $\vdash_{\mathbf{IK_t}} \varphi \to (\psi \to (\varphi \wedge \psi))$ A1
2. $\vdash_{\mathbf{IK_t}} G\varphi \to G(\psi \to (\varphi \wedge \psi))$ 1, RRG
3. $\vdash_{\mathbf{IK_t}} G(\psi \to (\varphi \wedge \psi)) \to (F\psi \to F(\varphi \wedge \psi))$ A4
4. $\vdash_{\mathbf{IK_t}} G\varphi \to (F\psi \to F(\varphi \wedge \psi))$ 2,3, SYLL
5. $\vdash_{\mathbf{IK_t}} (G\varphi \to (F\psi \to F(\varphi \wedge \psi))) \to ((G\varphi \wedge F\psi) \to F(\varphi \wedge \psi))$ A1
6. $\vdash_{\mathbf{IK_t}} (G\varphi \wedge F\psi) \to F(\varphi \wedge \psi)$ 4,5, MP

□

Lemma 7. $\vdash_{\mathbf{IK_t}} (H\varphi \wedge P\psi) \to P(\varphi \wedge \psi)$

Proof analogous to the proof of the previous lemma.

Lemma 8. $\vdash_{\mathbf{IK_t}} G\neg\varphi \to F\neg\varphi$

Proof.

1. $\vdash_{\mathbf{IK_t}} F\varphi \to \neg G\neg\varphi$ A5
2. $\vdash_{\mathbf{IK_t}} (F\varphi \to \neg G\neg\varphi) \to (G\neg\varphi \to \neg F\varphi)$ A1
3. $\vdash_{\mathbf{IK_t}} (G\neg\varphi \to \neg F\varphi)$ 1,2,MP

□

Lemma 9. $\vdash_{\mathbf{IK_t}} H\neg\varphi \to \neg P\varphi$

Proof analogous to the proof of the previous lemma.

Lemma 10. $\vdash_{\mathbf{IK_t}} F(\varphi \to \psi) \to (G\varphi \to F\psi)$

Proof.

1. $\vdash_{\mathbf{IK_t}} \varphi \to ((\varphi \to \psi) \to \psi)$ A1
2. $\vdash_{\mathbf{IK_t}} G\varphi \to G((\varphi \to \psi) \to \psi)$ 1,RRG
3. $\vdash_{\mathbf{IK_t}} G((\varphi \to \psi) \to \psi) \to (F(\varphi \to \psi) \to F\psi)$ A4
4. $\vdash_{\mathbf{IK_t}} G\varphi \to (F(\varphi \to \psi) \to F\psi)$ 2,3, SYLL

5. $\vdash_{IK_t} \left(G\varphi \to (F(\varphi \to \psi) \to F\psi) \right) \to \left(F(\varphi \to \psi) \to (G\varphi \to F\psi) \right)$ A1
6. $\vdash_{IK_t} F(\varphi \to \psi) \to (G\varphi \to F\psi)$ 4,5, MP

□

Lemma 11. $\vdash_{IK_t} P(\varphi \to \psi) \to (H\varphi \to P\psi)$

Proof analogous to the proof of the previous lemma.
We will show that the " new " A5 and A5′ axioms are we can derive from the' 'old' '8 and 8′ axioms.

Lemma 12. $G\neg\varphi \to \neg F\varphi \vdash_{IK_t} F\varphi \to \neg G\neg\varphi$

Proof.

1. $\vdash_{IK_t} G\neg\varphi \to \neg F\varphi$ assumption
2. $\vdash_{IK_t} (G\neg\varphi \to \neg F\varphi) \to (F\varphi \to \neg G\neg\varphi)$ axiom 1
3. $\vdash_{IK_t} (F\varphi \to \neg G\neg\varphi)$ 1,2,MP

□

It is likewise proved that:

Lemma 13. $H\neg\varphi \to \neg P\varphi \vdash_{IK_t} P\varphi \to \neg H\neg\varphi$

Thus, we have shown that the given axioms are equivalent. In further considerations we will use "new" axiomatics of **IK$_t$**.

8. The Adequacy of IK$_t$ Relative to Modified Semantics

The natural question is the question about the relationship between modified semantics and the assumed set of axioms for **IK$_t$**.

Theorem 14. *The **IK$_t$** axioms are true in any model, and the **IK$_t$** inference rules are infallible.*

Proof. We will prove only $A2'$, $A4'$ axioms and RH rule. Proofs for the other rules and axioms is carried out in analogous manner.

A2′ For any $\mathfrak{M}, m_i (\in \mathfrak{M})$, and $t (\in T_i)$: $\mathfrak{M}, m_i, t \models H(\varphi \to \psi) \to (H\varphi \to H\psi)$.

Suppose for some $\mathfrak{M}, m_i (\in \mathfrak{M})$ and $t (\in T_i)$: $\mathfrak{M}, m_i, t \not\models H(\varphi \to \psi) \to (H\varphi \to H\psi)$.

Therefore, from the condition of the truth for the implications, there is a state of knowledge m_j, $m_i \leq m_j$, such that:

$$\mathfrak{M}, m_j, t \models H(\varphi \to \psi), \tag{23}$$

$$\mathfrak{M}, m_j, t \not\models H\varphi \to H\psi. \tag{24}$$

From (24) and the condition of the truth for the implications, in a certain state of knowledge m_k, with a level of knowledge not lesser than the level of knowledge of the state m_j, i.e., such that $m_j \leq m_k$:

$$\mathfrak{M}, m_k, t \models H\varphi, \tag{25}$$

$$\mathfrak{M}, m_k, t \not\models H\psi. \tag{26}$$

From (25) and the condition of the truth for the H operator we get:

for any state of knowledge m_l such that $m_k \leq m_l$ and

$$\text{for any } t_1 \in T_l \text{ such that } t_1 R_l t \text{ holds: } \mathfrak{M}, m_l, t_1 \models \varphi. \quad (27)$$

From (26) and the condition of the truth for the H operator, there is a state m_p sucht that $m_k \leq m_p$) and there is a moment $t_2 \in T_p$ such that $t_2 R_p t$, in which:

$$\mathfrak{M}, m_p, t_2 \nvDash \psi. \quad (28)$$

Because $m_k \leq m_p$ and $t_2 R_p t$ therefore from (27) we have that at the moment t_2 holds $\mathfrak{M}, m_p, t_2 \models \varphi$. Hence, from (28) and the condition of the truth of the implications we get:

$$\mathfrak{M}, m_p, t_2 \nvDash \varphi \to \psi. \quad (29)$$

From (23) and the condition of the truth of the operator H we have:

for any m_r such that $m_j \leq m_r$ and

$$\text{for any } t_3 \in T_r \text{ such that } t_3 R_r t \text{ holds : } \mathfrak{M}, m_r, t_3 \models \varphi \to \psi. \quad (30)$$

Because: $m_j \leq m_k$, $m_k \leq m_p$, so from the transitivity of the relationship $\leq$ we get $m_j \leq m_p$. The moment t_2 is such that $t_2 R_p t$. Therefore, from (30) we have:

$$\mathfrak{M}, m_p, t_2 \models \varphi \to \psi.$$

This is contrary to 29.

A4′ For any $\mathfrak{M}, m_i (\in \mathfrak{M})$ and $t (\in T_i)$: $\mathfrak{M}, m_i, t \models H(\varphi \to \psi) \to (P\varphi \to P\psi)$.

Suppose for some $\mathfrak{M}, m_i (\in \mathfrak{M})$ and $t (\in T_i)$ $\mathfrak{M}, m_i, t \nvDash H(\varphi \to \psi) \to (P\varphi \to P\psi)$.

Thus, from the condition of the truth of the implications, in a certain state of knowledge m_j, such that $m_i \leq m_j$ we have:

$$\mathfrak{M}, m_j, t \models H(\varphi \to \psi), \quad (31)$$

$$\mathfrak{M}, m_j, t \nvDash P\varphi \to P\psi. \quad (32)$$

From (32) and the condition of the truth of the implications, in some state of knowledge m_k, such that $m_j \leq m_k$:

$$\mathfrak{M}, m_k, t \models P\varphi, \quad (33)$$

and

$$\mathfrak{M}, m_k, t \nvDash P\psi. \quad (34)$$

From (33) and the condition of the truth of the P operator we have:

$$\text{there exists } t_1 (\in T_k), t_1 R_k t \text{ such that } \mathfrak{M}, m_k, t_1 \models \varphi. \quad (35)$$

From (34) and the condition of the truth of the P operator we obtain:

$$\text{does not exist moment of time } t_2 (\in T_k), t_2 R_k t, \text{ such that } \mathfrak{M}, m_k, t_2 \models \psi. \quad (36)$$

Let us consider the moment t_1 satisfying (35). Because $t_1 R_k t$, so from (36) we have:

$$\mathfrak{M}, m_k, t_1 \not\models \psi. \tag{37}$$

If $\mathfrak{M}, m_k, t_1 \models \psi$, it would be against (36).

From (35), (37) and the condition of the truth of the implications, we get that $\mathfrak{M}, m_k, t_1 \not\models \varphi \to \psi$. From (31) and condition the truth of the operator H we have:

for any m_l such that $m_j \leq m_l$ and for any $t_3 (\in T_l)$ such that $t_3 R_l t : \mathfrak{M}, m_l, t_3 \models \varphi \to \psi$. (38)

Because $m_j \leq m_k, t_1 R_k t$ and $\mathfrak{M}, m_k, t_1 \not\models (\varphi \to \psi)$, so we get a contradiction with (38).

RH If $\mathfrak{M} \models \varphi$, then $\mathfrak{M} \models H\varphi$.

Let us assume that $\mathfrak{M} \models \varphi$. So for any m_i and for any $t(\in T_i)$ holds $\mathfrak{M}, m_i, t \models \varphi$. So especially for any $t_1(\in T_i)$ such that $t_1 R_i t : \mathfrak{M}, m_i, t_1 \models \varphi$. So for any $t(\in T_i)$ holds $\mathfrak{M}, m_i, t \models H\varphi$. Because we were considering any m_i, therefore $\mathfrak{M} \models H\varphi$.

□

Adequacy **IK$_t$** with respect to modified semantics was demonstrated by Surowik [?].

Theorem 15. $\Sigma \vdash_{IK_t} \varphi$ iff $\Sigma \models_{IK_t} \varphi$.

The proof of this theorem is similar to the proof of the adequacy theorem demonstrated by Ewald in [?].

9. Mutual Undefinability in IK$_t$ Operators H, P and G, F

We will now prove theorems that show some special properties of the **IK$_t$** system, essentially distinguishing this system from systems built on the basis of classical logic. For the formula to be the tautology of the **IK$_t$** system, it needs to be true at any time, in any state of knowledge. To show that a formula is not true, it is enough to indicate the state of knowledge and the moment in which this formula is not true.

We will show that some relationships between the operators H and P and G and F holds in the system **K$_t$** but do not occur between the equivalents of these operators in the system **IK$_t$**.

Theorem 16.

(a) $\not\models_{IK_t} \neg P \neg p \to Hp$,
(b) $\not\models_{IK_t} \neg H \neg p \to Pp$,
(c) $\not\models_{IK_t} \neg Hp \to P \neg p$,
(d) $\not\models_{IK_t} \neg F \neg p \to Gp$,
(e) $\not\models_{IK_t} \neg G \neg p \to Fp$

Proof.

(a) Let $T = \{t_1, t_2\}$, $R = \{(t_1, t_2)\}$. Let I be a set of indexes. For any i: $T_i = T$, $R_i = R$. Let $k, k > 1$, be a certain index of state of knowledge. Let $\mathcal{F} = \{V_i\}_{i \in I}$ be a class of functions satisfied the following conditions:

$$\text{for any } i \text{ such that } i \leq k \text{ holds } p \notin V_i(t_1), \tag{39}$$

and

$$\text{for any } i \text{ such that } k < i \text{ holds } p \in V_i(t_1). \tag{40}$$

The V_i valuations are therefore selected so that the sentence p is true at the time of t_1 in the states of knowledge with index not greater than k and at the same time it was not true at the time of t_1 in the states of knowledge with index greater than k.

Let $\mathcal{T} = \langle T, R \rangle$. Let $\mathfrak{M} = \{m_i : i \in I\}$. From the construction of the $\mathfrak{M}$ model, we get that there are states of knowledge in the $\mathfrak{M}$ which level of knowledge is not lesser than the level of knowledge of m_1 in which at the moment t_1 p is true and there are states of knowledge with a level of knowledge not lesser than the level of knowledge of m_1, in which at the moment t_1 is not true that p. Therefore, it is not true that in any state of knowledge $m_1^* (\in \mathfrak{M})$ holds $\mathfrak{M}, m_1^*, t_1 \not\models p$. Therefore, by Definition ?? we get $\mathfrak{M}, m_1, t_1 \not\models \neg p$. From the construction of the $\mathfrak{M}$ model we get that in any state of knowledge $m_1^* (\in \mathfrak{M})$ holds $\mathfrak{M}, m_1^*, t_1 \not\models \neg p$. Because $t_1 R_1^* t_2$, therefore, by the Definition ?? we have $\mathfrak{M}, m_1^*, t_2 \not\models P\neg p$. By the Definition ?? we get

$$\mathfrak{M}, m_1, t_2 \models \neg P \neg p. \tag{41}$$

Because the moment t_1 is such that $t_1 R_1 t_2$ and $\mathfrak{M}, m_1, t_1 \not\models p$ so by the Definition ??

$$\mathfrak{M}, m_1, t_2 \not\models Hp. \tag{42}$$

From (41), (42) and the Definition ??: we have $\mathfrak{M}, m_1, t_2 \not\models \neg P\neg p \to Hp$. Therefore $\not\models_{\mathbf{IK_t}} \neg P\neg p \to Hp$.

(b) The $\mathfrak{M}$ model proposed in the proof of a) will be used to prove that $\neg H\neg p \to Pp$ is not a tautology of $\mathbf{IK_t}$. Please note that from the construction of the model and by Definition ?? we have $\mathfrak{M}, m_1^*, t_1 \not\models \neg p$. Because in any state of knowledge $m_1^* (\in \mathfrak{M})$ the only time before t_2 is the time t_1, so by the Definition ?? for any m_1^* holds $\mathfrak{M}, m_1^*, t_2 \not\models H\neg p$. Hence, by the Definition ??

$$\mathfrak{M}, m_1, t_2 \models \neg H \neg p. \tag{43}$$

From the construction of the model $\mathfrak{M}$ we have $\mathfrak{M}, m_1, t_1 \not\models p$. Because $t_1 R_1 t_2$, therefore by the Definition ??

$$\mathfrak{M}, m_1, t_2 \not\models Pp. \tag{44}$$

From (43), (44) and the Definition of ?? we obtain: $\mathfrak{M}, m_1, t_2 \not\models \neg H\neg p \to Pp$. Therefore $\not\models \mathbf{IK_t} \neg H\neg p \to Pp$.

(c) We will now show that $\neg Hp \to P\neg p$ is not a tautology of $\mathbf{IK_t}$. Let $T_1 = \{t_1, t_2\}$, $R_1 = \{(t_2, t_1)\}$. Let the function V_1 be such that $p \notin V_1(t_2)$. States of knowledge in which the level of knowledge is not lower than the level of m_1 we construct as follows:

$$T_{i+1} = T_i \cup \{t_{i+2}\}, \tag{45}$$

$$R_{i+1} = R_i \cup \{(t_{i+2}, t_1)\}. \tag{46}$$

V_{i+1} is such that for $t \neq t_{i+2} : p \in V_{i+1}(t)$, and for $t = t_{i+2} : p \notin V_{i+1}(t)$. $\tag{47}$

State of knowledge m_{i+1} is an ordered triple $\langle T_{i+1}, R_{i+1}, V_{i+1} \rangle$. Let $\mathcal{F} = \{V_i\}_{i \in I}$ will be a class of functions satisfying the condition (47), $\mathcal{T} = \langle \bigcup_{i \in I} T_i, \bigcup_{i \in I} R_i \rangle$, $\mathfrak{M} = \{m_i : i \in I\}$, the states of knowledge m_i are constructed in accordance with conditions (45), (46) and (47). From the construction of the $\mathfrak{M}$ model we get that in every state of knowledge $m_1^* (\in \mathfrak{M})$, there is a moment t, earlier than t_1 such that $tR_1^* t_1$ in which such that $\mathfrak{M}, m_1^*, t \not\models p$. So by the Definition ?? for any state of knowledge m_1^* we have $\mathfrak{M}, m_1^*, t_1 \not\models Hp$. From the definition of ?? we have that:

$$\mathfrak{M}, m_1, t_2 \models \neg Hp. \tag{48}$$

From the construction of the model we have that if at some moment of time t, in any state of knowledge $m_i (\in \mathfrak{M})$ is that $\mathfrak{M}, m_i, t \not\models p$, then in every state of knowledge $m_i^* (\in \mathfrak{M})$ holds $\mathfrak{M}, m_i^*, t \models p$. In the state of knowledge $m_1 (\in \mathfrak{M})$ the only time before t_1 is the moment t_2. The moment t_2 is such that $p \notin V_1(t_2)$. In the classical model, this would suffice to say that $\mathfrak{M}, m_1, t_2 \models \neg p$. This is not the case in the temporal logic model built upon intuitionistic logic. From the way of constructing states of knowledge with no lower level of knowledge than the level of knowledge in the state m_1 we have $p \in V_2(t_2)$. Therefore, by the Definition ??

$$\mathfrak{M}, m_1, t_2 \not\models \neg p. \tag{49}$$

By the Definition ?? we have $\mathfrak{M}, m_1, t_1 \not\models \neg Hp \rightarrow P \neg p$.

We construct counter-examples for d), e) and f) in an analogous way. □

In the **IK**$_t$ system, between the G and F and H and P operators there are no relationships usually found in temporal logic systems that are based on classical logic. However, the above conclusion is not sufficient to state that the operators G and F as well as H and P are not mutually definable in **IK**$_t$. The conclusion is only that they do not occur between these operators definition relationships the same as those in *classical* tense logics. We will show that in intuitionistic temporal logic, temporal operators are not definable as they are in temporal logics based on classical propositional logic. We will show that intuitionistic temporal operators are not definable in any other way using sentence connectives and other intuitionistic temporal operators.

To show that a temporal operator is not definable in the **IK**$_t$, two structures should be indicated such that the sentence with the considered operator at a moment t in one structure is true, and it is false in the other. On the other hand, all sentences in which the operator does not appear have the same logical value in both structures at the moment t.

Theorem 17 ([?]). *The intuitionistic temporal operators F and G as well as P and H are not each other definable in the **IK**$_t$.*

Proof. We will show first that the operator F is not definable if we use of intuitionistic sentence connectives and other temporal operators. We will show that Fp is not equivalent to any temporal formula in which the F operator does not occur.

F: Let $T_1 = \{t_1, t_2\}$, $T_2 = \{t_1, t_2, t_3\}$, $R_1 = \{(t_1, t_2)\}$, $R_2 = \{(t_1, t_2), (t_1, t_3)\}$, $T = T_1 \cup T_2$. Let $V_1 : T_1 \rightarrow 2^{\mathcal{AP}}$ be such that $p \notin V_1(t_2)$ while $V_2 : T_2 \rightarrow 2^{\mathcal{AP}}$ will be such a function that: $V_2(t_1) = V_1(t_1)$, $V_2(t_2) = V_1(t_2) \cup \{p\}$, $V_2(t_3) = V_1(t_2)$. Let $\mathcal{F} = \{V_1, V_2\}$, $m_1 = \langle T_1, R_1, V_1 \rangle$, $m_2 = \langle T_2, R_2, V_2 \rangle$, $\mathfrak{M} = \{m_1, m_2\}$.

By means of structural induction, it can be shown that for any φ without the F operator we have

$$\mathfrak{M}, m_1, t_1 \vDash \varphi \text{ iff } \mathfrak{M}, m_2, t_1 \vDash \varphi. \tag{50}$$

At the same time, $\mathfrak{M}, m_2, t_1 \vDash Fp$ and $\mathfrak{M}, m_1, t_1 \nvDash Fp$. Therefore, the F operator is not definable in $\mathbf{IK_t}$.

G: We will now show that the operator G is not definable if we use of intuitionistic sentence connectives and other temporal operators. We will show that Gp is not equivalent to any temporal formula in which the G operator is not present.

Let $T_1 = \{t_1, t_2, t_3\}$, $T_2 = \{t_1, t_2, t_3\}$, $R_1 = \{(t_1, t_2)\}$, $R_2 = \{(t_1, t_2), (t_1, t_3)\}$, $T = T_1 \cup T_2$. Let $V_1 : T_1 \to 2^{AP}$ will be such a function that $p \notin V_1(t_2)$. Let $V_2 : T_2 \to 2^{AP}$ will be such a function that: $V_2(t_1) = V_1(t_1)$, $V_2(t_2) = V_1(t_2) \cup \{p\}$, $V_2(t_3) = V_1(t_3)$, $p \in V_1(t_3)$. Let $\mathcal{F} = \{V_1, V_2\}$. Let $m_1 = \langle T_1, R_1, V_1 \rangle$, $m_2 = \langle T_2, R_2, V_2 \rangle$. Let $\mathfrak{M} = \{m_1, m_2\}$. By means of structural induction, it can be shown that for any φ sentence without the G operator we have:

$$\mathfrak{M}, m_1, t_1 \vDash \varphi \text{ iff } \mathfrak{M}, m_2, t_1 \vDash \varphi. \tag{51}$$

At the same time, $\mathfrak{M}, m_2, t_1 \vDash Gp$ and $\mathfrak{M}, m_1, t_1 \nvDash Gp$. So the G operator is not definable in $\mathbf{IK_t}$.

Similarly, we can to show that P and H are not each other definable in $\mathbf{IK_t}$.

□

It is not, however, that the operators G, F, H, P are completely independent of each other. Certain relationships between the operators H and P and G and F occur in $\mathbf{IK_t}$. We will prove some of them:

Theorem 18.

(a) $\vdash_{\mathbf{IK_t}} H\neg\varphi \to \neg P\varphi$,
(b) $\vdash_{\mathbf{IK_t}} H\varphi \to \neg P\neg\varphi$,
(c) $\vdash_{\mathbf{IK_t}} P\neg\varphi \to \neg H\varphi$,
(d) $\vdash_{\mathbf{IK_t}} G\neg\varphi \to \neg F\varphi$,
(e) $\vdash_{\mathbf{IK_t}} G\varphi \to \neg F\neg\varphi$,
(f) $\vdash_{\mathbf{IK_t}} F\neg\varphi \to \neg G\varphi$.

Proof.

(a) $\vdash_{\mathbf{IK_t}} (H\neg\varphi \to \neg P\varphi)$

1.	$\vdash_{\mathbf{IK_t}} (P\varphi \to \neg H\neg\varphi) \to (H\neg\varphi \to \neg P\varphi)$	axiom 1,
2.	$\vdash_{\mathbf{IK_t}} H\neg\varphi \to \neg P\varphi$	A5',1,MP.

(b) $\vdash_{\mathbf{IK_t}} H\varphi \to \neg P\neg\varphi$

1.	$\vdash_{\mathbf{IK_t}} \varphi \to \neg\neg\varphi$	axiom 1,
2.	$\vdash_{\mathbf{IK_t}} H(\varphi \to \neg\neg\varphi)$	1,RH,
3.	$\vdash_{\mathbf{IK_t}} H(\varphi \to \neg\neg\varphi) \to (H\varphi \to H\neg\neg\varphi)$	A2',
4.	$\vdash_{\mathbf{IK_t}} H\varphi \to H\neg\neg\varphi$	2,3,MP,
5.	$\vdash_{\mathbf{IK_t}} H\neg\neg\varphi \to \neg P\neg\varphi$	case (a),
6.	$\vdash_{\mathbf{IK_t}} H\varphi \to \neg P\neg\varphi$	4,5, SYLL.

(c) $\vdash_{\mathbf{IK_t}} (P\neg\varphi \to \neg H\varphi)$

1. $\vdash_{\mathbf{IK_t}} (H\varphi \to \neg P\neg\varphi) \to (P\neg\varphi \to \neg H\varphi)$ axiom 1,
2. $\vdash_{\mathbf{IK_t}} P\neg\varphi \to \neg H\varphi$ 1, case (b),MP.

The proofs of the cases (d), (e) and (f) are similar, so we skip them. □

10. Summary

Temporal logic systems can be built in a variety of ways. They can be based on classical logic, but also, as we presented in this article, based on intuitionistic logic. The discussed systems are minimal systems, which means that no properties have been imposed on the time structure. One can, however, enrich these systems with additional specific axioms, build a temporal logic systems adequate to various time structures, e.g., reflexive, symmetrical, transitive, linear or branched. However, while in tense logic systems based on classical logic, the thesis of logical determinism can be rejected by modifying the structure of time and assuming, as a semantic time, a branching time into the future, in tense logics based on intuitionistic logic, modification of the time structure is not necessary. Formulas expressing the thesis of logical determinism are not theses of the minimal system because of its basic properties, no matter what time structure is adopted as a semantic time.

There is a relationship between the systems being discussed. Each thesis of the $\mathbf{IK_t}$ system is also the thesis of $\mathbf{K_t}$, so:

$$\mathbf{IK_t} \subset \mathbf{K_t}.$$

In addition, as we have shown in this article, intuitionistic temporal logic can be used to represent knowledge that changes over time. Intuitionistic logic and knowledge are closely related. This epistemic approach is the epicenter of Brouwer's intuitionistic explanation of truth as provability by an ideal mathematician, or more generally by an ideal cognitive subject. Kripke's intuitionistic models are good tools for modelling the evolutionary learning process of the cognitive subject.

The intuitionistic temporal logic $\mathbf{IK_t}$ has many advantages when we understand it as a formal tool for the logical representation of knowledge changing over time. Knowledge is implemented in this system on a semantic level in a natural way. In a natural way, by means of a set of partially ordered states of knowledge, the way of acquiring knowledge is also modeled. However, this system has some imperfections and limitations. The first is the limited applicability of this system. Due to the adopted monotonicity of knowledge, i.e., a fact recognized in a given state of knowledge is known in all states of knowledge with a not lower level of knowledge, this system is a good tool for a modelling of mathematical or logical knowledge that changes over time.

Conflicts of Interest: The authors declare no conflict of interest.

References

- Van Benthem, J.F.A.K. *The Logic of Time: A Model-Theoretic Investigation into the Varietes of Temporal Ontology and Temporal Discourse*, 2nd ed.; LII 874. Synthese library; Kluwer Academic Publishers: Dordrecht, The Netherlands; Boston, MA, USA; London, UK, 1991; Volume 156.
- Van Benthem, J.F.A.K. Reflections on epistemic logic. *Logique et Analyse* **1991**, *34*, 5–14.
- Ewald, W.B. Intuitionistic tense and modal logic. *J. Symb. Log.* **1986**, *51*, 166–179. [CrossRef]
- Surowik, D. Intuitionistic Tense Logic and Indeterminism. Ph.D. Thesis, Lodz University, Łódź, Poland, 2001. (In Polish)

- Trzęsicki, K. Logic of Grammar Time Operators and the Problem of Determinism. Habilitation Thesis, University of Warsaw, Warsaw, Poland, 1986. (In Polish)
- Surowik, D. Some thechnical results in a certain intuitionistic tense logic. In *Topics in Logic Informatics and Philosophy of Science*; University of Białystok: Białystok, Poland, 1999.

 © 2020 by the authors. Licensee MDPI, Basel, Switzerland. This article is an open access article distributed under the terms and conditions of the Creative Commons Attribution (CC BY) license (http://creativecommons.org/licenses/by/4.0/).

Article

Deduction in Non-Fregean Propositional Logic SCI

Joanna Golińska-Pilarek and Magdalena Welle

Institute of Philosophy, University of Warsaw, 00–927 Warsaw, Poland; m.welle@uw.edu.pl
* Correspondence: j.golinska@uw.edu.pl

Received: 5 September 2019; Accepted: 14 October 2019; Published: 17 October 2019

Abstract: We study deduction systems for the weakest, extensional and two-valued non-Fregean propositional logic SCI. The language of SCI is obtained by expanding the language of classical propositional logic with a new binary connective ≡ that expresses the identity of two statements; that is, it connects two statements and forms a new one, which is true whenever the semantic correlates of the arguments are the same. On the formal side, SCI is an extension of classical propositional logic with axioms characterizing the identity connective, postulating that identity must be an equivalence and obey an extensionality principle. First, we present and discuss two types of systems for SCI known from the literature, namely sequent calculus and a dual tableau-like system. Then, we present a new dual tableau system for SCI and prove its soundness and completeness. Finally, we discuss and compare the systems presented in the paper.

Keywords: non-Fregean logic; identity connective; sentential calculus with identity; situational semantics; deduction; (dual) tableau; Gentzen system

MSC: 03A05; 03B60; 03B65; 03B80; 03F99

---

1. Introduction

One of the tasks of formal logic is to provide adequate tools for the formal analysis of certain fragments of natural language, as well as for the languages of particular fields of science. It is commonly accepted that the theory of interpretation of a language is semantics. The choice of semantics determines how we think about a given language and what meaning we assign to its components. It is often acknowledged that the first precisely formulated semantic principles—that serves as a foundation for contemporary formal logic and have determined its development—were presented by Frege in his *Begriffsschrift*. According to Frege, a correct and adequate formal system of a given language should meet the following conditions:

F1 All names and all sentences have meaning and denotation. Meaning is not the same as denotation.
F2 A name and a sentence are the proper names of their denotations.
F3 Only one logical value can be assigned to each sentence: true or false.
F4 If two expressions have the same denotation, then they are exchangeable in any propositional context of a sentence without changing the logical value of that sentence.
F5 If two sentences are exchangeable in any propositional context of a sentence without changing its logical value, then they have the same denotation.

Note that crucial notions for Frege's account are the following: meaning, denotation, and logical valuation. Frege admits that the meaning of a sentence is not the same as its denotation. Indeed, the same holds for names: 'Evening Star' and 'Morning Star' denote the same object, but they have different meanings. The meaning of a sentence should be understood as its sense that the sentence expresses. Thus, we should ask what a sentence is referring to. From the formal point of view, the answer to this question requires, in particular, to decide what the denotation of a propositional variable is.

Propositional variables have an unusual character due to the fact that, at the same time, they are formulas. In classical logic, this ambiguity is removed as it identifies denotations of sentences with their truth values. In other words, in classical logic, propositional variables occur only at the metalogical level (they are not treated as real variables) or, as it holds in propositional calculus, variables range over the two element set of logical values. Within such an account, only terms have a meaningful ontological reference. Thus, the advocate of classical logic, if he is also a proponent of the Fregean semantic principles, must accept that all true sentences have one and the same denotation, namely the logical value Truth, and denotation of all false sentences is the logical value False.

The consequence of Frege's semantic principles, according to which denotations of sentences are logical values, is usually called the *Fregean principle*, and every logical system that adopts this principle is called a *Fregean system*. The Fregean approach can be considered as philosophically justified in the study of mathematical languages, but it is not obvious that such an approach is justified as a foundation of a philosophically adequate semantics. The debatable cases of the applicability of the Fregean account in the formalization of any language are, for instance, theories of meaning or ontology. If we admit that the references of sentences are situations (states of affairs) that these sentences describe (in analogy with the assumption that the semantic references of terms are objects named by these terms), then the Fregean principle is not only a semantic principle, but also an ontological one which imposes a quantitative restriction on the universe of situations: there are at most two situations described by sentences. This is an extremely strong assumption. Note that the Fregean account does not impose an upper limit on the universe of objects.

Roman Suszko in 1968 [1] proposed to change the Fregean paradigm and introduced the so-called *non-Fregean logic* (see also [2]). The basic philosophical assumption underlying non-Fregean logic is the thesis in reality to which the language is referring, and there exist semantic correlates of all expressions that are not purely syntactic. Therefore, in the non-Fregean approach, it is assumed that the semantic correlates of names are objects from the universe of objects, the semantic correlates of predicates are appropriate sets, and the semantic correlates of sentences are situations described by these sentences. Furthermore, a universe of situations cannot be quantitatively restricted, except that there are at least two situations.

The construction of a propositional non-Fregean logic, which at the same time preserves all properties of classical logic with respect to the classical connectives, is relatively simple. To make it possible to express statements on situations and interactions between them, propositional calculus is extended with the additional connective, named the *identity connective* and denoted by $\equiv$. The intended interpretation of the identity connective is the following: $\phi \equiv \psi$ is true if and only if the semantic correlates of ϕ and ψ are the same, that is, sentences φ and ψ describe the same situation (the same state of affairs). Note that, in a general case, the identity connective is different than the classical equivalence. Equivalent sentences, that is, sentences that do not have the same logical value, do not have to refer to the same situation, and so they do not have to be identical. In other words, the logical value of a sentence and the situation described by this sentence are two different things.

Suszko states that the identity connective is more primitive than other non-truth functional connectives such as modal connectives of possibility and necessity. The identity connective is more primitive in the sense that it cannot be eliminated without identifying it with the equivalence connective. However, if we add the identity connective to classical logic, we do not lose two-valuedness. If a non-Fregean system of logic would be constructed based on classical logic, extending its language with the identity connective and constructing semantics in which classical connectives preserve their classical meaning, we will obtain a two-valued system, in which each sentence is either true or false. On the other hand, being logically two-valued, non-Fregean logics can be seen as systems that are ontologically (referentially) many-valued.

The weakest extensional and two-valued propositional non-Fregean logic is SCI (*Sentential Calculus with Identity*). A detailed description of the philosophical assumptions of SCI can be found in [2]. Originally, logic SCI was defined as an extension of classical propositional logic with four axioms expressing the following properties of the identity connective: (1) any sentence is identical with itself, (2) if two sentences are identical, then so are their negations, (3) identical sentences are equivalent, and (4) sentences that are identical are interchangeable in any propositional context. A sound and complete semantics for SCI was designed by Suszko and Bloom in [3]. A models for SCI is a structure that consists of a universe of situations, a distinguished subset of facts (situations that actually hold), and operations that represent all the connectives. Furthermore, it is assumed that the operations satisfy certain conditions with respect to the distinguished set of situations so that the classical connectives gain their classical meaning, and the operation corresponding to the identity connective represents the identity between situations. For details, see Section 2.

It is established that SCI has the finite model property and is decidable. Moreover, as shown in [4], the class of different axiomatic extensions of SCI is uncountable. There is also some research on non-standard (deviant) modifications of SCI obtained by rejecting some of its fundamental assumptions or extending its language with additional operators. Recently, the most studied modifications of SCI are Grzegorczyk's non-Fregean logics, which are paraconsistent non-Fregean logics ([5–7]). Some research has been also focused on first-order non-Fregean logics, in particular SCI_Q. In [8], it has been proved that the logic SCI_Q, obtained by extending SCI with propositional quantifiers, is able to express infiniteness and many well-known mathematical theories (e.g., the theories of groups and fields, Peano arithmetic). Furthermore, SCI_Q does not have the finite model property, is undecidable, satisfies the Löwenheim-Skolem Theorem, and is an analog of Fagin's Theorem (the class of sets of natural numbers that are expressible in SCI_Q is precisely the complexity class NP).

The non-Fregean approach has many philosophical and logical advantages as it offers a relatively simple, natural and intuitive basis for exploring fundamental relationships between language and situations. Moreover, non-Fregean logics can be seen as a general framework for representing and comparing logics with different languages or semantics. Indeed, it turned out that many non-classical logics can be equivalently translated into some extensions of SCI. For instance, modal logics S3, S4, and S5 are equivalent to some extensions of SCI, that is, there are translations—from a non-Fregean language to a modal one and the other way round—that preserve the satisfaction of formulas with respect to the appropriate class of structures (see [9]). It has been also proved in [10] that SCI can serve as a basis for expressing many-valued logics. Furthermore, it has been shown that the weakest non-Fregean logic MGL (*Minimal Grzegorczyk Logic*) introduced in [11] is able to express most non-classical logics, including uncountably many extensions of SCI and paraconsistent non-Fregean Grzegorczyk's logics. Thus, MGL can be treated as a generic non-Fregean logic.

The non-Fregean approach could be relevant in cognitive science applications as well as in natural language processing. Last but not least, research on non-Fregean logics could lead to a better understanding of the capabilities and relationships of logics with mutually incompatible languages and semantics. Studies on various versions of non-Fregean logic may also shed light on which class of logics offers the most adequate account of logical symbols from point of view of natural language.

In this paper, we focus on deduction systems for the logic SCI. The deduction system for SCI was originally defined in a Hilbert-style. Since then, some other systems for SCI have been proposed: Gentzen sequent calculus ([12–14]) and a dual tableau system ([15,16]). The aim of the paper is to present a new dual tableau system for SCI, which is suitable for automated reasoning in SCI. The main advantage of the new system is that, contrary to previously known systems, it is more efficient: it does not involve any substitution rule, its rules for the identity connective do not branch a proof tree, and it generates shorter and simpler proof trees.

The paper consists of five sections: in Section 2, we present the basics of the non-Fregean propositional logic SCI, that is, its language, semantics, and axiomatization. In Sections 3 and 4, we briefly survey sequent calculus and a dual tableau system for SCI, respectively. In Section 5, we present a new dual tableau system for SCI and prove its soundness and completeness. Finally, in Section 6, we discuss and compare the systems presented in the paper.

2. The Non-Fregean Propositional Logic SCI

The vocabulary of the language of the non-fregean propositional logic, SCI, consists of the symbols from the following pairwise disjoint sets:

- $\mathbb{V} = \{p_1, p_2, p_3, \ldots\}$—a countable infinite set of propositional variables;
- $\{\neg, \vee, \wedge, \rightarrow, \leftrightarrow, \equiv\}$—the set of propositional operations of negation $\neg$, disjunction $\vee$, conjunction $\wedge$, implication $\rightarrow$, equivalence $\leftrightarrow$, and identity $\equiv$.

The set $\mathbb{FOR}$ of all SCI-formulas is the smallest set including $\mathbb{V}$ and closed with respect to all propositional operations.

An SCI-*model* is a structure $\mathcal{M} = (U, \sim, \sqcup, \sqcap, \Rightarrow, \Leftrightarrow, \circ, D)$, where U is a non-empty set referred to as a *universe*, D is any non-empty proper subset of U, and $\sim, \sqcup, \sqcap, \Rightarrow, \Leftrightarrow, \circ$ are operations on U with arities 1, 2, 2, 2, 2, 2, respectively, such that, for all $a, b \in U$ the following hold:

(SCI1) $\sim a \in D$ iff $(a \notin D)$;
(SCI2) $a \sqcup b \in D$ iff $(a \in D$ or $b \in D)$;
(SCI3) $a \sqcap b \in D$ iff $(a \in D$ and $b \in D)$;
(SCI4) $a \Rightarrow b \in D$ iff $(a \notin D$ or $b \in D)$;
(SCI5) $a \Leftrightarrow b \in D$ iff $(a \in D$ if and only if $b \in D)$;
(SCI6) $a \circ b \in D$ iff $a = b$.

Let $\mathcal{M}$ be an SCI-model. A valuation in $\mathcal{M}$ is any mapping $v \colon \mathbb{FOR} \to U$ such that $v(p) \in D$, for every $p \in \mathbb{V}$, and the following conditions hold for all SCI-formulas:

$v(\neg \varphi) = \sim v(\varphi)$ \qquad\qquad $v(\varphi \to \psi) = v(\varphi) \Rightarrow v(\psi)$
$v(\varphi \vee \psi) = v(\varphi) \sqcup v(\psi)$ \qquad\qquad $v(\varphi \wedge \psi) = v(\varphi) \sqcap v(\psi)$
$v(\varphi \leftrightarrow \psi) = v(\varphi) \Leftrightarrow v(\psi)$ \qquad\qquad $v(\varphi \equiv \psi) = v(\varphi) \circ v(\psi).$

Given an SCI-model $\mathcal{M}$ and a valuation v in $\mathcal{M}$, an SCI-formula φ is said to be *satisfied in $\mathcal{M}$ by v* (in short $\mathcal{M}, v \models \varphi$) whenever $v(\varphi) \in D$. An SCI-formula φ is *true* in $\mathcal{M}$ if and only if for every v in $\mathcal{M}$, $\mathcal{M}, v \models \varphi$. A formula is SCI-*valid* if and only if it is true in all SCI-models. An SCI-formula φ is said to be *satisfiable in an* SCI-*model* $\mathcal{M}$ whenever there exists a valuation v on $\mathcal{M}$ such that $\mathcal{M}, v \models \varphi$. A model is referred to as *finite* if its universe is finite.

The intended philosophical interpretation of an SCI-model $\mathcal{M} = (U, \sim, \sqcup, \sqcap, \Rightarrow, \Leftrightarrow, \circ, D)$ is as follows: U is the set of *situations* (denotations of sentences); D is the set of *facts*, that is, it consists of those situations that correspond to true sentences; the operations correspond to the formation of new formulas with connectives.

The logic SCI is two-valued. We may define the logical value of a formula φ in a model $\mathcal{M}$ as:

$$val_{\mathcal{M}}(\varphi) \stackrel{df}{=} \begin{cases} \text{true}, & \text{if for every } v \text{ in } \mathcal{M}, v(\varphi) \in D, \\ \text{false}, & \text{otherwise.} \end{cases}$$

The following proposition shows that SCI is extensional in the sense that any subformula ψ of an SCI-formula φ can be replaced with another formula ϑ denoting the same as ψ without affecting the denotation of φ.

Proposition 1. *Let $\mathcal{M}$ be an SCI-model, let v be a valuation in $\mathcal{M}$, let φ be an SCI-formula containing a subformula ψ, and let φ' be the result of replacing some occurrences of ψ in φ by a formula ϑ. Then, $\mathcal{M}, v \models \psi \equiv \vartheta$ implies $\mathcal{M}, v \models \varphi \equiv \varphi'$.*

The proof of Proposition 1 is presented in [16]. It should be emphasized that two-valuedness and extensionality concern different levels. Two-valuedness is a property of truth values, while extensionality holds for denotations.

As shown in [3], the logic SCI has the finite model property and is decidable:

Theorem 1 (Finite model property and decidability of SCI). *The logic SCI has the finite model property, i.e., every satisfiable SCI-formula is satisfiable in a finite SCI-model. Furthermore, the logic SCI is decidable.*

Corollary 1. *Let T be a set of SCI-formulas such that T is true in all finite SCI-models. Then, T is true in all infinite SCI-models as well.*

The proof of the above corollary can be found in [8].

A Hilbert-style axiomatization of SCI consists of axiom schemas of classical propositional logic PC, which characterize the operations $\neg, \vee, \wedge, \rightarrow, \leftrightarrow$, and the following axiom schemas for the identity operation $\equiv$:

($\equiv_1$) $\varphi \equiv \varphi$;
($\equiv_2$) $(\varphi \equiv \psi) \rightarrow (\neg\varphi \equiv \neg\psi)$;
($\equiv_3$) $(\varphi \equiv \psi) \rightarrow (\varphi \rightarrow \psi)$;
($\equiv_4$) $[(\varphi \equiv \psi) \wedge (\vartheta \equiv \zeta)] \rightarrow [(\varphi \# \vartheta) \equiv (\psi \# \zeta)]$, for $\# \in \{\wedge, \vee, \rightarrow, \leftrightarrow, \equiv\}$.

The only rule of inference is modus ponens. The notion of provability of a formula is defined as usual. Thus, an SCI-formula φ is said to be SCI-*provable* whenever there exists a finite sequence $\varphi_1, \ldots, \varphi_n$ of SCI-formulas, $n \geq 1$, such that $\varphi_n = \varphi$ and each $\varphi_i, i \in \{1, \ldots, n\}$, is an SCI-axiom or follows from earlier formulas in the sequence by the modus ponens rule. It is easy to see that all theorems of classical propositional logic are SCI-provable formulas.

Fact 2. *For every PC-formula φ, the following conditions are equivalent:*

1. *φ is provable in the classical propositional logic.*
2. *φ is SCI-provable.*

Soundness and completeness of SCI with respect to the class of SCI-models was proved in [3].

Theorem 2 (Soundness and Completeness of SCI). *For every SCI-formula φ, the following conditions are equivalent:*

1. *φ is SCI-provable.*
2. *φ is SCI-valid.*

The logic SCI is very weak as it does not impose any specific assumptions on the cardinality of the universe of situations (except that it has at least two elements). Furthermore, it does not assume any specific assumptions on the identities of equivalent formulas—for instance, the formula $(\varphi \wedge \psi) \equiv (\psi \wedge \varphi)$ is not SCI-valid. Indeed, the reduct $(U, \sim, \sqcup, \sqcap)$ of an SCI-model is not necessarily a Boolean algebra, since, for example, $a \sqcap b = b \sqcap a$ is not true in all SCI-models. Consider an SCI-model $\mathcal{M} = (U, \sim, \sqcup, \sqcap, \Rightarrow, \Leftrightarrow, \circ, D)$, where $U = \{0, 1, 2\}$, $D = \{1, 2\}$, and the operations $\sim, \sqcup, \sqcap, \Rightarrow, \Leftrightarrow, \circ$ are defined by:

$$\sim a \stackrel{df}{=} \begin{cases} 0, & \text{if } a \neq 0, \\ 1, & \text{otherwise,} \end{cases}$$

$$a \sqcup b \stackrel{df}{=} \begin{cases} 0, & \text{if } a = 0 \text{ and } b = 0, \\ 1, & \text{otherwise,} \end{cases}$$

$$a \Rightarrow b \stackrel{df}{=} \begin{cases} 0, & \text{if } a \neq 0 \text{ and } b = 0, \\ 1, & \text{otherwise,} \end{cases}$$

$$a \sqcap b \stackrel{df}{=} \begin{cases} 0, & \text{if } a = 0, \text{ or } b = 0, \\ 1, & \text{if } b = 2 \text{ and } a \neq 0, \\ 2, & \text{otherwise,} \end{cases}$$

$$a \Leftrightarrow b \stackrel{\mathrm{df}}{=} \begin{cases} 0, & \text{if } a \neq 0, b = 0 \text{ or } a = 0, b \neq 0, \\ 1, & \text{otherwise,} \end{cases}$$

$$a \circ b \stackrel{\mathrm{df}}{=} \begin{cases} 0, & \text{if } a \neq b, \\ a, & \text{if } a = b \text{ and } a \neq 0, \\ 1, & \text{otherwise.} \end{cases}$$

It is easy to verify that such a structure is an SCI-model. Then, the following hold in the model $\mathcal{M}\colon 2 \sqcap 1 = 2$ and $1 \sqcap 2 = 1$. Hence, $a \sqcap b = b \sqcap a$ is not true in this model.

Therefore, if we remove from the SCI-language some of the classical propositional connectives and define them equationally as usual, then the logic obtained in this way would not be a notational variant of SCI. Indeed, suppose that we remove the connective $\vee$ from SCI-language, and we add the definition $\varphi \vee \psi \stackrel{\mathrm{df}}{=} \neg \varphi \to \psi$. In such a logic, the formula $(\varphi \vee \psi) \equiv (\neg \varphi \to \psi)$ would be valid, while it is not an SCI-valid formula.

However, in some applications, there may be a need to impose some specific properties of situations or interactions between them. If we add additional assumptions on the universe of situations, we will obtain an extension of SCI. For example, if we add to the set of SCI-axioms the so-called Fregean axiom

$$\text{(Fregean Axiom)} \quad (\varphi \leftrightarrow \psi) \to (\varphi \equiv \psi),$$

which identifies the denotations of sentences with their truth values, then we get classical propositional logic. It is easy to see that classical propositional logic is the strongest among all propositional extensions of SCI. As shown in [4], there are uncountably many different non-Fregean theories stronger than SCI and weaker than classical propositional logic.

In the rest of the paper, we present and discuss two types of deduction systems for SCI: a sequent-style and a tableau-like systems. Although, as mentioned above, any restriction of the SCI-language leads to a different logic than the original SCI, to make the presentation more readable, we will assume that SCI-language consists of three connectives: $\neg$, $\to$, and $\equiv$. In the context of deduction systems, this restriction is minor because each of the presented system can be easily extended to the full SCI-language without loss of soundness and completeness.

3. Sequent-Style Formalizations for SCI

Sequent calculi constitute an important type of deduction systems. They were designed by Gerhard Gentzen for purely theoretical reasons, mainly as a theoretical framework for investigations properties of logical consequence. However, it turned out that Gentzen sequent calculus is not only another way of axiomatization of classical logic, but also a good alternative to Hilbert (axiomatic) systems: it is much easier and more convenient to use in practice. Anyone who has tried to construct an axiomatic proof of even a very simple formula knows that such a proof construction requires a lot of effort and creativity. The reason lies in the very nature of Hilbert systems: to prove a formula we must construct a sequence of formulas with this formula as the last element of the sequence. Thus, the main challenge in building axiomatic proofs is: How can we find the way to the formula in question? The Hilbert system itself does not provide any strategy on how to find proofs; it only says which sequences of formulas are proofs. In Gentzen systems, this weakness is mitigated by changing the notion of a proof: in order to build a proof of a formula, we start with that formula and decompose it according to the rules of the system; if the last formulas satisfy certain conditions, then we can conclude that a formula is a theorem. Thus, in each step of decomposition, a reasoner knows the given formulas and can analyze their possible derivations. This means that sequent calculus is a goal-oriented tool, and so it could be more easily implemented than Hilbert-style systems. In recent decades, systems that could be easily automated have become increasingly important, mainly due to growing interests in applications of logic and the rapid development of information technologies.

Gentzen sequent calculus provides—among other systems like tableaux or resolution—a good tool for automated theorem proving.

The first sequent calculus for the logic SCI was built by Michaels (see [12]); then, it has been simplified by Wasilewska in [13]) and modified by Chlebowski in [14]. Below, we present the basics of a sequent calculus for SCI, which is a version of systems from [12] and [13] adjusted to the well known sequent axiomatization of classical propositional logic.

By Γ, Δ, Σ, with indices if necessary, we will denote finite (possibly empty) sequences of SCI-formulas. If Γ and Δ are sequences $\varphi_1, \ldots, \varphi_n$ and $\psi_1, \ldots, \psi_m$, respectively, then Γ, Δ denotes a sequence $\varphi_1, \ldots, \varphi_n, \psi_1, \ldots, \psi_m$. Similarly, if φ is a formula and $\Gamma = (\varphi_1, \ldots, \varphi_n)$ is a sequence of formulas, then φ, Γ (resp. Γ, φ) denotes the sequence $\varphi, \varphi_1, \ldots, \varphi_n$ (resp. $\varphi_1, \ldots, \varphi_n, \varphi$). A sequence that contains only propositional variables (resp. identities of the form $\varphi \equiv \psi$) will be referred to as an *atomic sequence* (resp. an *identities sequence*). If φ, ψ, ϑ are SCI-formulas, then, by $\varphi[\psi/\vartheta]$, we denote any sequence consisting of all formulas, including φ, obtained from φ by replacing at least one occurrence of ψ by ϑ. Clearly, given formulas φ, ψ, ϑ, a sequence $\varphi[\psi/\vartheta]$ is finite. If Γ is a finite sequence of SCI-formulas and ψ, ϑ are SCI-formulas, then $\Gamma[\psi/\vartheta] \stackrel{df}{=} \{\varphi[\psi/\vartheta] : \varphi \in \Gamma\}$.

A *sequent* is an expression of the form $\Gamma \vdash \Delta$, where Γ (Δ) is referred to as *antecedent* (resp. *succedent*) of a sequent. Validity of a sequent $\Gamma \vdash \Delta$, for $\Gamma = (\varphi_1, \ldots, \varphi_n)$ and $\Delta = (\psi_1, \ldots, \psi_m)$ is equivalent with validity of a formula $(\varphi_1 \wedge \ldots \wedge \varphi_n) \to (\psi_1 \vee \ldots \vee \psi_m)$. Thus, if a sequence of formulas is on the left of the $\vdash$, then it is considered *conjunctively*, while, if it is on the right of the $\vdash$, the sequence of formulas is considered *disjunctively*. Sequent rules can be divided on the *left* and *right* rules, which in general correspond to valid formulas of the form $(\varphi_1 \wedge \ldots \wedge \varphi_n) \to (\psi_1 \vee \ldots \vee \psi_m)$. Sequent rules have the following general forms:

(left rule) $\quad \dfrac{\psi_1, \varphi_1, \ldots, \varphi_n, \Gamma \vdash \Delta \quad | \quad \ldots \quad | \quad \psi_m, \varphi_1, \ldots, \varphi_n, \Gamma \vdash \Delta}{\varphi_1, \ldots, \varphi_n, \Gamma \vdash \Delta},$

(right rule) $\quad \dfrac{\Gamma \vdash \Delta, \psi_1, \ldots, \psi_m, \varphi_1 \quad | \quad \ldots \quad | \quad \Gamma \vdash \Delta, \psi_1, \ldots, \psi_m, \varphi_n}{\Gamma \vdash \Delta, \psi_1, \ldots, \psi_m}.$

There are two major groups of sequent rules: logical and structural. A logical rule introduces a new formula either on the left or on the right of the $\vdash$. A structural rule operates on the structure of the sequents, ignoring the exact shape of the formulas. Some sequents are distinguished as axioms. In order to prove a sequent $\Gamma \vdash \Delta$, we write the sequent and then proceed to construct a tree in an upward direction. In each step, we follow the rules of a sequent calculus until we reach a closing sequent that is a sequent that is an axiom. If we apply a rule in which the symbol '|' occurs, then the tree splits. Each split in a tree adds a new branch. If a given branch has at its top an axiom, then it is called closed; otherwise, it is open. If all branches are closed, then a derivation of a sequent is its proof.

The sequent calculus for SCI, denoted by G_{SCI}, consists of logical rules for the classical connectives from Table 1, the rule for the identity connective depicted in Table 2, and structural rules given in Table 3. G_{SCI}-axioms are sequents of either of the following forms, for any SCI-formula φ and any finite sequences $\Gamma, \Gamma', \Delta, \Delta'$ of SCI-formulas: $\Gamma, \varphi, \Gamma' \vdash \Delta, \varphi, \Delta'$ or $\Gamma, \vdash \Delta, \varphi \equiv \varphi, \Delta'$.

Table 1. G_{SCI}-rules for classical connectives.

$(\neg L)$	$\dfrac{\Gamma \vdash \varphi, \Delta}{\Gamma, \neg \varphi \vdash \Delta}$	$(\neg R)$	$\dfrac{\Gamma, \varphi \vdash \Delta}{\Gamma \vdash \neg \varphi, \Delta},$	
$(\to L)$	$\dfrac{\Gamma \vdash \varphi, \Delta \quad | \quad \Gamma', \psi \vdash \Delta'}{\Gamma, \Gamma', \varphi \to \psi \vdash \Delta, \Delta'}$	$(\to R)$	$\dfrac{\Gamma, \varphi \vdash \psi, \Delta}{\Gamma \vdash \varphi \to \psi, \Delta},$	

where φ, ψ are any SCI-formulas
$\Gamma, \Gamma', \Delta, \Delta'$ are any finite (possibly empty) sequences of SCI-formulas

Table 2. G_{SCI}-rule for the identity connective.

$$(\equiv_{G_{SCI}}) \quad \frac{\Gamma, \Sigma, \Gamma'[\varphi/\psi] \vdash \Delta, \Delta'[\varphi/\psi] \quad | \quad \Gamma, \Gamma'[\varphi/\psi] \vdash \Delta, \Sigma, \Delta'[\varphi/\psi]}{\Gamma, \varphi \equiv \psi, \Gamma' \vdash \Delta, \Delta'},$$

where φ, ψ are any SCI-formulas and Σ is the sequence $\varphi, \psi, \varphi \equiv \psi, \psi \equiv \varphi$
Γ, Δ are atomic sequences and Γ', Δ' are identities sequences

Table 3. Structural rules of G_{SCI}-calculus.

$$(W_L) \quad \frac{\Gamma \vdash \Delta}{\Gamma, \varphi \vdash \Delta} \qquad (W_R) \quad \frac{\Gamma \vdash \Delta}{\Gamma \vdash \varphi, \Delta}$$

$$(C_L) \quad \frac{\Gamma, \varphi, \varphi \vdash \Delta}{\Gamma, \varphi \vdash \Delta} \qquad (C_R) \quad \frac{\Gamma \vdash \varphi, \varphi, \Delta}{\Gamma \vdash \varphi, \Delta}$$

$$(P_L) \quad \frac{\Gamma, \varphi, \psi, \Gamma' \vdash \Delta}{\Gamma, \psi, \varphi, \Gamma' \vdash \Delta} \qquad (P_R) \quad \frac{\Gamma \vdash \Delta, \varphi, \psi, \Delta'}{\Gamma \vdash \Delta, \psi, \varphi, \Delta'}$$

$$(\text{cut}) \quad \frac{\Gamma \vdash \varphi, \Delta \quad | \quad \varphi, \Gamma' \vdash \Delta'}{\Gamma, \Gamma' \vdash \Delta, \Delta'}$$

where φ, ψ are any SCI-formulas
$\Gamma, \Gamma', \Delta, \Delta'$ are any finite (possibly empty) sequences of SCI-formulas

An SCI-formula φ is said to be G_{SCI}-provable if and only if there is a G_{SCI}-proof for the sequent $\vdash \varphi$. As proved in [12] (cf. [13]), the system G_{SCI} is sound and complete:

Theorem 3 (Soundness and Completeness of G_{SCI}). *Let φ be an SCI-formula. Then, the following conditions are equivalent:*

1. *φ is SCI-valid;*
2. *φ is G_{SCI}-provable.*

Note that the rule ($\equiv_{G_{SCI}}$) for the identity connective is a branching rule, and it can be applied only if no other logical rule can be applied that is all formulas in sequents are either propositional variables or identities. Furthermore, observe that the rule ($\equiv_{G_{SCI}}$) corresponds to the extensionality property as stated in Proposition 1. This means that the rule ($\equiv_{G_{SCI}}$) reflects the following property of the logic SCI: if $((\varphi \equiv \psi) \wedge \vartheta) \to \chi$ is SCI-valid, then $((\varphi \equiv \psi) \wedge \vartheta') \to \chi'$ is SCI-valid, where $\varphi, \psi, \vartheta, \chi$ are any SCI-formulas and ϑ' (resp. χ') is the result of replacing some occurrences of φ in ϑ (resp. χ) by a formula ψ.

Clearly, the rule ($\equiv_{G_{SCI}}$) uses substitution, thus it may produce many formulas which are not necessary to close a tree. Chlebowski in 2018 proposed two sound and complete sequent calculi for SCI whose rules for identity do not make use of substitution. The idea of Chlebowski's systems is to translate each of the SCI-axiom schemas ($\equiv_3$)–($\equiv_4$) to sequent rules. For instance, a left rule corresponding to the axiom schema ($\equiv_4$) for $\to$ has the following form:

$$\frac{(\varphi \to \vartheta) \equiv (\psi \to \chi), \varphi \equiv \psi, \vartheta \equiv \chi, \Gamma \vdash \Delta}{\varphi \equiv \psi, \vartheta \equiv \chi, \Gamma \vdash \Delta}.$$

For details of Chlebowski's systems, we refer the reader to [14].

Figure 1 presents a closed G_{SCI}-derivation of the formula $(p_1 \equiv p_2) \to (p_1 \to p_2)$, which is an instance of the axiom schema ($\equiv_3$), while, in Figure 2, we show how to prove in G_{SCI} the formula $(p_1 \equiv p_2) \to [(p_2 \equiv p_3) \to (p_1 \equiv p_3)]$, which expresses the fact that the identity connective represents a transitive relation.

$$
\frac{\frac{\frac{\frac{\overbrace{p_1,p_1,p_2,p_1 \equiv p_2,p_2 \equiv p_1 \vdash p_2}^{\text{closed}} \mid \overbrace{p_1 \vdash p_2,p_1,p_2,p_1 \equiv p_2,p_2 \equiv p_1}^{\text{closed}}}{p_1,p_1 \equiv p_2 \vdash p_2}(\equiv_{\mathsf{G_{SCI}}})}{\frac{p_1 \equiv p_2, p_1 \vdash p_2}{p_1 \equiv p_2 \vdash p_1 \to p_2}(\to_R)}(P_L)}{\vdash (p_1 \equiv p_2) \to (p_1 \to p_2)}(\to_R)
$$

Figure 1. A $\mathsf{G_{SCI}}$-proof of a formula $(p_1 \equiv p_2) \to (p_1 \to p_2)$.

$$
\frac{\frac{\frac{\frac{\overbrace{p_2,p_3,p_1 \equiv p_3,\ldots \vdash p_1 \equiv p_3}^{\text{closed}} \mid \overbrace{p_1 \equiv p_3,\ldots \vdash p_2,p_3,p_1 \equiv p_3,\ldots}^{\text{closed}}}{p_2 \equiv p_3, p_1 \equiv p_2 \vdash p_1 \equiv p_3}(\equiv_{\mathsf{G_{SCI}}})}{\frac{p_1 \equiv p_2, p_2 \equiv p_3 \vdash p_1 \equiv p_3}{p_1 \equiv p_2 \vdash (p_2 \equiv p_3) \to (p_1 \equiv p_3)}(\to_R)}(P_L)}{\vdash (p_1 \equiv p_2) \to [(p_2 \equiv p_3) \to (p_1 \equiv p_3)]}(\to_R)
$$

Figure 2. A $\mathsf{G_{SCI}}$-proof of a formula $(p_1 \equiv p_2) \to [(p_2 \equiv p_3) \to (p_1 \equiv p_3)]$.

4. Dual Tableau System $\mathsf{DT_{SCI}}$

Dual tableau systems are based on Rasiowa–Sikorski diagrams for classical predicate logic (see [17]). They are top–down systems determined by the rules of inferences and axioms. Rules have the following form (rule) $\frac{\Phi}{\Phi_1 | \ldots | \Phi_n}$, $n \geq 1$, where $\Phi, \Phi_1, \ldots, \Phi_n$ are finite sets of formulas. The set Φ is called the *premise* of the rule. Sets $\Phi_1, \ldots, \Phi_n$ are said to be *conclusions*. Some systems allow infinitary rules with infinitely countable many conclusions. The rules are supposed to preserve the validity of the sets of formulas to which they are applied, where the validity of a finite set of formulas is understood as the validity of the disjunction of its elements. Thus, a comma in the sets of a rule (rule) can be interpreted as the meta-disjunction, while branching '|' as the meta-conjunction. The rules apply to finite sets of formulas. A rule (rule) is applicable to a finite set X, whenever $X = \Phi$, and there is $i \in \{1, \ldots, n\}$ such that $X \neq \Phi_i$. Axioms are distinguished valid sets of formulas, also referred to as *axiomatic sets*. The key notion in the methodology of dual tableau systems is a *proof tree*. A proof tree for a formula φ is a (finitely) branching tree whose root consists of the set $\{\varphi\}$ and each node of the tree, except the root, is obtained by an application of a rule to its predecessor node. A formula φ is said to be *provable*, whenever there is a proof tree for φ such that all of its branches ends with an axiom.

Dual tableau systems are *validity checkers* that are in order to prove a formula φ we build a proof tree directly for that formula. It distinguishes dual tableaux from tableau systems, which are *unsatisfiability checkers*, as in tableau systems in order to prove a formula a proof tree for its negation is constructed.

Over the years, dual tableaux have been constructed for a great variety of logics, in particular for modal, intuitionistic, relevant, many-valued, temporal, spatial, fuzzy, dynamic programming logics, among others. A very recent comprehensive survey on the foundations and applications of dual tableaux is the book [16].

The first sound and complete dual tableau for the fragment of SCI-language was presented in [15]. A dual tableau for the full SCI-language is described in [16]. The system presented in [15] (resp. [16]) was defined for SCI-language which among its classical connectives contains $\neg, \wedge$, and $\vee$ (resp. $\neg, \wedge, \vee, \to, \leftrightarrow$). In this section, we present the basics of a dual tableau from [16] adjusted to SCI-language that contains only three connectives $\neg, \to$, and $\equiv$. For the simplicity of our presentation, we will write $\varphi \not\equiv \psi$ instead of $\neg(\varphi \equiv \psi)$.

The dual tableau system for SCI, denoted by $\mathsf{DT_{SCI}}$, consists of $\mathsf{DT_{SCI}}$-axiomatic sets, decomposition rules $(\neg), (\to), (\neg \to)$ presented in Table 4, and the specific rule $(\equiv)$ depicted in Table 5.

Decomposition rules enable us to decompose formulas built by means of the classical connectives $\neg$ and $\rightarrow$, while the specific rule reflects properties of the identity connective $\equiv$. DT_{SCI}-axiomatic sets have either of the following forms, where φ is an SCI-formula and X is a finite (possible empty) set of SCI-formulas: $\{\varphi \equiv \varphi\} \cup X$ or $\{\varphi, \neg\varphi\} \cup X$.

Table 4. DT_{SCI}-decomposition rules.

$$(\rightarrow) \quad \frac{\{\varphi \rightarrow \psi\} \cup X}{\{\neg\varphi, \psi\} \cup X} \qquad (\neg\rightarrow) \quad \frac{\{\neg(\varphi \rightarrow \psi)\} \cup X}{\{\varphi\} \cup X \mid \{\neg\psi\} \cup X}$$

$$(\neg) \quad \frac{\{\neg\neg\varphi\} \cup X}{\{\varphi\} \cup X}$$

where φ and ψ are any SCI-formulas,
and X is a finite (possibly empty) set of SCI-formulas.

Table 5. DT_{SCI}-specific rule for the identity connective $\equiv$.

$$(\equiv) \quad \frac{\{\varphi(\psi)\} \cup X}{\{\psi \equiv \vartheta, \varphi(\psi)\} \cup X \mid \{\varphi(\psi/\vartheta), \varphi(\psi)\} \cup X}$$

where φ and ϑ are any SCI-formulas, ψ is any subformula of φ,
$\varphi(\vartheta)$ is obtained from $\varphi(\psi)$ by replacing some occurrences of ψ with ϑ
and X is a finite (possibly empty) set of SCI-formulas.

A finite set of SCI-formulas $\{\varphi_1, \ldots, \varphi_n\}$ is said to be SCI-*valid* whenever the disjunction of its elements is SCI-valid that is for every SCI-model $\mathcal{M}$ and for every valuation v in $\mathcal{M}$ there exists $i \in \{1, \ldots, n\}$ such that $\mathcal{M}, v \models \varphi_i$. A DT_{SCI}-*proof tree* for an SCI-formula φ is a finitely branching tree whose nodes are sets of formulas satisfying the following conditions:

- the formula φ is at the root of this tree,
- each node except the root is obtained by an application of a DT_{SCI}-rule to its predecessor node,
- a node does not have successors whenever it is a DT_{SCI}-axiomatic set or none of the rules applies to its set of formulas.

A branch of a DT_{SCI}-proof tree is said to be *closed* whenever it contains a node with a DT_{SCI}-axiomatic set of formulas. A DT_{SCI}-proof tree is *closed* whenever all of its branches are closed. A formula φ is DT_{SCI}-*provable* whenever there is a closed DT_{SCI}-proof tree for φ, which is then referred to as its DT_{SCI}-proof.

Figure 3 presents a DT_{SCI}-proof for the formula $(p_1 \equiv p_2) \rightarrow (p_1 \rightarrow p_2)$, which is an instance of the axiom schema $(\equiv_3)$. Figure 4 presents a closed DT_{SCI}-proof tree for the formula $(p_1 \equiv p_2) \rightarrow [(p_2 \equiv p_3) \rightarrow (p_1 \equiv p_3)]$. In each node of the proof tree, we underline the formula to which a rule has been applied during the construction of the proof tree, and we indicate only those formulas in a node which are essential for this construction.

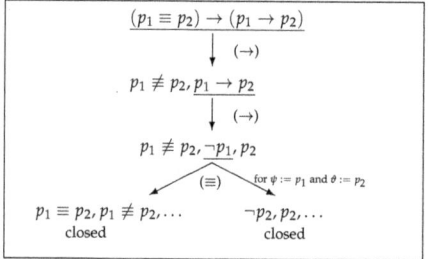

Figure 3. A DT_{SCI}-proof for the formula $(p_1 \equiv p_2) \rightarrow (p_1 \rightarrow p_2)$.

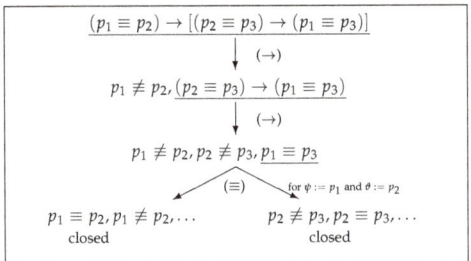

Figure 4. A DT_{SCI}-proof for $(p_1 \equiv p_2) \rightarrow [(p_2 \equiv p_3) \rightarrow (p_1 \equiv p_3)]$.

The proof of soundness and completeness of DT_{SCI}-system is presented in [16] (cf. [15]):

Theorem 4 (Soundness and Completeness of DT_{SCI}). *Let φ be an SCI-formula. Then, the following conditions are equivalent:*

1. *φ is SCI-valid;*
2. *φ is DT_{SCI}-provable.*

As in G_{SCI}-system, the rule for the identity connective of the system DT_{SCI} branches a tree and involves use of substitution. In the next section, we present a new dual tableau that has no such disadvantages.

5. A Substitution-Free Dual Tableau for SCI

In this section, we present the system DT^*_{SCI}, which is a modification of DT_{SCI}-system by replacing the rule ($\equiv$) with several rules for the identity connective that do not involve substitutions. The system DT^*_{SCI} consists of DT_{SCI}-decomposition rules presented in Table 4 (see Section 4) and the specific rules presented in Table 6. The specific rule (ref) (resp. (sym), (tran)) expresses reflexivity (resp. symmetry, transitivity) of a relation which in SCI-models corresponds to the identity connective. Thus, specific rules (ref), (sym), and (tran) reflect the fact that the relation corresponding to the identity connective is an equivalence relation. The specific rule ($\equiv_\neg$) (resp. ($\equiv_\rightarrow$), ($\equiv_\equiv$)) expresses the instance of SCI-axiom ($\equiv_2$) (resp. ($\equiv_4$) for $\rightarrow$ and ($\equiv_4$) for $\equiv$) presented in Section 2. Therefore, specific rules ($\equiv_\neg$), ($\equiv_\rightarrow$), and ($\equiv_\equiv$) correspond to the extensionality property for connectives $\neg$, $\rightarrow$, and $\equiv$ (see Proposition 1). As the identity connective can be characterized as an operation that satisfies the extensionality property and represents an equivalence relation, we will show that specific rules (ref), (sym), (tran), ($\equiv_\neg$), ($\equiv_\rightarrow$), and ($\equiv_\equiv$) are sufficient to prove completeness of the system DT^*_{SCI}.

The axiomatic sets of DT^*_{SCI} have either of the following forms, where φ, ψ are any SCI-formulas and X is a finite (possible empty) set of SCI-formulas:

$(Ax^1_{DT^*_{SCI}})$ $\{\varphi \equiv \varphi\} \cup X$, $\qquad$ $(Ax^2_{DT^*_{SCI}})$ $\{\varphi, \neg\varphi\} \cup X$,
$(Ax^3_{DT^*_{SCI}})$ $\{\varphi, \neg\psi, \varphi \not\equiv \psi\} \cup X$, $\qquad$ $(Ax^4_{DT^*_{SCI}})$ $\{\neg\varphi, \psi, \varphi \not\equiv \psi\} \cup X$.

The notions of an SCI-valid set of formulas, a DT^*_{SCI}-proof tree, a closed branch of such a tree, a closed DT^*_{SCI}-proof tree, and DT^*_{SCI}-provability are defined in a similar way as for DT_{SCI}-system in Section 4. Observe that none of the specific rules split a tree or use substitutions.

Now, we will prove the soundness and completeness of DT^*_{SCI}-system.

Table 6. DT^*_{SCI}-specific rules for the identity connective.

(ref)	$\dfrac{X}{\{\varphi \not\equiv \varphi\} \cup X}$	(sym)	$\dfrac{\{\varphi \not\equiv \psi\} \cup X}{\{\psi \not\equiv \varphi, \varphi \not\equiv \psi\} \cup X}$
(tran)	$\dfrac{\{\varphi \not\equiv \psi, \psi \not\equiv \vartheta\} \cup X}{\{\varphi \not\equiv \vartheta, \varphi \not\equiv \psi, \psi \not\equiv \vartheta\} \cup X}$		
$(\equiv_\neg)$	$\dfrac{\{\varphi \not\equiv \psi\} \cup X}{\{\neg\varphi \not\equiv \neg\psi, \varphi \not\equiv \psi\} \cup X}$		
$(\equiv_\rightarrow)$	$\dfrac{\{\varphi \not\equiv \psi, \vartheta \not\equiv \chi\} \cup X}{\{(\varphi \rightarrow \vartheta) \not\equiv (\psi \rightarrow \chi), \varphi \not\equiv \psi, \vartheta \not\equiv \chi\} \cup X}$		
$(\equiv_\equiv)$	$\dfrac{\{\varphi \not\equiv \psi, \vartheta \not\equiv \chi\} \cup X}{\{(\varphi \equiv \vartheta) \not\equiv (\psi \equiv \chi), \varphi \not\equiv \psi, \vartheta \not\equiv \chi\} \cup X}$		

where $\varphi, \psi, \vartheta, \chi$ are any SCI-formulas,
and X is a finite (possibly empty) set of SCI-formulas.

Proposition 3 (Correctness of DT^*_{SCI}-rules). *For every DT^*_{SCI}-rule (rule), the premise of rule is SCI-valid if and only if all of its conclusions are SCI-valid.*

Proof. The proof for decomposition rules is straightforward, so we will prove the proposition for the specific rules of DT^*_{SCI}: (ref), (sym), (tran), $(\equiv_\neg)$, $(\equiv_\rightarrow)$, $(\equiv_\equiv)$. Let $\varphi, \psi, \vartheta, \chi$ be any SCI-formulas and let X be a finite (possibly empty) set of SCI-formulas. Observe that, in each of the specific DT^*_{SCI}-rules, the premise of a rule is a subset of its conclusion. Thus, if the premise is SCI-valid, then so is its conclusion. Therefore, it suffices to show that SCI-validity of the conclusion of a rule implies SCI-validity of its premise.

Correctness of the rule (ref)

Assume $\{\varphi \not\equiv \varphi\} \cup X$ is SCI-valid and suppose X is not SCI-valid. Then, there are an SCI-model $\mathcal{M}$ and a valuation v in $\mathcal{M}$ such that, for every $\xi \in X$, $\mathcal{M}, v \not\models \xi$. Thus, since $\{\varphi \not\equiv \varphi\} \cup X$ is SCI-valid and $\mathcal{M}, v \not\models \xi$ for every $\xi \in X$, we obtain $\mathcal{M}, v \models \varphi \not\equiv \varphi$, so $\mathcal{M}, v \not\models \varphi \equiv \varphi$. However, all SCI-formulas, SCI-models $\mathcal{M}$, and valuations satisfy $\mathcal{M}, v \models \varphi \equiv \varphi$, a contradiction.

Correctness of the rule (sym)

Assume that $\{\psi \not\equiv \varphi, \varphi \not\equiv \psi\} \cup X$ is SCI-valid and suppose $\{\varphi \not\equiv \psi\} \cup X$ is not SCI-valid. Then, there exist an SCI-model $\mathcal{M}$ and a valuation v in $\mathcal{M}$ such that $\mathcal{M}, v \not\models \varphi \not\equiv \psi$ and, for every $\xi \in X$, $\mathcal{M}, v \not\models \xi$. Thus, $\mathcal{M}, v \models \varphi \equiv \psi$, which means that $v(\varphi) = v(\psi)$. Furthermore, by the assumption, the model $\mathcal{M}$ and the valuation v must satisfy the formula $\psi \not\equiv \varphi$, so $v(\psi) \neq v(\varphi)$, a contradiction.

Correctness of the rule (tran)

Assume $\{\varphi \not\equiv \vartheta, \varphi \not\equiv \psi, \psi \not\equiv \vartheta\} \cup X$ is SCI-valid and suppose $\{\varphi \not\equiv \psi, \psi \not\equiv \vartheta\} \cup X$ is not SCI-valid. Then, there exists an SCI-model $\mathcal{M}$ and a valuation v in $\mathcal{M}$ that do not satisfy any formula from the set $\{\varphi \not\equiv \psi, \psi \not\equiv \vartheta\} \cup X$, so $\mathcal{M}, v \models \varphi \equiv \psi$ and $\mathcal{M}, v \models \psi \equiv \vartheta$. Hence, we obtain $v(\varphi) = v(\psi) = v(\vartheta)$. However, by the assumption, it must hold that $\mathcal{M}, v \models \varphi \not\equiv \vartheta$, which imply $v(\varphi) \neq v(\vartheta)$, a contradiction.

Correctness of the rule $(\equiv_\neg)$

Assume $\{\neg\varphi \not\equiv \neg\psi, \varphi \not\equiv \psi\} \cup X$ is SCI-valid and suppose $\{\varphi \not\equiv \psi\} \cup X$ is not SCI-valid. Then, there exist an SCI-model $\mathcal{M} = (U, \sim, \Rightarrow, \circ, D)$ and a valuation v such that $\mathcal{M}, v \models \varphi \equiv \psi$ that is $v(\varphi) = v(\psi)$. On the other hand, by the assumption, $\mathcal{M}, v \models \neg\varphi \not\equiv \neg\psi$, which imply $v(\neg\varphi) \neq v(\neg\psi)$. However, if $v(\varphi) = v(\psi)$, then clearly $v(\neg\varphi) = \sim v(\varphi) = \sim v(\psi) = v(\neg\psi)$, which contradicts $v(\neg\varphi) \neq v(\neg\psi)$.

Correctness of the rule $(\equiv_\rightarrow)$

Assume $\{(\varphi \rightarrow \vartheta) \not\equiv (\psi \rightarrow \chi), \varphi \not\equiv \psi, \vartheta \not\equiv \chi\} \cup X$ is SCI-valid and suppose $\{\varphi \not\equiv \psi, \vartheta \not\equiv \chi\} \cup X$ is not SCI-valid. Then, there exist an SCI-model $\mathcal{M} = (U, \sim, \Rightarrow, \circ, D)$ and a valuation v such that $\mathcal{M}, v \models \varphi \equiv \psi$ and $\mathcal{M}, v \models \vartheta \equiv \chi$, that is, $v(\varphi) = v(\psi)$ and $v(\vartheta) = v(\chi)$. Hence, due to the definition of an SCI-model, we obtain: $v(\varphi \rightarrow \vartheta) = v(\varphi) \Rightarrow v(\vartheta) = v(\psi) \Rightarrow v(\chi) = v(\psi \rightarrow \chi)$. Therefore, $\mathcal{M}, v \models (\varphi \rightarrow \vartheta) \equiv (\psi \rightarrow \chi)$. However, by the assumption, $\mathcal{M}, v \models (\varphi \rightarrow \vartheta) \not\equiv (\psi \rightarrow \chi)$, a contradiction.

Correctness of the rule $(\equiv_\equiv)$

Assume $\{(\varphi \equiv \vartheta) \not\equiv (\psi \equiv \chi), \varphi \not\equiv \psi, \vartheta \not\equiv \chi\} \cup X$ is SCI-valid and suppose $\{\varphi \not\equiv \psi, \vartheta \not\equiv \chi\} \cup X$ is not SCI-valid. Then, there exist an SCI-model $\mathcal{M} = (U, \sim, \Rightarrow, \circ, D)$ and a valuation v such that $\mathcal{M}, v \models \varphi \equiv \psi$ and $\mathcal{M}, v \models \vartheta \equiv \chi$ that is $v(\varphi) = v(\psi)$ and $v(\vartheta) = v(\chi)$. Thus, by the definition of an SCI-model, we obtain that $v(\varphi \equiv \vartheta) = v(\varphi) \circ v(\vartheta) = v(\psi) \circ v(\chi) = v(\psi \equiv \chi)$, and so $\mathcal{M}, v \models (\varphi \equiv \vartheta) \equiv (\psi \equiv \chi)$. By the assumption, $\mathcal{M}, v \models (\varphi \equiv \vartheta) \not\equiv (\psi \equiv \chi)$, a contradiction. □

Proposition 4 (Validity of $\mathsf{DT}^*_{\mathsf{SCI}}$-axiomatic sets). *All the $\mathsf{DT}^*_{\mathsf{SCI}}$-axiomatic sets are SCI-valid.*

Proof. Let φ, ψ be any SCI-formulas and let X be any finite (possibly empty) set of SCI-formulas. The proof of validity of sets $(\mathsf{Ax}^1_{\mathsf{DT}^*_{\mathsf{SCI}}})$ and $(\mathsf{Ax}^2_{\mathsf{DT}^*_{\mathsf{SCI}}})$ is obvious. By way of example, we will prove validity of sets $(\mathsf{Ax}^3_{\mathsf{DT}^*_{\mathsf{SCI}}})$, since the proof for $(\mathsf{Ax}^4_{\mathsf{DT}^*_{\mathsf{SCI}}})$ is similar. Suppose a set $\{\varphi, \neg\psi, \varphi \not\equiv \psi\} \cup X$ is not SCI-valid. Then, there exist an SCI-model $\mathcal{M} = (U, \sim, \Rightarrow, \circ, D)$ and a valuation v such that $\mathcal{M}, v \not\models \varphi$, $\mathcal{M}, v \not\models \neg\psi$, and $\mathcal{M}, v \not\models \varphi \equiv \psi$. Thus, by the definition of an SCI-model, $v(\varphi) \notin D$, $\sim v(\psi) \notin D$, and $\sim(v(\varphi) \circ v(\psi)) \notin D$. However, this means that $v(\varphi) \notin D$, $v(\psi) \in D$, and $v(\varphi) = v(\psi)$, which is not possible since the latter implies $v(\varphi) \in D$ iff $v(\psi) \in D$. □

Due to Propositions 3 and 4, the soundness of $\mathsf{DT}^*_{\mathsf{SCI}}$ can be easily proved:

Theorem 5 (Soundness of $\mathsf{DT}^*_{\mathsf{SCI}}$). *If an SCI-formula is $\mathsf{DT}^*_{\mathsf{SCI}}$-provable, then it is SCI-valid.*

Proof. Let φ be a $\mathsf{DT}^*_{\mathsf{SCI}}$-provable formula. Then, there exists a closed $\mathsf{DT}^*_{\mathsf{SCI}}$-proof tree for φ that is all of its branches end with $\mathsf{DT}^*_{\mathsf{SCI}}$-axiomatic sets. By Proposition 4, all $\mathsf{DT}^*_{\mathsf{SCI}}$-axiomatic sets are SCI-valid, so the leaves in the closed $\mathsf{DT}^*_{\mathsf{SCI}}$-proof tree for φ are SCI-valid sets of formulas. Moreover, by Proposition 3, if conclusions of a rule are SCI-valid, then so is its premise. Therefore, going from the leaves to the root of the tree, in each step, we obtain nodes that are SCI-valid. Hence, the root $\{\varphi\}$ is SCI-valid, and so we conclude that the formula φ is SCI-valid. □

In order to prove completeness of $\mathsf{DT}^*_{\mathsf{SCI}}$, we will construct an SCI-model and a valuation that do not satisfy a formula, which is not $\mathsf{DT}^*_{\mathsf{SCI}}$-provable. We call a branch b of a $\mathsf{DT}^*_{\mathsf{SCI}}$-proof tree $\mathsf{DT}^*_{\mathsf{SCI}}$-*complete* whenever it satisfies the following completion conditions for all SCI-formulas $\varphi, \psi, \vartheta, \chi$:

Cpl($\neg$) If $\neg\neg\varphi \in b$, then $\varphi \in b$.

Cpl($\rightarrow$) If $\varphi \rightarrow \psi \in b$, then both $\neg\varphi \in b$ and $\psi \in b$.

Cpl($\neg\rightarrow$) If $\neg(\varphi \rightarrow \psi) \in b$, then either $\varphi \in b$ or $\neg\psi \in b$.

Cpl(ref) For every SCI-formula φ, $\varphi \not\equiv \varphi \in b$.

Cpl(sym) If $\varphi \not\equiv \psi \in b$, then $\psi \not\equiv \varphi \in b$.

Cpl(tran) If $\varphi \not\equiv \psi \in b$ and $\psi \not\equiv \vartheta \in b$, then $\varphi \not\equiv \vartheta \in b$.

Cpl($\equiv_\neg$) If $\varphi \not\equiv \psi \in b$, then $\neg\varphi \not\equiv \neg\psi \in b$.

Cpl($\equiv_\rightarrow$) If $\varphi \not\equiv \psi \in b$ and $\vartheta \not\equiv \chi \in b$, then $(\varphi \rightarrow \vartheta) \not\equiv (\psi \rightarrow \chi) \in b$.

Cpl($\equiv_\equiv$) If $\varphi \not\equiv \psi \in b$ and $\vartheta \not\equiv \chi \in b$, then $(\varphi \equiv \vartheta) \not\equiv (\psi \equiv \chi) \in b$.

A DT^*_{SCI}-proof tree is said to be DT^*_{SCI}-*complete* whenever each of its branches is either closed or DT^*_{SCI}-complete. The rules of DT^*_{SCI}-system guarantee that, for every SCI-formula, there is a complete DT^*_{SCI}-proof tree. A non-closed branch that is DT^*_{SCI}-complete will be referred to as *open*. The following property can be easily proved:

Proposition 5 (Closed Branch Property). *For every complete branch b of a DT^*_{SCI}-proof tree and for every SCI-formula φ, if both $\varphi \in b$ and $\neg\varphi \in b$, then the branch b is closed.*

Proof. Let b be a complete branch of DT^*_{SCI}-proof tree and let φ be an SCI-formula such that both $\varphi \in b$ and $\neg\varphi \in b$. Suppose b is not closed. We will prove the proposition by the induction on complexity of formulas. First, observe that all the DT^*_{SCI}-rules preserves propositional variables, negations of propositional variables, identities and negations of identities, that is, if a node contains φ or $\neg\varphi$, for $\varphi \in \mathbb{V} \cup \{\psi \equiv \vartheta; \psi, \vartheta \in \mathbb{FOR}\}$, then all of its successors contain these formulas. Thus, if $\varphi \in b$ and $\neg\varphi \in b$, then there exists a node t in branch b such that both $\varphi \in t$ and $\neg\varphi \in t$, which means that a node t is DT^*_{SCI}-axiomatic, and branch b is closed. Hence, the proposition holds for formulas from the set $\mathbb{V} \cup \{\psi \equiv \vartheta; \psi, \vartheta \in \mathbb{FOR}\}$. Assume the proposition holds for φ and ψ. We will show that it holds for formulas $\neg\varphi$ and $\varphi \to \psi$. Assume $\neg\varphi \in b$ and $\neg\neg\varphi \in b$. Then, as b is a non-closed complete branch and $\neg\neg\varphi \in b$, by the completion condition Cpl($\neg$), $\varphi \in b$. Thus, we have $\neg\varphi \in b$ and $\varphi \in b$, so by the inductive hypothesis, b is closed. Now, let $\varphi \to \psi \in b$ and $\neg(\varphi \to \psi) \in b$. Since b is a non-closed complete branch and $\varphi \to \psi \in b$, by the completion condition Cpl($\to$), we obtain that both $\neg\varphi \in b$ and $\psi \in b$. Similarly, by the completion condition Cpl($\neg \to$), we have that either $\varphi \in b$ or $\neg\psi \in b$. Therefore, either both $\varphi \in b$ and $\neg\varphi \in b$ or both $\psi \in b$ and $\neg\psi \in b$. Hence, by the inductive hypothesis, the branch b must be closed, which ends the proof. □

Let b be an open branch of a DT^*_{SCI}-proof tree and let $R_\circ$ be defined on the set of all SCI-formulas as follows:

$$(\varphi, \psi) \in R_\circ \stackrel{\mathrm{df}}{\iff} (\varphi \not\equiv \psi) \in b.$$

Proposition 6. *For every open branch b of a DT^*_{SCI}-proof tree, $R_\circ$ is an equivalence relation on the set of all SCI-formulas.*

Proof. Let b be an open branch of a DT^*_{SCI}-proof tree and let φ be an SCI-formula. Then, by the completion condition Cpl(ref), $\varphi \not\equiv \varphi$ belongs to the branch b, and so $(\varphi, \varphi) \in R_\circ$, that is, $R_\circ$ is reflexive. Assume φ and ψ are SCI-formulas such that $(\varphi, \psi) \in R_\circ$. Then, $\varphi \not\equiv \psi \in b$, and by the completion condition Cpl(sym), $\psi \not\equiv \varphi \in b$. Thus, $(\psi, \varphi) \in R_\circ$, which means that $R_\circ$ is symmetric. Now, assume that $(\varphi, \psi) \in R_\circ$ and $(\psi, \vartheta) \in R_\circ$, that is, both formulas $\varphi \not\equiv \psi$ and $\psi \not\equiv \vartheta$ are in b. Therefore, by the completion condition Cpl(tran), $\varphi \not\equiv \vartheta \in b$, so $(\varphi, \vartheta) \in R_\circ$. Thus, the relation $R_\circ$ is transitive. Hence, we have proved that $R_\circ$ is an equivalence relation. □

Proposition 7. *For every open branch b of a DT^*_{SCI}-proof tree, the relation $R_\circ$ is compatible with all the connectives of SCI.*

Proof. Let b be an open branch of a DT^*_{SCI}-proof tree and let $\varphi, \psi, \vartheta, \chi$ be SCI-formulas. Assume $(\varphi, \psi) \in R_\circ$, that is, $\varphi \not\equiv \psi \in b$. Then, due to the completion condition Cpl($\equiv_\neg$), $\neg\varphi \not\equiv \neg\psi \in b$, so $(\neg\varphi, \neg\psi) \in R_\circ$. Thus, $R_\circ$ is compatible with $\neg$. Now, let $(\varphi, \psi) \in R_\circ$ and $(\vartheta, \chi) \in R_\circ$, that is, $\varphi \not\equiv \psi \in b$ and $\vartheta \not\equiv \chi \in b$. Then, by the completion condition Cpl($\equiv_\to$), we obtain $(\varphi \to \vartheta) \not\equiv (\psi \to \chi) \in b$, that is, $((\varphi \to \vartheta), (\psi \to \chi)) \in R_\circ$, so $R_\circ$ is compatible with $\to$. Finally, assume that $(\varphi, \psi) \in R_\circ$ and $(\vartheta, \chi) \in R_\circ$, that is, $\varphi \not\equiv \psi \in b$ and $\vartheta \not\equiv \chi \in b$. Then, by the completion condition Cpl($\equiv_\equiv$), we have $(\varphi \equiv \vartheta) \not\equiv (\psi \equiv \chi) \in b$, that is, $((\varphi \equiv \vartheta), (\psi \equiv \chi)) \in R_\circ$, so $R_\circ$ is compatible with $\equiv$, which ends the proof. □

Let $p \in \mathbb{V}$ and let φ, ψ be SCI-formulas. We define the *depth* of an SCI-formula as follows:

$$d(p) = d(\varphi \equiv \psi) = 0 \qquad d(\neg\varphi) = d(\varphi) + 1 \qquad d(\varphi \to \psi) = \max(d(\varphi), d(\psi)) + 1.$$

By $\mathbb{FOR}^n$, we denote the set of all SCI-formulas of depth n. Given an SCI-formula φ, by $[\varphi]_{R_\circ}$ we denote the equivalence class of $R_\circ$ determined by φ. Let b be an open branch of a $\mathsf{DT}^*_{\mathsf{SCI}}$-proof tree and let $\mathcal{M}^b = (U^b, \sim^b, \Rightarrow^b, \circ^b, D^b)$ be the *branch structure* defined as follows:

$U^b = \{[\varphi]_{R_\circ} : \varphi \in \mathbb{FOR}\}$,
$D^b = \{[\varphi]_{R_\circ} : \varphi \in \bigcup_{n \in \mathbb{N}} D_n\}$, where

$$D_0 = \{\varphi \in \mathbb{FOR}^0 : \neg\varphi \in b\},$$
$$D_{n+1} = D^1_{n+1} \cup D^2_{n+1}, \text{ for}$$
$$D^1_{n+1} = \{\neg\varphi \in \mathbb{FOR}^{n+1} : \varphi \notin D_n\}$$
$$D^2_{n+1} = \{\varphi \to \psi \in \mathbb{FOR}^{n+1} : \varphi \notin \bigcup_{k \leq n} D_k \text{ or } \psi \in \bigcup_{k \leq n} D_k)\},$$

operations $\sim^b, \Rightarrow^b, \circ^b$ are defined as:

$$\sim^b [\varphi]_{R_\circ} \stackrel{\mathrm{df}}{=} [\neg\varphi]_{R_\circ} \quad [\varphi]_{R_\circ} \Rightarrow^b [\psi]_{R_\circ} \stackrel{\mathrm{df}}{=} [\varphi \to \psi]_{R_\circ} \quad [\varphi]_{R_\circ} \circ^b [\psi]_{R_\circ} \stackrel{\mathrm{df}}{=} [\varphi \equiv \psi]_{R_\circ}.$$

Proposition 8 (Branch Model Property). *For every open branch b of a $\mathsf{DT}^*_{\mathsf{SCI}}$-proof tree, the branch structure $\mathcal{M}^b$ is an SCI-model.*

Proof. Let b be an open branch of a $\mathsf{DT}^*_{\mathsf{SCI}}$-proof tree. Clearly, U^b is not empty. Observe also that, for every formula $\varphi \in \mathbb{FOR}^n$, it holds that $\varphi \in \bigcup_{n \in \mathbb{N}} D_n$ iff $\varphi \in D_n$. Hence, for every formula $\varphi \in \mathbb{FOR}^n$, $[\varphi]_{R_\circ} \in D^b$ iff $\varphi \in D_n$. Moreover, by the completion condition Cpl(ref), for every SCI-formula φ, $\varphi \equiv \varphi \in b$, which means that $\varphi \equiv \varphi \in D_0$, and so $[\varphi \equiv \varphi]_{R_\circ} \in D^b$. Thus, $D^b \neq \emptyset$. Now, as $\varphi \equiv \varphi \in D_0$ and $\varphi \not\equiv \varphi \in \mathbb{FOR}^1$, by the definition of D_n, $\varphi \not\equiv \varphi \notin D_1$, and so $[\varphi \not\equiv \varphi]_{R_\circ} \notin D^b$. Thus, $U^b \setminus D^b \neq \emptyset$.

Due to Proposition 7, operations $\sim^b, \Rightarrow^b$, and $\circ^b$ are well defined. Indeed, assume $[\varphi]_{R_\circ} = [\psi]_{R_\circ}$, that is, $(\varphi, \psi) \in R_\circ$. Then, by Proposition 7, $(\neg\varphi, \neg\psi) \in R_\circ$, and so $[\neg\varphi]_{R_\circ} = [\neg\psi]_{R_\circ}$. Thus, since $\sim^b [\varphi]_{R_\circ} = [\neg\varphi]_{R_\circ}$ and $\sim^b [\psi]_{R_\circ} = [\neg\psi]_{R_\circ}$, we get $\sim^b [\varphi]_{R_\circ} = \sim^b [\psi]_{R_\circ}$. Therefore, if $[\varphi]_{R_\circ} = [\psi]_{R_\circ}$, then $\sim^b [\varphi]_{R_\circ} = \sim^b [\psi]_{R_\circ}$. Now, let $\# \in \{\to, \equiv\}$ and let $\#^b$ be defined as: $\#^b = \Rightarrow^b$, if $\# = \to$; and $\#^b = \circ^b$ otherwise. Assume $[\varphi]_{R_\circ} = [\psi]_{R_\circ}$ and $[\vartheta]_{R_\circ} = [\chi]_{R_\circ}$, that is, $(\varphi, \psi) \in R_\circ$ and $(\vartheta, \chi) \in R_\circ$. By Proposition 7, we obtain that $((\varphi\#\vartheta), (\psi\#\chi)) \in R_\circ$, and so $[\varphi\#\vartheta]_{R_\circ} = [\psi\#\chi]_{R_\circ}$. Therefore, we have:

$$[\varphi]_{R_\circ} \#^b [\vartheta]_{R_\circ} = [\varphi\#\vartheta]_{R_\circ} = [\psi\#\chi]_{R_\circ} = [\psi]_{R_\circ} \#^b [\chi]_{R_\circ}.$$

Hence, operations $\sim^b, \Rightarrow^b$, and $\circ^b$ are well defined. Now, we will prove that they satisfy semantic conditions with respect to D^b. Note that D^b satisfy the following properties for every SCI-formula φ and for all $n, k \in \mathbb{N}$:

(*)　If $d(\varphi) = n$, then $[\varphi]_{R_\circ} \in D^b$ iff $\varphi \in D_n$.
(**)　If $\varphi \in D_n$, then $d(\varphi) = n$.
(***)　If $d(\varphi) = n$ and $k \neq n$, then $\varphi \notin D_k$.

Let $[\varphi]_{R_\circ} \in U^b$ be such that $d(\varphi) = n$, for some $n \in \mathbb{N}$. Assume $\sim^b [\varphi]_{R_\circ} \in D^b$, which by the definition of the operation $\sim^b$ means that $[\neg\varphi]_{R_\circ} \in D^b$. Thus, since $d(\neg\varphi) = n+1$, by (*), we have $\neg\varphi \in D_{n+1}$. Then, by the definition of D^b, it holds that $\varphi \notin D_n$, and thus due to (*) we obtain that $[\varphi]_{R_\circ} \notin D^b$. Now, assume that $[\varphi]_{R_\circ} \notin D^b$, that is, by (*), we obtain $\varphi \notin D_n$. Thus, by the definition of D^b, we get $\neg\varphi \in D_{n+1}$, which due to (*) means that $[\neg\varphi]_{R_\circ} \in D^b$. Hence, by the definition of the operation $\sim^b$, we have $\sim^b [\varphi]_{R_\circ} \in D^b$. Therefore, we have proved that $\sim^b [\varphi]_{R_\circ} \in D^b$ iff $[\varphi]_{R_\circ} \notin D^b$.

Let $[\varphi]_{R_\circ}, [\psi]_{R_\circ} \in U^b$. Assume $[\varphi]_{R_\circ} \Rightarrow^b [\psi]_{R_\circ} \in D^b$. Then, by the definition of $\Rightarrow^b$, $[\varphi \to \psi]_{R_\circ} \in D^b$. By the definition of D^b, there exists $n \in \mathbb{N}$ such that $\varphi \to \psi \in D_n$, which, by (**), implies $d(\varphi \to \psi) = n$, and clearly $n \geq 1$. Since $\varphi \to \psi \in D_n$, by the definition of D^b, we obtain that either $\varphi \notin \bigcup_{k<n} D_k$ or $\psi \in \bigcup_{k<n} D_k$. Clearly, $d(\varphi) < n$, so, if $\varphi \notin \bigcup_{k<n} D_k$, then, due to (*), we get $[\varphi]_{R_\circ} \notin D^b$.

Moreover, $d(\psi) < n$, so, if $\psi \in \bigcup_{k<n} D_k$, then, by (*), it holds that $[\psi]_{R_\circ} \in D^b$. Hence, we have proved that, if $[\varphi]_{R_\circ} \Rightarrow^b [\psi]_{R_\circ} \in D^b$, then either $[\varphi]_{R_\circ} \notin D^b$ or $[\psi]_{R_\circ} \in D^b$. Now, let us assume that $d(\varphi) = i$, $d(\psi) = j$, for some $i, j \in \mathbb{N}$, and either $[\varphi]_{R_\circ} \notin D^b$ or $[\psi]_{R_\circ} \in D^b$. Thus, by (**), $\varphi \notin D_i$ or $\psi \in D_j$. Let $n = \max(i, j) + 1$. If $\varphi \notin D_i$, then, by (***), it can be easily proved that $\varphi \notin \bigcup_{k<n} D_k$. If $\psi \in D_j$, then, by (*), $\psi \in \bigcup_{k<n} D_k$. Therefore, either $\varphi \notin \bigcup_{k<n} D_k$ or $\psi \in \bigcup_{k<n} D_k$. Then, by the definition of D^b, it follows that $\varphi \to \psi \in D_n$, and, by (*), we have $[\varphi \to \psi]_{R_\circ} \in D^b$. Thus, by the definition of the operation $\Rightarrow^b$, we obtain that $[\varphi]_{R_\circ} \Rightarrow^b [\psi]_{R_\circ} \in D^b$. Hence, we have proved that $[\varphi]_{R_\circ} \Rightarrow^b [\psi]_{R_\circ} \in D^b$ iff either $[\varphi]_{R_\circ} \notin D^b$ or $[\psi]_{R_\circ} \in D^b$.

Now, let $[\varphi]_{R_\circ}, [\psi]_{R_\circ} \in U^b$. Clearly, $d(\varphi \equiv \psi) = 0$. Then, the following can be easily shown: $[\varphi]_{R_\circ} \circ^b [\psi]_{R_\circ} \in D^b$ iff $[\varphi \equiv \psi]_{R_\circ} \in D^b$ iff $\varphi \equiv \psi \in D_0$ iff $\varphi \not\equiv \psi \in b$ iff $(\varphi, \psi) \in R_\circ$ iff $[\varphi]_{R_\circ} = [\psi]_{R_\circ}$.

Hence, we have shown that $[\varphi]_{R_\circ} \circ^b [\psi]_{R_\circ} \in D^b$ iff $[\varphi]_{R_\circ} = [\psi]_{R_\circ}$. Therefore, we have proved that the branch structure $\mathcal{M}^b$ is an SCI-model. □

Let $\mathcal{M}^b = (U^b, \sim^b, \Rightarrow^b, \circ^b, D^b)$ be the branch structure for an open branch b of a $\mathsf{DT}^*_{\mathsf{SCI}}$-proof tree. Let $v^b : \mathbb{FOR} \to U^b$ be a function such that $v^b(\varphi) = [\varphi]_{R_\circ}$, for all $\varphi \in \mathbb{FOR}$. Due to the definition of $\mathcal{M}^b$, the following can be easily proved:

Proposition 9. *Let b be an open branch of a $\mathsf{DT}^*_{\mathsf{SCI}}$-proof tree and let $\mathcal{M}^b = (U^b, \sim^b, \Rightarrow^b, \circ^b, D^b)$ be the branch structure. Then, the function $v^b : \mathbb{FOR} \to U^b$ such that $v^b(\varphi) = [\varphi]_{R_\circ}$, for all $\varphi \in \mathbb{FOR}$, is an SCI-valuation in $\mathcal{M}^b$, that is, for all SCI-formulas φ and ψ, the following hold:*

$$v^b(\neg \varphi) = \sim^b [\varphi]_{R_\circ} \quad v^b(\varphi \to \psi) = [\varphi]_{R_\circ} \Rightarrow^b [\psi]_{R_\circ} \quad v^b(\varphi \equiv \psi) = [\varphi]_{R_\circ} \circ^b [\psi]_{R_\circ}.$$

The valuation v^b will be referred to as the *branch valuation*. Now, we will prove the property that will enable us to prove the completeness theorem.

Proposition 10 (Satisfaction in Branch Model Property). *Let $\mathcal{M}^b = (U^b, \sim^b, \Rightarrow^b, \circ^b, D^b)$ be the branch structure for an open branch b of a $\mathsf{DT}^*_{\mathsf{SCI}}$-proof tree and let v^b be the branch valuation in $\mathcal{M}^b$. Then, for every SCI-formula φ, if $\mathcal{M}^b, v^b \models \varphi$, then $\varphi \notin b$.*

Proof. Let $\mathcal{M}^b = (U^b, \sim^b, \Rightarrow^b, \circ^b, D^b)$ be the branch structure for an open branch b of a $\mathsf{DT}^*_{\mathsf{SCI}}$-proof tree and v^b the branch valuation in $\mathcal{M}^b$. We will prove the proposition by the induction on the depth of SCI-formulas. Let φ be an SCI-formula such that $d(\varphi) = 0$.

Assume $\mathcal{M}^b, v^b \models \varphi$. Note that the following holds: $\mathcal{M}^b, v^b \models \varphi$ iff $v^b(\varphi) = [\varphi]_{R_\circ} \in D^b$ iff $\varphi \in D_0$ iff $\neg \varphi \in b$. Thus, by the assumption, we obtain $\neg \varphi \in b$, which, by Proposition 5, implies $\varphi \notin b$.

Assume $\mathcal{M}^b, v^b \models \neg \varphi$. Then, $\mathcal{M}^b, v^b \models \neg \varphi$ iff $v^b(\neg \varphi) = [\neg \varphi]_{R_\circ} \in D^b$ iff $\neg \varphi \in D_1$. Suppose $\neg \varphi \in b$. Then, $\varphi \in D_0$, so, by the definition of D_1, we have $\neg \varphi \notin D_1$, a contradiction.

Assume that the proposition holds for SCI-formulas φ and ψ and their negations. We will show that it holds for formulas $\neg \neg \varphi$, $\varphi \to \psi$, and $\neg(\varphi \to \psi)$.

Let $\mathcal{M}^b, v^b \models \neg \neg \varphi$. Since $\mathcal{M}^b$ is an SCI-model, by the assumption $\mathcal{M}^b, v^b \models \varphi$. Thus, by the inductive hypothesis, $\varphi \notin b$. Suppose $\neg \neg \varphi \in b$. Then, by the completion condition Cpl($\neg$), $\varphi \in b$, a contradiction.

Let $\mathcal{M}^b, v^b \models \varphi \to \psi$. Then, either $\mathcal{M}^b, v^b \models \neg \varphi$ or $\mathcal{M}^b, v^b \models \psi$. Then, by the inductive hypothesis, either $\neg \varphi \notin b$ or $\psi \notin b$. Suppose $\varphi \to \psi \in b$. Then, by the completion condition Cpl($\to$), both $\neg \varphi \in b$ and $\psi \in b$, a contradiction.

Let $\mathcal{M}^b, v^b \models \neg(\varphi \to \psi)$. Then, both $\mathcal{M}^b, v^b \models \varphi$ and $\mathcal{M}^b, v^b \models \neg \psi$. Then, by the inductive hypothesis, both $\varphi \notin b$ and $\neg \psi \notin b$. Suppose $\neg(\varphi \to \psi) \in b$. Then, by the completion condition Cpl($\neg \to$), either $\varphi \in b$ or $\neg \psi \in b$, a contradiction. □

Now, we will prove completeness of an $\mathsf{DT}^*_{\mathsf{SCI}}$-system:

Theorem 6 (Completeness of DT^*_{SCI}). *If an SCI-formula is SCI-valid, then it is DT^*_{SCI}-provable.*

Proof. Let φ be SCI-valid and suppose that a closed DT^*_{SCI}-proof tree for φ does not exist. Then, there exists a complete DT^*_{SCI}-proof tree for φ with an open branch, say b. Clearly, $\varphi \in b$, so by Proposition 10, the branch structure $\mathcal{M}^b$ and the branch valuation v^b do not satisfy φ. However, by Proposition 8, $\mathcal{M}^b$ is an SCI-model. Thus, φ is not true in some SCI-model, and hence φ is not SCI-valid, a contradiction. □

Theorems 5 and 6 imply:

Theorem 7 (Soundness and Completeness of DT^*_{SCI}). *Let φ be an SCI-formula. Then, the following conditions are equivalent:*

1. φ is SCI-valid;
2. φ is DT^*_{SCI}-provable.

Below, we present examples of DT^*_{SCI}-proofs, namely DT^*_{SCI}-proofs of $(p_1 \equiv p_2) \to (p_1 \to p_2)$ and $(p_1 \equiv p_2) \to [(p_2 \equiv p_3) \to (p_1 \equiv p_3)]$ are presented in Figures 5 and 6, respectively. Note that DT^*_{SCI}-proofs are much shorter than the corresponding proofs of these formulas in the systems G_{SCI} and DT_{SCI}. Furthermore, contrary to the proofs in G_{SCI} and DT_{SCI}, DT^*_{SCI}-proofs of formulas in question are one-branching proofs.

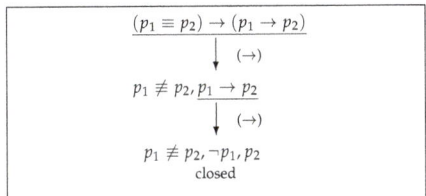

Figure 5. A DT^*_{SCI}-proof for the formula $(p_1 \equiv p_2) \to (p_1 \to p_2)$.

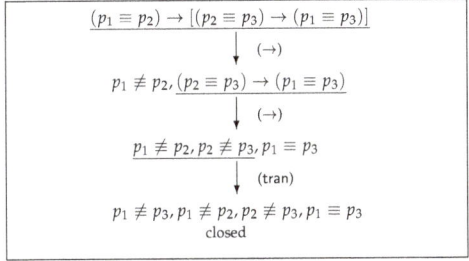

Figure 6. A DT^*_{SCI}-proof for $(p_1 \equiv p_2) \to [(p_2 \equiv p_3) \to (p_1 \equiv p_3)]$.

6. Discussion

All the systems presented in the previous sections are sound and complete deduction systems for SCI. Comparing with systems G_{SCI} and DT_{SCI}, the system DT^*_{SCI} seems to be simpler, more intuitive, and more effective. Its rules for the identity connective do not split a branch of a tree and do not make use of substitution. It should also be emphasized that the only rule of DT^*_{SCI}-system that may introduce branching is the rule $(\neg \to)$. Furthermore, although DT^*_{SCI} contains nine rules, while G_{SCI}-system has 12 rules, DT^*_{SCI}-system generates proofs that are much simpler and shorter than corresponding proofs in G_{SCI}.

However, all the systems presented in this paper have one important disadvantage. The logic SCI is decidable, while the systems in question are not decision procedures for SCI as, in particular, they may generate infinite trees. Although there is a decision procedure for SCI based on G_{SCI}-system, as shown in [13], but a procedure described in [13] contains external machinery that is not a part of the system itself, so it provides rather another proof for decidability of SCI than a decision procedure itself. Hence, further research on deduction systems for SCI should focus on seeking its decision procedure. The system DT^*_{SCI} seems to have a significant advantage over other systems G_{SCI} and DT_{SCI}, as its relatively simple modification could provide a decision procedure for SCI. A possible modification of DT^*_{SCI} should restrict applicability of the rules for the identity connective as follows: (1) the rule (ref) can be applied only for φ that are subformulas or negated subformulas of the initial formula; (2) given the formulas φ, ψ, ϑ, the rules (sym) and (tran) can be applied only once; (3) the rules $(\equiv_\neg)$, $(\equiv_\to)$, and $(\equiv_\equiv)$ can be applied to a finite set of formulas provided that the length of new formulas introduced by rules is not greater than the length of the initial formula plus 1. Additionally, we should also impose a general restriction on closeness of a branch, namely that, if a node is a 'copy' of some earlier node, then the branch is closed. It seems that such a modification could guarantee termination of proof trees, and thus it could provide a decision procedure for SCI.

7. Conclusions

We have presented and discussed two types of systems for SCI known from the literature: sequent calculus G_{SCI} and a dual tableau-like system DT_{SCI}. Then, we presented the system DT^*_{SCI}, which is a new dual tableau system for the logic SCI. We proved soundness and completeness of DT^*_{SCI} and we showed that it is more efficient than G_{SCI} and DT_{SCI}: it does not involve any substitution rule, its rules for the identity connective do not branch a proof tree, and it generates shorter and simpler proof trees. Further research on deduction systems for non-Fregean logics should concentrate on decision procedures for SCI and a methodology of designing deduction systems in tableuax style for non-Fregean logics which are extensions and modifications of SCI.

Author Contributions: Defining the general research problem and the scientific ideas for its solution: J.G.-P.; Elaboration of the results known from the literature included in the thematic scope of the research problem studied in the paper: J.G.-P. and M.W.; Construction of the new system DT^*_{SCI}: J.G.-P.; Proving soundness and completeness of the system: J.G.-P. and M.W.; Checking correctness of proofs: J.G.-P.; Writing the manuscript: J.G.-P. and M.W.; Editing and proofreading: J.G.-P. and M.W.; Funding: J.G.-P.; Supervision: J.G.-P.

Funding: The research presented in the paper was funded by the National Science Centre, Poland, research project No. 2017/25/B/HS1/00503.

Conflicts of Interest: The authors declare no conflict of interest.

References

1. Suszko, R. Non-Fregean logic and theories. *Analele Univ. Bucur. Acta Log.* **1968**, *11*, 105–125.
2. Suszko, R. Abolition of the Fregean axiom. In *Logic Colloquium: Symposium on Logic Held at Boston, 1972–73*; Parikh, R., Ed.; Lecture Notes in Mathematics; Springer: Heidelberg, Germany, 1975; Volume 453, pp. 169–239.
3. Bloom, S.; Suszko, R. Investigation into the sentential calculus with identity. *Notre Dame J. Form. Log.* **1972**, *13*, 289–308. [CrossRef]
4. Golińska-Pilarek, J.; Huuskonen, T. Number of extensions of non-Fregean logics. *J. Philos. Log.* **2005**, *34*, 193–206. [CrossRef]
5. Golińska-Pilarek, J.; Huuskonen, T. Logic of descriptions. A new approach to the foundations of mathematics and science. *Stud. Log. Gramm. Rhetor.* **2012**, *40*, 63–94.
6. Golińska-Pilarek, J.; Huuskonen, T. Grzegorczyk's non-Fregean logics and their formal properties. In *Applications of Formal Philosophy*; Urbaniak, R., Payette, G., Eds.; Logic, Argumentation and Reasoning; Springer International Publishing: New York, NY, USA, 2017; Volume 14, pp. 243–263.
7. Golińska-Pilarek, J.; Huuskonen, T. A mystery of Grzegorczyk's logic of descriptions. In *The Lvov-Warsaw School. Past and Present*; Garrido, A., Wybraniec-Skardowska, U., Eds.; Studies in Universal Logic; Springer Nature: Stuttgart, Germany, 2018; pp. 731–745.

8. Golińska-Pilarek, J.; Huuskonen, T. Non-Fregean propositional logic with quantifiers. *Notre Dame J. Form. Log.* **2016**, *57*, 249–279. [CrossRef]
9. Suszko, R. Identity connective and modality. *Stud. Log.* **1971**, *27*, 7–39. [CrossRef]
10. Malinowski, G. Identity, many-valuedness and referentiality. *Log. Log. Philos.* **2013**, *22*, 375–387. [CrossRef]
11. Golińska-Pilarek, J. On the minimal non-Fregean Grzegorczyk's logic. *Stud. Log.* **2016**, *104*, 209–234. [CrossRef]
12. Michaels, A. A uniform proof proceduree for SCI tautologies. *Stud. Log.* **1974**, *33*, 299–310. [CrossRef]
13. Wasilewska, A. A sequence formalization for SCI. *Stud. Log.* **1976**, *35*, 213–217. [CrossRef]
14. Chlebowski, S. Sequent calculi for SCI. *Stud. Log.* **2018**, *106*, 541–563. [CrossRef]
15. Golińska-Pilarek, J. Rasiowa-Sikorski proof system for the non-Fregean sentential logic SCI. *J. Appl.-Non-Class. Log.* **2007**, *17*, 511–519. [CrossRef]
16. Orłowska, E.; Golińska-Pilarek, J. *Dual Tableaux: Foundations, Methodology, Case Studies*; Trends in Logic; Springer: Dordrecht Heidelberg London New York, NY, USA, 2011; Volume 33.
17. Rasiowa, H.; Sikorski, R. On Gentzen theorem. *Fundam. Math.* **1960**, *48*, 57–69. [CrossRef]

© 2019 by the authors. Licensee MDPI, Basel, Switzerland. This article is an open access article distributed under the terms and conditions of the Creative Commons Attribution (CC BY) license (http://creativecommons.org/licenses/by/4.0/).

Article

Sequent-Type Calculi for Three-Valued and Disjunctive Default Logic

Sopo Pkhakadze [†,‡] and Hans Tompits [*,†,‡]

Knowledge-Based System Group, Institut für Logic and Computation, Technische Universität Wien, Favoritenstraße 9-11, 1040 Vienna, Austria; pkhakadze@kr.tuwien.ac.at
* Correspondence: tompits@kr.tuwien.ac.at
† This paper is an extended version of our paper published in the proceedings of the 15th International Conference on Logic Programming and Non-monotonic Reasoning (LPNMR 2019) as well as of an abstract published in the proceedings of the conference "Kurt Gödel's Legacy: Does Future lie in the Past?" held 2019 in Vienna. This paper is dedicated to the memory of Khimuri Rukhaia, logician, professor, and a kind man, who was the teacher of the first author during her Bachelor studies and who sadly passed away during the preparation of this work.
‡ These authors contributed equally to this work.

Received: 9 June 2020; Accepted: 15 July 2020; Published: 21 July 2020

Abstract: Default logic is one of the basic formalisms for nonmonotonic reasoning, a well-established area from logic-based artificial intelligence dealing with the representation of *rational conclusions*, which are characterised by the feature that the inference process may require to retract prior conclusions given additional premises. This nonmonotonic aspect is in contrast to *valid inference relations*, which are monotonic. Although nonmonotonic reasoning has been extensively studied in the literature, only few works exist dealing with a proper proof theory for specific logics. In this paper, we introduce sequent-type calculi for two variants of default logic, viz., on the one hand, for *three-valued default logic* due to Radzikowska, and on the other hand, for *disjunctive default logic*, due to Gelfond, Lifschitz, Przymusinska, and Truszczyński. The first variant of default logic employs Łukasiewicz's three-valued logic as the underlying base logic and the second variant generalises defaults by allowing a selection of consequents in defaults. Both versions have been introduced to address certain representational shortcomings of standard default logic. The calculi we introduce axiomatise *brave reasoning* for these versions of default logic, which is the task of determining whether a given formula is contained in some extension of a given default theory. Our approach follows the sequent method first introduced in the context of nonmonotonic reasoning by Bonatti, which employs a *rejection calculus* for axiomatising invalid formulas, taking care of expressing the consistency condition of defaults.

Keywords: sequent-type calculi; nonmonotonic logics; default logic; rejection systems

1. Introduction

Most formal logics studied in the literature are *monotonic* in the sense that an increased set of premisses never yields a reduced set of conclusions. An important class of logics, closely related to the formalisation of human common-sense reasoning and important in the area of logic-based artificial intelligence (AI), however, do not enjoy this property—they are *nonmonotonic*. A central nonmonotonic-reasoning formalism is *default logic*, introduced by Raymond Reiter in 1980 [1]. In default logic, conclusions may be asserted on the basis of having no evidence, making such inferences unjustified. A typical argument schema along these lines is to assume a certain statement given no evidence to the contrary. Such nonmonotonic conclusions are *defeasible* as they may be invalidated by additional information. In general, nonmonotonic logics deal with the representation of *rational arguments* while traditional logics formalise *valid conclusions*. Other important nonmonotonic-reasoning formalisms that have been introduced in the literature, besides default logic, are e.g., *autoepistemic logic* [2], *circumscription* [3], *logic programming under the answer-set semantics* [4], and *equilibrium logic* [5]. The term of referring to a logical system as being "nonmonotonic" was first introduced by Marvin Minsky in 1975 [6].

Given the large body of works devoted to nonmonotonic reasoning, only few investigations exist dealing with concrete proof systems for it. Prominent among these are the sequent-type calculi for default logic and autoepistemic logic introduced by Bonatti [7] and those for default logic, autoepistemic logic, and circumscription by Bonatti and Olivetti [8]. In this paper, we introduce sequent-type calculi for brave reasoning in the style of Bonatti [7] for two variants of default logic, viz., on the one hand, for *three-valued default logic*, due to Radzikowska [9], and on the other hand, for *disjunctive default logic*, due to Gelfond, Lifschitz, Przymusinska, and Truszczyński [10]. The first variant of default logic employs Łukasiewicz's three-valued logic [11] as the underlying base logic and the second variant generalises default rules by allowing a selection of consequents in defaults, closely related to the answer-set semantics of disjunctive logic programs [4]. Both versions have been introduced to address certain representational shortcomings of standard default logic. Other variants of default logic include, e.g., *justified default logic* [12], *constrained default logic* [13,14], *rational default logic* [15], *general default logic* [16], and *four-valued default logic* [17] (an overview about different versions of default logic is given by Antoniou and Wang [18]).

A distinguishing feature of the approach of Bonatti and Olivetti is the usage of a *rejection calculus* for axiomatising invalid formulas, i.e., of non-theorems, taking care of formalising consistency conditions, which makes these calculi arguably particularly elegant and suitable for proof-complexity elaborations as, e.g., , recently undertaken by Beyersdorff, Meier, Thomas, and Vollmer [19]. In a rejection calculus, the inference rules formalise the propagation of refutability instead of validity and establish invalidity by deduction. Rejection calculi are also referred to in the literature as *complementary calculi* or *refutation calculi*, and the first axiomatic treatment of rejection was done by Łukasiewicz in his formalisation of Aristotle's syllogistic [20].

Since a sound and complete axiomatisation of non-theorems is only possible for logics that are decidable (or at least where the set of non-theorems is semi-decidable), Bonatti [7] considered only propositional versions of the nonmonotonic logics for which he developed sequent calculi. The same holds also for the subsequent calculi introduced by Bonatti and Olivetti [8], and this is what we follow here too.

Analogous to the method of Bonatti [7], our calculi comprise three kinds of sequents each:

(i) assertional sequents for axiomatising validity in the respective underlying monotonic base logic;
(ii) *anti-sequents* for axiomatising invalidity for the underlying monotonic logics, taking care of the consistency check of defaults; and
(iii) proper default sequents, for representing nonmonotonic conclusions.

Although it would be possible to use just one kind of sequents, this would be at the expense of losing clarity of the structure of sequents. In addition, the usage of different types of sequents also reflects the interactions between the underlying monotonic proof machinery and nonmonotonic inferences in a much clearer manner.

As far as three-valued logics are concerned, different kinds of sequent-style systems exist in the literature, like systems based on (two-sided) sequents [21,22] in the style of Gentzen's original work [23] and employing additional non-standard rules, or using *hypersequents* [24], which are tuples of Gentzen-style sequents. In our sequent and anti-sequent calculi for Łukasiewicz's three-valued logic, we adopt the approach of Rousseau [25], which is a natural generalisation for many-valued logics of the classical two-sided sequent formulation of Gentzen. The respective calculi are obtained from a systematic construction for many-valued logics as described by Zach [26] and by Tompits and Bogojeski [27].

For the case of disjunctive default logic, the calculus we define employs the well-known sequent-type calculus following Gentzen [23] and an anti-sequent calculus due to Bonatti [7].

Concerning rejection systems in general, its history goes back already to Aristotle who not only analysed correct reasoning in his system of syllogisms but also studied invalid arguments, where in particular he rejected arguments by reducing them to other already rejected ones. The first usage of the term "rejection" in modern logic was done by Jan Łukasiewicz in his 1921 paper *Logika dwuwartościowa* ("Two-valued logic") in which he states that by doing so he follows Brentano [28]. An axiomatic treatment of rejection was then discussed in Łukasiewicz's treatment of Aristotle's syllogistic [20,29] where he introduced a Hilbert-type rejection system. This was then further elaborated by his student Jerzy Słupecki [30] and eventually extended to a theory of rejected propositions [31–35]. In general, work about axiomatic rejection methods comprise of not only investigations about classical logic [36–38] but also for varieties of other logics, like intuitionistic logic [39–43], modal logics [38,44], or description logics [45]. For an excellent survey on the development of rejection systems, we refer to a paper by Wybraniec-Skardowska [46].

The paper is organised as follows. In the next section, we present the background on the formalisms employed in our work, that is, on the underlying monotonic logics (Section 2.1) and the two variants of default logic (Section 2.2). Afterwards, in Section 3, we introduce our sequent calculus for three-valued default logic, and in Section 4, we discuss our calculus for disjunctive default logic. The paper concludes with Section 5, providing a brief summary and an outlook for future work.

2. Background

2.1. Underlying Monotonic Logics

We start with setting down the basic definitions and notation for classical propositional logic and Łukasiewicz's three-valued logic [11], which are required for our subsequent elaborations.

2.1.1. Classical Propositional Logic

The alphabet of classical propositional logic, **PL**, consists of (i) a countable set $\mathcal{P}$ of *propositional constants*, (ii) the *truth constants* "⊤" ("truth") and "⊥" ("falsehood"), (iii) the *primitive logical connectives* "¬" ("negation") and "⊃" ("implication"), and (iv) the punctuation symbols "(" ("and") ")". The class of *formulas* is built from elements of the alphabet of **PL** in the usual inductive fashion, whereby the propositional constants and truth constants constitute the *atomic formulas*. Formulas which are non-atomic are referred to as *composite formulas*.

Besides the primitive connectives ¬ and ⊃, we also make use of the standard connectives "∨" ("disjunction"), "∧" ("conjunction"), and "≡" ("equivalence"), defined in the usual way: $(A \vee B) := ((\neg A) \supset B)), (A \wedge B) := \neg(\neg A \vee \neg B)$, and $(A \equiv B) := ((A \supset B) \wedge (B \supset A))$.

In what follows, we will use the letters "P", "Q", "R", ... (possibly appended with subscripts and/or with primes) or words from everyday English to refer to propositional constants, and use the letters "A", "B", "C", ... (again possibly appended with subscripts and/or with primes) to refer to arbitrary formulas (distinct such letters need not represent distinct formulas).

A *(two-valued) interpretation* is a mapping I assigning each propositional constant from $\mathcal{P}$ an element from the set $\{\mathbf{t}, \mathbf{f}\}$, whose elements are referred to as *truth values*, where **t** represents *truth* and **f** represents *falsity*. The truth value of a composite formula A under an interpretation I, denoted by $V^I(A)$, is defined in terms of the usual truth-table conditions of classical propositional logic. Accordingly, a formula A is *true under I* iff $V^I(A) = \mathbf{t}$, and *false under I* if $V^I(A) = \mathbf{f}$. If A is true under I, then I is said to be a *model* of A, and if A is false under I, then I is a *countermodel* of A. If I is a countermodel of A, then we also say that I *refutes* A. We call A *satisfiable* (*in* **PL**) if it has some model, and *falsifiable* (*in* **PL**), or *refutable* (*in* **PL**), if it has some countermodel. Moreover, A is *unsatisfiable* (*in* **PL**) if it has no model. Finally, A is a *tautology*, symbolically $\models_2 A$, if it is true in every interpretation, and *refutable* (*in* **PL**), symbolically $\not\models_2 A$, otherwise.

A set of formulas is also referred to as a *theory*. An interpretation I is a model of a theory T if I is a model of all elements of T, otherwise I is a countermodel of T. If a theory T has a model, then T is satisfiable, and if T has a countermodel, then T is falsifiable. A theory is unsatisfiable if it has no model.

A formula A is a *valid consequence* of a theory T (*in* **PL**), or T entails A (*in* **PL**), in symbols $T \models_2 A$, iff A is true in any model of T. Two formulas, A and B, are (*logically*) *equivalent* (*in* **PL**) iff $\models_2 (A \equiv B)$. In general, two theories are (logically) equivalent iff they have the same models.

As customary, we will write expressions like "$T \cup \{A\} \models_2 B$" as "$T, A \models_2 B$", and similarly for finite sets of form $\{A_1, \ldots, A_n\}$ instead of a singleton set $\{A\}$.

We denote by $\vdash_2$ the usual derivability operator of **PL** with respect to some fixed sound and complete Hilbert-type system. The *deductive closure operator* of **PL** is given by:

$$\mathrm{Th}_2(T) := \{A \mid T \vdash_2 A\},$$

where T is a theory. A theory T is *deductively closed* iff $T = \mathrm{Th}_2(T)$. As well known, the operator $\mathrm{Th}_2(\cdot)$ enjoys the following properties (for any theory T and T'):

1. $T \subseteq \mathrm{Th}_2(T)$. ("Inflationaryness".)
2. $\mathrm{Th}_2(\mathrm{Th}_2(T)) = \mathrm{Th}_2(T)$. ("Idempotency".)
3. $T \subseteq T'$ implies $\mathrm{Th}_2(T) \subseteq \mathrm{Th}_2(T')$. ("Monotonicity".)

If A is not derivable from T, then we indicate this by writing $T \nvdash_2 A$. Later on, we will define proof systems axiomatising formulas that are *not* derivable from a given theory. Such axiom systems are accordingly also referred to as *complementary calculi* as they axiomatise the *complement* of the provable formulas of a logic.

We say that a theory T is *consistent* iff there is a formula A such that $T \nvdash_2 A$. Clearly, T is consistent iff it is satisfiable. Moreover, a formula A is *consistent with* T iff $T \nvdash_2 \neg A$.

2.1.2. Łukasiewicz's Three-Valued Logic

We now turn to the three-valued logic of Łukasiewicz [11] for the propositional case, henceforth denoted by $Ł_3$. Our presentation follows the one given by Radzikowska [9].

The alphabet of $Ł_3$ consists of the alphabet of **PL** along with the additional truth constant $\sqcup$ ("undetermined"). Again, we assume $\mathcal{P}$ as a countable set of propositional constants. The class of formulas of $Ł_3$ is built similarly to the formulas of **PL**, except that $\sqcup$ is counted as an additional atomic formula.

A difference to the syntax of the logic **PL** concerns the defined connectives; while conjunction, $\wedge$, and material equivalence, $\equiv$, are defined as in propositional logic, disjunction in $Ł_3$ is defined differently:

$$(A \vee_3 B) := ((A \supset B) \supset B).$$

Furthermore, there are also additional unary defined operators, viz.

- the connective "$\sim$" ("weak negation"), given by

$$\sim A := (A \supset \neg A);$$

- the unary operators "L" ("certainty operator") and "M" ("possibility operator"), defined by

$$LA := \neg(A \supset \neg A) \quad \text{and} \quad MA := (\neg A \supset A),$$

which, according to Łukasiewicz [11], were first formalised in 1921 by Tarski; and
- the operator "I", given by

$$IA := (MA \wedge \neg LA).$$

Intuitively, LA expresses that A is certain, whilst MA means that A is possible. These operators will be used subsequently to distinguish between *certain knowledge* and *defeasible conclusions*. Furthermore, IA expresses that A is *contingent* or *modally indifferent*.

A (*three-valued*) *interpretation* is a mapping m assigning to each propositional constant from $\mathcal{P}$ an element from $\{\mathbf{t}, \mathbf{f}, \mathbf{u}\}$. Here, besides the truth values $\mathbf{t}$ and $\mathbf{f}$, the symbol $\mathbf{u}$ represents a truth value standing for "undetermined" or "indeterminacy". As usual, $m(P)$ is the *truth value of P under* m, where now P is true under an interpretation m if $m(P) = \mathbf{t}$, false under m if $m(P) = \mathbf{f}$, and has undetermined truth value if $m(P) = \mathbf{u}$.

The truth value, $V^m(A)$, of an arbitrary formula A under an interpretation m is given subject to the following conditions:

1. If $A = \top$, then $V^m(A) = \mathbf{t}$.
2. If $A = \sqcup$, then $V^m(A) = \mathbf{u}$.
3. If $A = \bot$, then $V^m(A) = \mathbf{f}$.
4. If A is an atomic formula, then $V^m(A) = m(A)$.
5. If $A = \neg B$, for some formula B, or $A = (C \supset D)$, for some formulas C and D, then $V^m(A)$ is determined according to the truth tables given in Figure 1 (there, the corresponding truth conditions for the defined connectives are also given).

$\neg$			$\supset$	t	u	f		$\vee_3$	t	u	f		$\wedge$	t	u	f
t	f		t	t	u	f		t	t	t	t		t	t	u	f
u	u		u	t	t	u		u	t	u	u		u	u	u	f
f	t		f	t	t	t		f	t	u	f		f	f	f	f

$\equiv$	t	u	f		$\sim$			L			M			I	
t	t	u	f		t	f		t	t		t	t		t	f
u	u	t	u		u	t		u	f		u	t		u	t
f	f	u	t		f	t		f	f		f	f		f	f

Figure 1. Truth tables for the connectives of Ł$_3$.

If $V^m(A) = \mathbf{t}$, then A is *true under* m, if $V^m(A) = \mathbf{u}$, then A is *undetermined under* m, and if $V^m(A) = \mathbf{f}$, then A is *false under m*. If A is true under m, then m is a *model* of A. If A is true in every interpretation, then A is *valid* (in Ł$_3$), written $\models_3 A$.

Clearly, the classically valid principle of *tertium non datur*, i.e., the law of excluded middle, $A \vee \neg A$, as well as the corresponding law of non-contradiction, $\neg(A \wedge \neg A)$, are not valid in Ł$_3$. However, their three-valued pendants, viz., the principle of *quartum non datur*, i.e., the law of excluded fourth, $A \vee IA \vee \neg A$, and the corresponding extended non-contradiction principle, $\neg(A \wedge \neg IA \wedge \neg A)$, are valid in Ł$_3$.

In classical logic, two formulas are logically equivalent if and only if, they have the same models, where logical equivalence between formulas A and B is defined by the condition that $\models_2 (A \equiv B)$ holds. However, such a relation between logical equivalence and equality of models does not hold in general in the three-valued logic case. Indeed, following Radzikowska [9], let us define that two formulas A and B are *strongly equivalent*, symbolically $A \Leftrightarrow_s B$, iff $\models_3 (A \equiv B)$. That is, A and B are strongly equivalent iff, for any three-valued interpretation m, $V^m(A) = V^m(B)$. Furthermore, let us call A and B *equivalent* (*in* Ł$_3$), symbolically $A \Leftrightarrow B$, iff A and B have the same models. Clearly, strong equivalence implies equivalence, but in general not vice versa. For instance, P and LP, for an atom P, are equivalent but not strongly equivalent. In addition, strong equivalence is an equivalence relation (i.e., reflexive, symmetric, and transitive) and enjoys a substitution principle, similar to the one of classical logic, i.e., if a formula C_A contains a subformula A, and C_B is the result of substituting at least one occurrence of A in C_A by a formula B, then $A \Leftrightarrow_s B$ implies $C_A \Leftrightarrow_s C_B$.

Let us also note some strong equivalences which hold in $Ł_3$:

1. $(A \supset B) \Leftrightarrow_s (M\neg A \vee_3 B) \wedge (MB \vee_3 \neg A)$.
2. $O(A \circ B) \Leftrightarrow_s (OA \circ OB)$, for $O \in \{L, M\}$ and $\circ \in \{\wedge, \vee\}$.
3. $OO'A \Leftrightarrow_s O'A$, for $O, O' \in \{L, M\}$.
4. $\sim A \Leftrightarrow_s M\neg A$.
5. $\neg LA \Leftrightarrow_s M\neg A$.
6. $\neg MA \Leftrightarrow_s L\neg A$.
7. $((A \wedge B) \vee_3 C \Leftrightarrow_s (A \vee_3 C) \wedge (B \vee_3 C)$.
8. $((A \vee_3 B) \wedge C \Leftrightarrow_s (A \wedge C) \vee_3 (B \wedge C)$.

The notion of a *theory* in $Ł_3$ is defined as in **PL**, i.e., a theory is a set of formulas. Likewise, the notion of a *model* or of a *countermodel* of a theory, and of a theory being *satisfiable*, *falsifiable* (or *refutable*), or *unsatisfiable* are defined in $Ł_3$ mutatis mutandis as in **PL**. A theory T is said to *entail* a formula A (in $Ł_3$), or A is a *valid consequence* of T (in $Ł_3$), symbolically $T \models_3 A$, iff every model (in $Ł_3$) of T is also a model (in $Ł_3$) of A.

Sound and complete Hilbert-style axiomatisations of the logic $Ł_3$ can be readily found in the literature [47,48]; the first one was introduced by Wajsberg in 1931 [49]. We write $T \vdash_3 A$ if A has a derivation (in some fixed Hilbert-style calculus) from T in $Ł_3$. As well, the *deductive closure operator* of $Ł_3$ is given by

$$\text{Th}_3(T) := \{A \mid T \vdash_3 A\},$$

where T is a theory. The notions of a theory being *deductively closed* and of being *consistent*, as well as of a formula being *consistent with* a theory, are defined similarly as in **PL**. Moreover, the properties of inflationaryness, idempotency, and monotonicity hold for $\text{Th}_3(\cdot)$ like for $\text{Th}_2(\cdot)$, and consistency of a theory T in $Ł_3$ is equivalent to the satisfiability of T in $Ł_3$.

While in **PL** we have the well-known properties that (i) $T \vdash_2 A$ iff $T \cup \{\neg A\}$ is inconsistent and (ii) $T, A \vdash_2 B$ iff $T \vdash_2 (A \supset B)$ (the "only if" part of the latter is generally referred to as the *deduction theorem*), for a theory T and formulas A and B, in $Ł_3$ sight variations thereof hold:

Proposition 1. *Let T be a theory, and A and B formulas.*

1. $T \vdash_3 A$ iff $T \cup \{M\neg A\}$ is inconsistent (in $Ł_3$).
2. $T, A \vdash_3 B$ iff $T \vdash_3 (LA \supset B)$.

Note that, as a consequence, the consistency of a formula A with a theory T implies the consistency of the theory $T \cup \{MA\}$, but it does not necessarily imply the consistency of $T \cup \{A\}$. For instance, $\neg P$ is consistent with $\{MP\}$, for an atomic formula P, so $\{MP, M\neg P\}$ is consistent, but $\{MP, \neg P\}$ is not.

Furthermore, although in $Ł_3$ it always holds that $T \vdash_3 A \supset B$ implies $T, A \vdash_3 B$, it is the converse direction (i.e., the classical version of the deduction theorem) that fails in general.

2.2. Two Variants of Default Logic

We continue with the basic elements of *three-valued default logic*, due to Radzikowska [9], and of *disjunctive default logic*, introduced by Gelfond, Lifschitz, Przymusinska, and Truszczyński [10]. Note that we deal here with propositional versions of the formalisms as our subsequent calculi are defined for the propositional case only, similar to the undertaking of Bonatti [7,37] and of Bonatti and Olivetti [8].

2.2.1. Three-Valued Default Logic

Radzikowska's three-valued default logic [9], which in what follows we will denote by $\mathbf{DL_3}$, differs from Reiter's standard default logic [1] (henceforth referred to as **DL**) in two aspects; not only is in **DL₃** the deductive machinery of classical logic replaced with Ł₃, but there is also a modified consistency check for default rules employed in which the consequent of a default is taken into account as well. The latter feature is somewhat reminiscent to the consistency checks used in *justified default logic* [12] and in *constrained default logic* [13,14], where a default may only be applied if it does not lead to a contradiction a posteriori.

Formally, a *default rule*, or simply a *default*, d, is an expression of the form

$$\frac{A : B_1, \ldots, B_n}{C},$$

where A is the *prerequisite*, $B_1, \ldots, B_n$ are the *justifications*, and C is the *consequent* of d. The intuitive meaning of such a default is:

if A is believed, and $B_1, \ldots, B_n$ and LC are consistent with what is believed, then MC is asserted.

Note that under this reading, by applying a default of the above form, it is assumed that C cannot be false, but it is not assumed that C is true in all situations. It is only assumed that C must be true in at least one such situation. This reflects the intuition that accepting a default conclusion, we are prepared to rule out all situations where it is false, but we can imagine at least one such situation in which it is true. As a consequence, we cannot conclude both MC and M¬C simultaneously.

In what follows, formulas of the form MC obtained by applying defaults will be referred to as *default assumptions*. For simplicity, defaults will also be written in the form $(A : B_1, \ldots, B_n/C)$.

A *default theory*, T, is a pair $\langle W, D \rangle$, where W is a set of formulas (i.e., a theory in Ł₃), called the *premises* of T, and D is a set of defaults. An *extension* of a default theory $T = \langle W, D \rangle$ in the three-valued default logic **DL₃** is defined thus: For a set S of formulas, let $\Gamma_T(S)$ be the smallest set K of formulas obeying the following conditions:

1. $K = \text{Th}_3(K)$.
2. $W \subseteq K$.
3. If $(A : B_1, \ldots, B_n/C) \in D$, $A \in K$, $\neg B_1 \notin S, \ldots, \neg B_n \notin S$, and $\neg LC \notin S$, then M$C \in K$.

Then, E is an extension of T iff $\Gamma_T(E) = E$.

Note that the criterion of the applicability of a default in **DL₃** makes the two defaults:

$$d = \frac{A : B_1, \ldots, B_n}{C} \quad \text{and} \quad d' = \frac{A : MB_1, \ldots, MB_n}{C}$$

equivalent in the sense that the application of d implies the application of d' and vice versa. Thus, in a default theory $T = \langle W, D \rangle$, we can replace every $d \in D$ with its corresponding version d' without changing extensions.

Note further that, for obtaining extensions in the sense of Reiter [1], in the above definition, instead of Th₃(·) we use Th₂(·), and the condition 3 is replaced by:

3′. If $(A : B_1, \ldots, B_n/C) \in D$, $A \in K$, and $\neg B_1 \notin S, \ldots, \neg B_n \notin S$, then $C \in K$.

There are two basic reasoning tasks in the context of default logic, viz., *brave reasoning* and *skeptical reasoning*. The former task is the problem of checking whether a formula A belongs to at least one extension of a given default theory T, whilst the latter task examines whether A belongs to all extensions of T. Our aim is to give a sequent-type axiomatisation of brave default reasoning, following the approach of Bonatti [7] for standard default logic.

To conclude our review of three-valued default logic, we give two examples, as discussed by Radzikowska [9], showing the representational advantages of **DL₃**.

Example 1 ([50]). *Consider the default theory $T = \langle W, D \rangle$, where*

$$W = \{Summer, \neg Sun\_Shining\} \quad \text{and} \quad D = \left\{ \frac{Summer : \neg Rain}{Sun\_Shining} \right\}.$$

The only default of this theory is inapplicable since $W \vdash_3 \neg LSun\_Shining$ holds. Hence, T has a single extension, viz. $E = Th_3(W)$. Note that T has no extension in Reiter's default logic due to the weaker consistency check which results in a vicious circle where the application of the default violates its justification for applying it.

Example 2 ([51]). *Consider the default rules*

$$d_1 = \frac{P : Q}{Q} \quad \text{and} \quad d_2 = \frac{Q : R}{R},$$

where P, Q, and R stand for the following propositions:

- P: "Tony recites passages from Shakespeare";
- Q: "Tony can read and write";
- R: "Tony is over seven years old".

Obviously, common sense suggests that, given P, there are perfect reasons to apply both defaults to infer that Tony is over seven years old. Suppose now that we add the default rule

$$d_3 = \frac{S : Q}{Q},$$

where S stands for "Tony is a child prodigy". Given S, it is reasonable to infer that Tony can read and write, but the inference of R that Tony is over seven years old seems to be unjustified.

In standard default logic **DL**, *a common way of suppressing R in the latter scenario would be to employ a default rule with exceptions of the form*

$$d_2' = \frac{Q : R \land \neg S}{R}.$$

However, this remedy is somewhat unsatisfactory as it requires that every default may possess a potentially large number of conceivable exceptions which, each time a new default is added, the previous ones must be revised, which is arguably ad hoc. In **DL₃**, *on the other hand, this can easily be accommodated by using the defaults*

$$\frac{P : LQ}{LQ} \quad \text{and} \quad \frac{Q : LR}{LR}$$

instead of d_1 and d_2, as well as

$$\frac{MS : Q}{Q}$$

instead of d_3.

Actually, the last example illustrates the difference between *causal rules* ("expectation-evoking rules") and *evidential rules* ("explanation-evoking rules") [51]. An example of the first kind of rules is "fire usually causes smoke" whilst "smoke usually suggests fire" is an instance of the second kind. As argued by Pearl [51], an evidential rule should not be applied if its prerequisite is derived by applying at least one causal rule. In **DL₃**, this can be taken into account by formalising causal default rules in the form of $(MA : B/B)$, $(LA : B/B)$, or $(A : B/B)$, whilst evidential rules are formalised by $(LA : LB/LB)$ or, equivalently, by $(A : LB/LB)$.

2.2.2. Disjunctive Default Logic

We now turn to the basics of disjunctive default logic [10], henceforth referred to as **DL_D**.

The main motivation for introducing disjunctive default logics was to address a difficulty encountered when using defaults in the presence of disjunctive information, a problem which was first observed by David Poole [52]. More specifically, the difficulty lies in the difference between a default theory having two extensions, one containing a formula A and the other a formula B, and a theory with a single extension, containing the disjunction $A \vee B$. This problem was also noted by Lin and Shoham [53], who gave an example of a theory in a modal-logic language, containing disjunctive information, and observed that no default theory exists which corresponds to this theory.

Another nice feature of disjunctive default logic is that it provides a one-to-one correspondence between answer-sets of disjunctive logic programs [4] and extensions of a corresponding disjunctive default theory. Such a correspondence does likewise not directly hold for standard default logic—and again the key problem lies in the presence of disjunctive information. More specifically, viewing $P \vee Q$ as a rule in a logic program under the answer-set semantics, the default naturally corresponding to this rule would be the default rule

$$d = \frac{\top :}{P \vee Q}.$$

Now, while the program consisting of the single rule $P \vee Q$ has two answer sets, viz. $\{P\}$ and $\{Q\}$, the default theory $\langle \emptyset, \{d\} \rangle$ has only one extension, $Th_2(\{P \vee Q\})$. As long as only programs without disjunctions are considered, such a natural translation of program rules into defaults gives rise to a one-to-one correspondence between answer sets of the given program and the extensions of its translation.

To formally introduce **DL_D**, by a *disjunctive default rule*, or simply a *disjunctive default*, d, we understand an expression of the form

$$\frac{A : B_1, \ldots, B_n}{C_1 | \cdots | C_m},$$

where $A, B_1, \ldots, B_n$, and $C_1, \ldots, C_m$ are formulas from **PL**. Similar to **DL_3**, we call A the *prerequisite*, $B_1, \ldots, B_n$ the *justifications*, and $C_1, \ldots, C_n$ the *consequents* of d. Furthermore, following Baumgartner and Gottlob [54], we refer to the symbol "|" as *effective disjunction*.

The intuitive meaning of such a default is:

if A is believed and $B_1, \ldots, B_n$ are consistent with what is believed, then one of $C_1, \ldots, C_m$ is asserted.

Similar to conventions in standard default logic, if the prerequisite of a default d is $\top$, then we will omit it from d. If, additionally, d has no justifications, then d is simply written as

$$C_1 | \cdots | C_m,$$

where $C_1, \ldots, C_m$ are the consequents of d. For convenience, disjunctive defaults will also be written in the form $(A : B_1, \ldots, B_n / C_1 | \cdots | C_m)$.

A *disjunctive default theory*, T, is a pair $\langle W, D \rangle$, where W is a set of formulas of **PL** (again referred to as the *premises* of T) and D is a set of disjunctive defaults.

For defining extensions of disjunctive default theories, we need some further notation: Let us call a set S of formulas *closed under propositional consequence* if, whenever $S \vdash_2 A$, then $A \in S$. Clearly, the deductive closure of a set S, $Th_2(S)$, is the smallest set of formulas closed under propositional consequence containing S. Moreover, for a family F of sets, let $min(F)$ denote the minimal elements of F, where minimality is defined with respect to set inclusion, i.e.,

$$min(F) = \{X \mid X \in F \text{ and there is no } Z \in F \text{ such that } Z \subset X\}.$$

Consider now a disjunctive default theory $T = \langle W, D \rangle$. Given a set S of formulas of **PL**, let $Cl_T(S)$ be the collection of all sets K satisfying the following conditions:

1. $K = \text{Th}_2(K)$.
2. $W \subseteq K$.
3. If $(A : B_1, \ldots, B_n \,/\, C_1 | \cdots | C_m) \in D$, $A \in K$ and $\{\neg B_1, \ldots, \neg B_n\} \cap S = \emptyset$, then $C_i \in K$, for some $i \in \{1, \ldots, m\}$.

Moreover, let $\Delta_T(S) = \min(Cl_T(S))$, i.e., $\Delta_T(S)$ consists of all minimal sets obeying conditions 1–3. Then, a set E of formulas of **PL** is an *extension* of T if $E \in \Delta_T(E)$.

The notion of a brave and a skeptical consequence given a disjunctive default theory is defined as before *mutatis mutandis*.

Let us now discuss some examples showing the differences between disjunctive default logic and standard default logic, following Gelfond, Lifschitz, Przymusinska, and Truszczyński [10].

Example 3 ([1]). *Consider the default theory* $T = \langle W, D \rangle$, *for*

$$W = \{P \vee Q\} \quad \text{and} \quad D = \left\{ \frac{P : R}{R}, \frac{Q : S}{S} \right\},$$

where P, Q, R, and S are atomic formulas. Intuitively, given the disjunctive information $P \vee Q$, *we would expect to derive* $R \vee S$, *because, in case P holds, we could apply the first default, and in case Q holds, we could accordingly apply the second default. However, in* **DL**, *neither of the two defaults is applicable and the single extension of T is* $\text{Th}_2(W)$.

Now, in disjunctive default logic, we can represent the information expressed by T in terms of a disjunctive default theory T' *containing the three defaults*

$$P|Q, \quad \frac{P : R}{R}, \quad \text{and} \quad \frac{Q : S}{S}.$$

In contrast to the situation in **DL**, T' *possesses two extensions in* **DL**$_\mathbf{D}$, *viz.* $\text{Th}_2(\{P, R\})$ *and* $\text{Th}_2(\{Q, S\})$, *and* $R \vee S$ *is contained in both, which is in accordance to our expectations.*

We next discuss the example by Poole [52].

Example 4. *Let us assume the following commonsense information: By default, a person's left arm is usable, the exception being when it is broken, and similarly for the right arm.*

In standard default logic, we can express this by the following two defaults:

$$d_1 := \frac{: U_l \wedge \neg B_l}{U_l} \quad \text{and} \quad d_2 := \frac{: U_r \wedge \neg B_r}{U_r},$$

where "U_l" and "U_r" stand for that the left arm is usable and that the right arm is usable, respectively, and, similarly, "B_l" and "B_r" mean that the left arm or the right arm is broken.

If there is no further information about one's hands, then one can conclude that both hands are usable. Indeed, the default theory $T = \langle \emptyset, \{d_1, d_2\} \rangle$ *has a single extension in* **DL**, *containing both* U_l *and* U_r.

However, if it is now known that the left arm is broken, i.e., B_l *is asserted, then the application of* d_1 *is blocked and the extended default theory*

$$T' = \langle \{B_l\}, \{d_1, d_2\} \rangle$$

has again one extension, containing U_r.

But let us assume now that we only know that one arm is broken, but we do not remember exactly which one. So, what we can assert now is the formula

$$B_l \vee B_r.$$

Considering now the extensions of the default theory

$$T'' = \langle \{B_l \vee B_r\}, \{d_1, d_2\} \rangle,$$

this default theory has still one extension, but unfortunately it contains both $U_l \vee U_r$, which is contrary to our intuition.

Using **DL**$_\mathbf{D}$, on the other hand, we can represent the information of T'' by a disjunctive default theory containing

$$B_l | B_r$$

together with the two defaults d_1 and d_2. The resulting theory has two extensions, viz.

$$\text{Th}_2(\{B_l, U_r\}) \quad \text{and} \quad \text{Th}_2(\{B_r, U_l\}),$$

both containing

$$U_l \vee U_r,$$

which corresponds with our intuition.

Note that the difference between a formula $A \vee B$ and a disjunctive default $A|B$ amounts to the difference between the assertions "A or B is known" and "A is known or B is known".

3. A Sequent Calculus for Three-Valued Default Logic

We now introduce our sequent calculus B$_3$ for brave reasoning in **DL**$_3$. Following the general design of the approach of Bonatti [7,55], B$_3$ involves three kinds of sequents, viz. assertional sequents for axiomatising validity in **Ł**$_3$, anti-sequents for axiomatising non-tautologies of **Ł**$_3$, and special default sequents representing brave reasoning in **DL**.

We start with laying down the postulates of B$_3$ and then, in Section 3.2, we show soundness and completeness.

3.1. Postulates of the Calculus

As far as sequent-type calculi for three-valued logics are concerned,—or, more generally, many-valued logics—different techniques have been discussed in the literature [21,24,26,56–58]. Here, we use an approach due to Rousseau [25], which is a natural generalisation for many-valued logics of the classical two-sided sequent formulation as pioneered by Gentzen [23]. In Rousseau's approach, a sequent for a three-valued logic is a triple of sets of formulas where each component of the sequent represents one of the three truth values.

3.1.1. A Sequent Calculus for **Ł**$_3$

Formally, we introduce sequents for **Ł**$_3$ as follows:

Definition 1. *A (three-valued) sequent is a triple of the form $\Gamma_1 \mid \Gamma_2 \mid \Gamma_3$, where each Γ_i, for $i \in \{1,2,3\}$, is a finite set of formulas, called a component of the sequent.*

For a (three-valued) interpretation m, a sequent $\Gamma_1 \mid \Gamma_2 \mid \Gamma_3$ is true under m if, for at least one $i \in \{1,2,3\}$, Γ_i contains some formula A such that $V^m(A) = v_i$, where $v_1 = \mathbf{f}$, $v_2 = \mathbf{u}$, and $v_3 = \mathbf{t}$. Furthermore, a sequent is valid if it is true under each interpretation.

Note that a standard classical sequent $\Gamma \vdash \Delta$ in the sense of Gentzen [23] corresponds to a pair $\Gamma \mid \Delta$ under the usual two-valued semantics of **PL**.

As customary for sequents, we write sequent components comprised of a singleton set $\{A\}$ simply as "A", and likewise we write $\Gamma \cup \{A\}$ as "Γ, A".

For obtaining the postulates of a many-valued logic in Rousseau's approach, the conditions of the logical connectives of a given logic are encoded in two-valued logic by means of a so-called *partial normal form* [47] and expressed by suitable inference rules.

The calculus we employ for $Ł_3$, which we denote by $SŁ_3$, is taken from Zach [26], which is obtained from a systematic construction of sequent-style calculi for many-valued logics and by applying some optimisations of the corresponding partial normal forms.

Definition 2. *The postulates of $SŁ_3$ are as follows:*

- axioms of $SŁ_3$ are sequents of the form

    - $\bot \mid \emptyset \mid \emptyset$,
    - $\emptyset \mid \sqcup \mid \emptyset$,
    - $\emptyset \mid \emptyset \mid \top$, and
    - $A \mid A \mid A$, where A is a formula;

and

- the inference rules of $SŁ_3$ are comprised of the rules depicted in Figure 2.

Note that from the inference rules of $SŁ_3$, we can easily obtain derived rules for the defined connectives of $Ł_3$. Furthermore, the last three rules in Figure 2 are also referred to as *weakening rules*.

$$\frac{\Gamma \mid \Delta \mid \Pi, A \quad \Gamma, B \mid \Delta \mid \Pi}{\Gamma, A \supset B \mid \Delta \mid \Pi} \; (\supset : \mathbf{f})$$

$$\frac{\Gamma \mid \Delta, A, B \mid \Pi \quad \Gamma, B \mid \Delta \mid \Pi, A}{\Gamma \mid \Delta, A \supset B \mid \Pi} \; (\supset : \mathbf{u})$$

$$\frac{\Gamma, A \mid \Delta, A \mid \Pi, B \quad \Gamma, A \mid \Delta, B \mid \Pi, B}{\Gamma \mid \Delta \mid \Pi, A \supset B} \; (\supset : \mathbf{t})$$

$$\frac{\Gamma \mid \Delta \mid \Pi, A}{\Gamma, \neg A \mid \Delta \mid \Pi} \; (\neg : \mathbf{f}) \qquad \frac{\Gamma \mid \Delta, A \mid \Pi}{\Gamma \mid \Delta, \neg A \mid \Pi} \; (\neg : \mathbf{u}) \qquad \frac{\Gamma, A \mid \Delta \mid \Pi}{\Gamma \mid \Delta \mid \Pi, \neg A} \; (\neg : \mathbf{t})$$

$$\frac{\Gamma \mid \Delta \mid \Pi}{\Gamma, A \mid \Delta \mid \Pi} \; (w : \mathbf{f}) \qquad \frac{\Gamma \mid \Delta \mid \Pi}{\Gamma \mid \Delta, A \mid \Pi} \; (w : \mathbf{u}) \qquad \frac{\Gamma \mid \Delta \mid \Pi}{\Gamma \mid \Delta \mid \Pi, A} \; (w : \mathbf{t})$$

Figure 2. Rules of the sequent calculus $SŁ_3$.

Soundness and completeness of $SŁ_3$ follows directly from the method AS described by Zach [26]:

Proposition 2. *A sequent $\Gamma \mid \Delta \mid \Pi$ is valid iff it is provable in $SŁ_3$.*

Note that sequents in the style of Rousseau are *truth functional* rather than formalising entailment directly, but, by a general result for many-valued logics as shown by Zach [26], the latter can be expressed simply as follows:

Proposition 3. *For a theory T and a formula A, $T \vdash_3 A$ iff the sequent $T \mid T \mid A$ is provable in $SŁ_3$.*

3.1.2. An Anti-Sequent Calculus for Ł₃

As for axiomatising non-theorems of **Ł₃**, a systematic construction of rejection calculi for many-valued logics has been developed by Bogojeski and Tompits [27], based on adapting the approach of Zach [26]. The refutation calculus we describe now for axiomatising invalid sequents in **Ł₃**, denoted by RŁ₃, is obtained from the method of Bogojeski and Tompits [27].

Definition 3. *A (three-valued) anti-sequent is a triple of form $\Gamma_1 \nmid \Gamma_2 \nmid \Gamma_3$, where each Γ_i, for $i \in \{1,2,3\}$, is a finite set of formulas, called a component of the anti-sequent.*

For a (three-valued) interpretation m, an anti-sequent $\Gamma_1 \nmid \Gamma_2 \nmid \Gamma_3$ is refuted by m, or m refutes $\Gamma_1 \nmid \Gamma_2 \nmid \Gamma_3$, if, for every $i \in \{1,2,3\}$ and every formula $A \in \Gamma_i$, $V^m(A) \neq v_i$, where v_i is defined as in Definition 1. An anti-sequent $\Gamma_1 \nmid \Gamma_2 \nmid \Gamma_3$ is refutable *if there is at least one interpretation that refutes $\Gamma_1 \nmid \Gamma_2 \nmid \Gamma_3$.*

Clearly, an anti-sequent $\Gamma_1 \nmid \Gamma_2 \nmid \Gamma_3$ is refutable iff the corresponding sequent $\Gamma_1 \mid \Gamma_2 \mid \Gamma_3$ is not valid.

Definition 4. *The postulates of RŁ₃ are as follows:*

- *the axioms of RŁ₃ are anti-sequents of the form $\Gamma_1 \nmid \Gamma_2 \nmid \Gamma_3$, where each Γ_i ($i \in \{1,2,3\}$) is a set of atomic formulas such that $\Gamma_1 \cap \Gamma_2 \cap \Gamma_3 = \emptyset$, $\top \notin \Gamma_1$, $\sqcup \notin \Gamma_2$, and $\bot \notin \Gamma_3$; and*
- *the inference rules of RŁ₃ are those given in Figure 3.*

Note that, in contrast to SŁ₃, the inference rules of RŁ₃ have only single premises. Indeed, this is a general pattern in sequent-style rejection calculi: If an inference rule for standard (assertional) sequents for a connective has n premises, then there are usually n corresponding unary inference rules in the associated rejection calculus. Intuitively, what is an exhaustive search in a standard sequent calculus becomes nondeterminism in a rejection calculus.

$$\frac{\Gamma \nmid \Delta \nmid \Pi, A}{\Gamma, A \supset B \nmid \Delta \nmid \Pi} \ (\supset : \mathbf{f}^1)^r \qquad \frac{\Gamma, B \nmid \Delta \nmid \Pi}{\Gamma, A \supset B \nmid \Delta \nmid \Pi} \ (\supset : \mathbf{f}^2)^r$$

$$\frac{\Gamma \nmid \Delta, A, B \nmid \Pi}{\Gamma \nmid \Delta, A \supset B \nmid \Pi} \ (\supset : \mathbf{u}^1)^r \qquad \frac{\Gamma, B \nmid \Delta \nmid \Pi, A}{\Gamma \nmid \Delta, A \supset B \nmid \Pi} \ (\supset : \mathbf{u}^2)^r$$

$$\frac{\Gamma, A \nmid \Delta, A \nmid \Pi, B}{\Gamma \nmid \Delta \nmid \Pi, A \supset B} \ (\supset : \mathbf{t}^1)^r \qquad \frac{\Gamma, A \nmid \Delta, B \nmid \Pi, B}{\Gamma \nmid \Delta \nmid \Pi, A \supset B} \ (\supset : \mathbf{t}^2)^r$$

$$\frac{\Gamma \nmid \Delta \nmid \Pi, A}{\Gamma, \neg A \nmid \Delta \nmid \Pi} \ (\neg : \mathbf{f})^r \qquad \frac{\Gamma \nmid \Delta, A \nmid \Pi}{\Gamma \nmid \Delta, \neg A \nmid \Pi} \ (\neg : \mathbf{u})^r \qquad \frac{\Gamma, A \nmid \Delta \nmid \Pi}{\Gamma \nmid \Delta \nmid \Pi, \neg A} \ (\neg : \mathbf{t})^r$$

Figure 3. Rules of the anti-sequent calculus RŁ₃.

Again, soundness and completeness of RŁ₃ follow from the systematic construction as described by Bogojeski and Tompits [27]. Likewise, non-entailment in **Ł₃** is expressed similarly as for SŁ₃.

Proposition 4. *An anti-sequent $\Gamma \nmid \Delta \nmid \Pi$ is refutable iff it is provable in RŁ₃.*

Proposition 5. *For a theory T and a formula A, $T \nvdash_3 A$ iff $T \nmid T \nmid A$ is provable in RŁ₃.*

3.1.3. The Default-Sequent Calculus B₃

We are now in a position to specify our calculus B₃ for brave reasoning in **DL₃**.

Definition 5. *A (brave) default sequent is an ordered quadruple of the form* $\Gamma; \Delta \Rightarrow \Sigma; \Theta$, *where* Γ, Σ, *and* Θ *are finite sets of formulas and* Δ *is a finite set of defaults.*

A default sequent $\Gamma; \Delta \Rightarrow \Sigma; \Theta$ *is true if there is an extension E of the default theory* $\langle \Gamma, \Delta \rangle$ *such that* $\Sigma \subseteq E$ *and* $\Theta \cap E = \emptyset$; *E is called a witness of* $\Gamma; \Delta \Rightarrow \Sigma; \Theta$.

The default sequent calculus B_3 consists of three-valued sequents, anti-sequents, and default sequents. It incorporates the systems $S\text{Ł}_3$ for three-valued sequents and $R\text{Ł}_3$ for anti-sequents, as well as additional axioms and inference rules for default sequents, described as follows:

Definition 6. *The postulates of* B_3 *comprise the following items:*
- *all axioms and inference rules of* $S\text{Ł}_3$ *and* $R\text{Ł}_3$;
- *axioms of the form* $\Gamma; \emptyset \Rightarrow \emptyset; \emptyset$, *where* Γ *is a finite set of formulas of* Ł_3; *and*
- *the inference rules depicted in Figure 4.*

$$\frac{\Gamma \mid \Gamma \mid A}{\Gamma; \emptyset \Rightarrow A; \emptyset} \ (l_1) \qquad \frac{\Gamma \dagger \Gamma \dagger A}{\Gamma; \emptyset \Rightarrow \emptyset; A} \ (l_2)$$

$$\frac{\Gamma; \emptyset \Rightarrow \Sigma_1; \Theta_1 \quad \Gamma; \emptyset \Rightarrow \Sigma_2; \Theta_2}{\Gamma; \emptyset \Rightarrow \Sigma_1, \Sigma_2; \Theta_1, \Theta_2} \ (mu)$$

$$\frac{\Gamma; \Delta \Rightarrow \Sigma; \Theta, A}{\Gamma; \Delta, (A : B_1, \ldots, B_n / C) \Rightarrow \Sigma; \Theta} \ (d_1) \qquad \frac{\Gamma; \Delta \Rightarrow \Sigma, \neg B_i; \Theta}{\Gamma; \Delta, (A : B_1, \ldots, B_i, \ldots, B_n / C) \Rightarrow \Sigma; \Theta} \ (d_2)$$

$$\frac{\Gamma; \Delta \Rightarrow \Sigma, \neg LC; \Theta}{\Gamma; \Delta, (A : B_1, \ldots, B_n / C) \Rightarrow \Sigma; \Theta} \ (d_3)$$

$$\frac{\Gamma; \emptyset \Rightarrow A; \emptyset \quad \Gamma, MC; \Delta \Rightarrow \Sigma; \Theta, \neg B_1, \ldots, \neg B_n, \neg LC}{\Gamma; \Delta, (A : B_1, \ldots, B_n / C) \Rightarrow \Sigma; \Theta} \ (d_4)$$

Figure 4. Rules for default sequents of the calculus B_3.

The informal meaning of the inference rules for the default sequents is the following:

(i) rules (l_1) and (l_2) combine three-valued sequents and anti-sequents with default sequents, respectively;
(ii) rule (mu) is the rule of "monotonic union"—it allows the joining of information in case that no default is present; and
(iii) rules (d_1)–(d_4) are the default introduction rules, where rules (d_1), (d_2), and (d_3) take care of introducing non-active defaults, whilst rule (d_4) allows to introduce an active default.

Let us give an example to illustrate the functioning of the calculus.

Example 5. *Consider the default theory* $T = \langle W, D \rangle$ *from Example 1, where*

$$W = \{Summer, \neg Sun\_Shining\} \quad \text{and} \quad D = \left\{ \frac{Summer : \neg Rain}{Sun\_Shining} \right\}.$$

As we saw, the single default of this theory is inapplicable since $W \vdash_3 \neg L Sun\_Shining$ *and* $E = Th_3(W)$ *is therefore the only extension of T. Consequently, $Sun\_Shining \notin E$ also holds. Hence, the default sequent*

$$Summer, \neg Sun\_Shining; (Summer : \neg Rain / Sun\_Shining) \Rightarrow \neg L Sun\_Shining; Sun\_Shining \quad (1)$$

is true. We will give a proof of (1) in B_3.

The proof of (1), depicted below and denoted by β, *uses the proof* α *as subproof. For brevity, we will use "S" for "Summer", "R" for "Rain", and "H" for Sun_Shining.*

- Proof α:

$$\cfrac{\cfrac{\cfrac{\cfrac{\cfrac{\cfrac{\cfrac{\cfrac{\cfrac{\neg H \mid \neg H \mid \neg H}{\neg H \mid S, \neg H \mid \neg H}\,(w:\mathbf{u})}{\neg H, H \mid S, \neg H \mid \neg H}\,(w:\mathbf{f})}{\neg H, H \mid S, \neg H, H \mid \neg H}\,(w:\mathbf{u})}{S, \neg H, H \mid S, \neg H, H \mid \neg H}\,(w:\mathbf{f})\qquad \cfrac{\cfrac{\cfrac{\cfrac{\neg H \mid \neg H \mid \neg H}{\neg H \mid S, \neg H \mid \neg H}\,(w:\mathbf{u})}{\neg H, H \mid S, \neg H \mid \neg H}\,(w:\mathbf{f})}{\neg H, H \mid S, \neg H, H \mid \neg H}\,(w:\mathbf{u})}{S, \neg H, H \mid S, \neg H \mid \neg H}\,(w:\mathbf{f})}{S, \neg H \mid S, \neg H \mid H \supset \neg H}\,(\supset:\mathbf{t})}{S, \neg H, \neg(H \supset \neg H) \mid S, \neg H \mid \emptyset}\,(\neg:\mathbf{f})}{S, \neg H \mid S, \neg H \mid \neg\neg(H \supset \neg H)}\,(\neg:\mathbf{t})}{S, \neg H; \emptyset \Rightarrow \neg LH; \emptyset}\,(l_1),\ \text{definition of L}$$

- Proof β:

$$\cfrac{\cfrac{\cfrac{\cfrac{\cfrac{S \dagger S, H \dagger H}{S \dagger S, \neg H \dagger H}\,(\neg:\mathbf{u})^r}{S, \neg H \dagger S, \neg H \dagger H}\,(\neg:\mathbf{f})^r}{S, \neg H; \emptyset \Rightarrow \emptyset; H}\,(l_2)}{S, \neg H; \emptyset \Rightarrow \neg LH; H}\ \alpha}{S, \neg H; (S:\neg R/H) \Rightarrow \neg LH; H}\,(d_3)\ (mu)$$

3.2. Adequacy of the Calculus

We now show soundness and completeness of B_3. To this end, we need some auxiliary results first, dealing with alternative characterisations and properties of extensions.

3.2.1. Preparatory Characterisations: Residues and Extensions

We start with some properties of extensions concerning adding defaults to default theories which provide the groundwork on which our adequacy proofs are built. In doing so, we first introduce an alternative formulation of **DL₃** extensions, adapting a proof-theoretical characterisation as described by Marek and Truszczyński [59] for standard default logic, and afterwards we provide results concerning so-called *residues*, which are inference rules resulting from defaults satisfying their consistency conditions. The latter endeavour generalises the approach of Bonatti [7] to the three-valued case.

Definition 7. *Let E be a set of formulas. A default $(A:B_1,\ldots,B_n/C)$ is $\mathbf{DL_3}$-active in E iff $E \vdash_3 A$ and $\{\neg B_1,\ldots,\neg B_n, \neg LC\} \cap E = \emptyset$.*

Definition 8. *Let D be a set of defaults and E a set of formulas. The $\mathbf{DL_3}$-reduct of D with respect to E, denoted by D_E, is the set consisting of the following inference rules:*

$$D_E := \left\{ \frac{A}{MC} \ \Big|\ \frac{A:B_1,\ldots,B_n}{C} \in D \text{ and } \{\neg B_1,\ldots,\neg B_n, \neg LC\} \cap E = \emptyset \right\}.$$

An inference rule A/MC is called $\mathbf{DL_3}$-residue of a default $(A:B_1,\ldots,B_n/C)$.

Whenever it is clear from the context, we will allow ourselves to drop the prefix "**DL₃**-" in "**DL₃**-active", "**DL₃**-reduct", and "**DL₃**-residue" to ease notation.

For a set R of inference rules, let $\vdash_3^R$ be the inference relation obtained from $\vdash_3$ by augmenting the postulates of the Hilbert-type calculus for **Ł₃** underlying the relation $\vdash_3$ with the inference rules from R. Let the corresponding deductive closure operator for $\vdash_3^R$ be given by

$$\text{Th}_3^R(W) := \{A \mid W \vdash_3^R A\}.$$

Clearly, $\text{Th}_3^\emptyset(W) = \text{Th}_3(W)$.

We then obtain the following characterisation of the operator Γ_T, mirroring the analogous property for standard default logic as discussed by Marek and Truszczyński [59]:

Theorem 1. *Let $T = \langle W, D \rangle$ be a three-valued default theory, E a set of formulas of $\textbf{Ł}_3$, and D_E the $\textbf{DL}_3$-reduct of D with respect to E. Then,*
$$\Gamma_T(E) = \text{Th}_3^{D_E}(W).$$

Proof. The result follows by a straightforward adaption of the proof of the analogous result for the case of standard default logic as given by Marek and Truszczyński [59]. □

By the definition of an extension, we thus obtain:

Corollary 1. *Let $T = \langle W, D \rangle$ be a three-valued default theory and E a set of formulas. Then,*
$$E \text{ is an extension of } T \text{ iff } \text{Th}_3^{D_E}(W) = E.$$

Next, we give some properties of extensions with respect to active and non-active defaults which underlay the construction of the default inference rules of B_3. We start with two lemmata whose proofs are obvious.

Lemma 1. *Let R and R' be sets of inference rules, and let W and W' be sets of formulas. Then, the following properties hold:*

1. $W \subseteq \text{Th}_3^R(W)$.
2. $\text{Th}_3^R(W) = \text{Th}_3^R(\text{Th}_3^R(W))$.
3. *If $R \subseteq R'$, then $\text{Th}_3^R(W) \subseteq \text{Th}_3^{R'}(W)$.*
4. *If $W \subseteq W'$, then $\text{Th}_3^R(W) \subseteq \text{Th}_3^R(W')$.*

Lemma 2. *Let A and B be formulas, W a set of formulas, and R a set of inference rules. Then:*

1. *If $A \notin \text{Th}_3^R(W)$, then $\text{Th}_3^R(W) = \text{Th}_3^{R \cup \{A/B\}}(W)$.*
2. *If $A \in \text{Th}_3^{R \cup \{A/B\}}(W)$, then $\text{Th}_3^{R \cup \{A/B\}}(W) = \text{Th}_3^R(W \cup \{B\})$.*

For convenience, we employ the following notation in what follows: For a default
$$d = \frac{A : B_1, \ldots, B_n}{C},$$
we write:

- $\text{p}(d) := A$;
- $\text{j}(d) := \{B_1, \ldots, B_n, \text{LC}\}$; and
- $\text{c}(d) := \text{MC}$.

Furthermore, for a set S of formulas, $\neg S$ stands for $\{\neg A \mid A \in S\}$.

Theorem 2. *Let $T = \langle W, D \rangle$ be a default theory, E a set of formulas, and d a default not active in E. Then,*
$$E \text{ is an extension of } \langle W, D \rangle \text{ iff } E \text{ is an extension of } \langle W, D \cup \{d\} \rangle.$$

Proof. If $\neg \text{j}(d) \cap E \neq \emptyset$, then $(D \cup \{d\})_E = D_E$. So,
$$\text{Th}_3^{(D \cup \{d\})_E}(W) = \text{Th}_3^{D_E}(W)$$

and the statement of the theorem holds quite trivially by Corollary 1.

For the rest of the proof, assume thus $\neg j(d) \cap E = \emptyset$. Since d is not active in E, $E \not\vdash_3 p(d)$ must then hold. Furthermore,

$$(D \cup \{d\})_E = D_E \cup \{p(d)/c(d)\} \tag{2}$$

holds.

Suppose E is an extension of $T = \langle W, D \rangle$, i.e., $E = \text{Th}_3^{D_E}(W)$. Since $E \not\vdash_3 p(d)$ and E is deductively closed, we obtain $p(d) \notin E$, and so $p(d) \notin \text{Th}_3^{D_E}(W)$. By part 1 of Lemma 2,

$$\text{Th}_3^{D_E}(W) = \text{Th}_3^{D_E \cup \{p(d)/c(d)\}}(W).$$

But in view of (2), we have that,

$$\text{Th}_3^{D_E}(W) = \text{Th}_3^{(D \cup \{d\})_E}(W).$$

Hence, since $E = \text{Th}_3^{D_E}(W)$, we obtain $E = \text{Th}_3^{(D \cup \{d\})_E}(W)$ and E is an extension of $\langle W, D \cup \{d\} \rangle$. This proves the "only if" direction.

For the "if" direction, assume now that E is an extension of $\langle W, D \cup \{d\} \rangle$. So, $E = \text{Th}_3^{(D \cup \{d\})_E}(W)$. Since we again have that $p(d) \notin E$ and $(D \cup \{d\})_E = D_E \cup \{p(d)/c(d)\}$ by (2), it follows that $p(d) \notin \text{Th}_3^{D_E \cup \{p(d)/c(d)\}}(W)$. Part 3 of Lemma 1 implies that $p(d) \notin \text{Th}_3^{D_E}(W)$ also holds, and thus, by part 1 of Lemma 2,

$$\text{Th}_3^{D_E}(W) = \text{Th}_3^{D_E \cup \{p(d)/c(d)\}}(W). \tag{3}$$

Since $E = \text{Th}_3^{(D \cup \{d\})_E}(W)$ by hypothesis and $\text{Th}_3^{(D \cup \{d\})_E}(W) = \text{Th}_3^{D_E \cup \{p(d)/c(d)\}}(W)$, by (3) we get that $E = \text{Th}_3^{D_E}(W)$, i.e., E is an extension of $T = \langle W, D \rangle$. □

Theorem 3. *Let E be a set of formulas and d a default. If E is an extension of $\langle W, D \cup \{d\} \rangle$ and d is active in E, then E is an extension of $\langle W \cup \{c(d)\}, D \rangle$.*

Proof. Suppose E is an extension of $\langle W, D \cup \{d\} \rangle$ and d is active in E. Then,

$$E = \text{Th}_3^{(D \cup \{d\})_E}(W)$$

and, since d is active in E, $\neg j(d) \cap E = \emptyset$. Therefore,

$$(D \cup \{d\})_E = D_E \cup \{p(d)/c(d)\}$$

and thus

$$E = \text{Th}_3^{D_E \cup \{p(d)/c(d)\}}(W).$$

But $E \vdash_3 p(d)$ also holds (since d is active in E), and so,

$$p(d) \in \text{Th}_3^{D_E \cup \{p(d)/c(d)\}}(W).$$

Therefore, by part 2 of Lemma 2,

$$\text{Th}_3^{D_E \cup \{p(d)/c(d)\}}(W) = \text{Th}_3^{D_E}(W \cup \{c(d)\}).$$

Thus, $E = \text{Th}_3^{D_E}(W \cup \{c(d)\})$, and so E is an extension of $\langle W \cup \{c(d)\}, D \rangle$. □

Theorem 4. *Let E be a set of formulas and d a default. If (i) E is an extension of the default theory $\langle W \cup \{c(d)\}, D \rangle$, (ii) $W \vdash_3 p(d)$, and (iii) $\neg j(d) \cap E = \emptyset$, then E is an extension of $\langle W, D \cup \{d\} \rangle$.*

Proof. Assume that the preconditions of the theorem hold. Since E is an extension of $\langle W \cup \{c(d)\}, D \rangle$,

$$E = \text{Th}_3^{D_E}(W \cup \{c(d)\}).$$

Furthermore, by the hypothesis $W \vdash_3 p(d)$, we have $p(d) \in \text{Th}_3^{D_E \cup \{p(d)/c(d)\}}(W)$. We thus get,

$$\text{Th}_3^{D_E \cup \{p(d)/c(d)\}}(W) = \text{Th}_3^{D_E}(W \cup \{c(d)\})$$

in view of part 2 of Lemma 2, and therefore,

$$E = \text{Th}_3^{D_E \cup \{p(d)/c(d)\}}(W).$$

By observing that the assumption $\neg j(d) \cap E = \varnothing$ implies $D_E \cup \{p(d)/c(d)\} = (D \cup \{d\})_E$, the result follows. □

3.2.2. Soundness and Completeness of B_3

We are now in a position to prove soundness and completeness of B_3.

Theorem 5 (Soundness). *If* $\Gamma; \Delta \Rightarrow \Sigma; \Theta$ *is provable in* B_3, *then it is true.*

Proof. We show that all axioms are true, and that the conclusions of all inference rules are true whenever its premisses are true (resp., valid or refutable in case of rules (l_1) and (l_2)).

First of all, an axiom $\Gamma; \varnothing \Rightarrow \varnothing; \varnothing$ is trivially true, because $\text{Th}_3(\Gamma)$ is the unique extension of the default theory $\langle \Gamma, \varnothing \rangle$ and hence the unique witness of $\Gamma; \varnothing \Rightarrow \varnothing; \varnothing$.

Assume that the premiss $\Gamma \mid \Gamma \mid A$ of rule (l_1) is valid. Then, $\Gamma \vdash_3 A$ holds and we therefore have $A \in \text{Th}_3(\Gamma)$. But $\text{Th}_3(\Gamma)$ is the unique extension of $\langle \Gamma, \varnothing \rangle$, so $\text{Th}_3(\Gamma)$ is the unique witness of $\Gamma; \varnothing \Rightarrow A; \varnothing$. Likewise, if the premiss $\Gamma \nmid \Gamma \nmid A$ of rule (l_2) is refutable, then $A \notin \text{Th}_3(\Gamma)$, and therefore $\text{Th}_3(\Gamma)$ is the (unique) witness of $\Gamma; \varnothing \Rightarrow \varnothing; A$.

If the two premisses $\Gamma; \varnothing \Rightarrow \Sigma_1; \Theta_1$ and $\Gamma; \varnothing \Rightarrow \Sigma_2; \Theta_2$ of rule (mu) are true, then they must have the same witness $E = \text{Th}_3(\Gamma)$. So, $\Sigma_i \subseteq E$ and $\Theta_i \cap E = \varnothing$, for $i = 1, 2$, holds, and hence $\Sigma_1 \cup \Sigma_2 \subseteq E$ and $(\Theta_1 \cup \Theta_2) \cap E = \varnothing$ holds too, which means that E is also a witness of $\Gamma; \varnothing \Rightarrow \Sigma_1, \Sigma_2; \Theta_1, \Theta_2$.

For showing the soundness of the rules d_1, d_2, and d_3, we only deal with the case for d_3; the other two cases are similar. So, let E be a witness of $\Gamma; \Delta \Rightarrow \Sigma, \neg LC; \Theta$. Then, E is an extension of $\langle \Gamma, \Delta \rangle$, $\Sigma \cup \{\neg LC\} \subseteq E$, and $\Theta \cap E = \varnothing$. Hence, $\neg LC \in E$ and thus the default $(A : B_1, \ldots, B_n/C)$ is not active in E. By Theorem 2, it follows that E is an extension of $\langle \Gamma, \Delta \cup \{(A : B_1, \ldots, B_n/C)\}\rangle$. Moreover, since $\Sigma \subseteq E$ and $\Theta \cap E = \varnothing$, E is a witness of $\Gamma; \Delta, (A : B_1, \ldots, B_n/C) \Rightarrow \Sigma; \Theta$.

Finally, assume that the premisses of rule (d_4) are true. Let E_1 be a witness of

$$\Gamma; \varnothing \Rightarrow A; \varnothing$$

and E_2 a witness of

$$\Gamma, MC; \Delta \Rightarrow \Sigma; \Theta, \neg B_1, \ldots, \neg B_n, \neg LC.$$

Thus, E_2 is an extension of $\langle \Gamma \cup \{MC\}, \Delta \rangle$ and $\{\neg B_1, \ldots, \neg B_n, \neg LC\} \cap E_2 = \varnothing$ holds. Moreover, E_1 is an extension of $\langle \Gamma, \varnothing \rangle$ with $A \in E_1$, and therefore $\Gamma \vdash_3 A$. Hence, by Theorem 4, E_2 is an extension of

$$\langle \Gamma, \Delta \cup \{(A : B_1, \ldots, B_n/C)\}\rangle.$$

Clearly, $\Sigma \subseteq E_2$ and $\Theta \cap E_2 = \varnothing$ holds, so E_2 is a witness of $\Gamma; \Delta, (A : B_1, \ldots, B_n/C) \Rightarrow \Sigma; \Theta$. □

Theorem 6 (Completeness). *If* $\Gamma; \Delta \Rightarrow \Sigma; \Theta$ *is true, then it is provable in* B_3.

Proof. Suppose $S = \Gamma; \Delta \Rightarrow \Sigma; \Theta$ is true, with E as its witness. The proof proceeds by induction on the cardinality $|\Delta|$ of Δ.

INDUCTION BASE. Assume $|\Delta| = 0$. If $\Sigma = \Theta = \emptyset$, then S is an axiom and hence provable in B_3. So suppose that either $\Sigma \neq \emptyset$ or $\Theta \neq \emptyset$. Since $Th_3(\Gamma)$ is the unique extension of $\langle \Gamma, \emptyset \rangle$, we have $E = Th_3(\Gamma)$. Furthermore, $\Sigma \subseteq E$ and $\Theta \cap E = \emptyset$ holds. It follows that for any $A \in \Sigma$, the sequent $\Gamma \mid \Gamma \mid A$ is provable in St_3, and for any $B \in \Theta$, the anti-sequent $\Gamma \nmid \Gamma \nmid B$ is provable in Rt_3. Repeated applications of rules (l_1), (l_2), and (mu) yield a proof of S in B_3.

INDUCTION STEP. Assume $|\Delta| > 0$, and let the statement hold for all default sequents $\Gamma'; \Delta' \Rightarrow \Sigma'; \Theta'$ such that $|\Delta'| < |\Delta|$. We distinguish two cases: (i) There is some default in Δ which is active in E, or (ii) none of the defaults in Δ is active in E.

If (i) holds, then there must be some default $d = (A : B_1, \ldots, B_n/C)$ in Δ such that d is active in E and $\Gamma \vdash_3 A$. Consider $\Delta_0 := \Delta \setminus \{d\}$. Then, $|\Delta_0| = |\Delta| - 1$ and $\Delta_0 \cup \{d\} = \Delta$. By Theorem 3, E is an extension of $\langle \Gamma \cup \{MC\}, \Delta_0 \rangle$. Since d is active in E, $\{\neg B_1, \ldots, \neg B_n, \neg LC\} \cap E = \emptyset$ holds and since E is a witness of $S = \Gamma; \Delta \Rightarrow \Sigma; \Theta$, we have that $\Sigma \subseteq E$ and $\Theta \cap E = \emptyset$. So, E is a witness of

$$S' = \Gamma, MC; \Delta_0 \Rightarrow \Sigma; \Theta, \neg B_1, \ldots, \neg B_n, \neg LC.$$

Since $|\Delta_0| < |\Delta|$, by induction hypothesis there is some proof α in B_3 of S'. Furthermore, $\Gamma \vdash_3 A$, so there is some proof β of the sequent $\Gamma \mid \Gamma \mid A$ in St_3. The following figure is a proof of S in B_3 (note that in this figure, the endsequents of α and β have been displayed explicitly for better clarity):

$$\dfrac{\dfrac{\overset{\beta}{\Gamma \mid \Gamma \mid A}}{\Gamma; \emptyset \Rightarrow A; \emptyset}\,(l_1) \qquad \overset{\alpha}{\Gamma, MC; \Delta_0 \Rightarrow \Sigma; \Theta, \neg B_1, \ldots, \neg B_n, \neg LC}}{\Gamma; \Delta \Rightarrow \Sigma; \Theta}\,(d_4)$$

Now assume that (ii) holds, i.e., no default in Δ is active in E. Since $|\Delta| > 0$, there is some default $d = (A : B_1, \ldots, B_n/C)$ in Δ such that $\Delta = \Delta_0 \cup \{d\}$ with $\Delta_0 := \Delta \setminus \{d\}$. Since d is not active in E, according to Theorem 2, E is an extension of $\langle \Gamma, \Delta_0 \rangle$. Furthermore, either:

- $E \nvdash_3 A$;
- there is some $B_{i_0} \in \{B_1, \ldots, B_n\}$ such that $\neg B_{i_0} \in E$; or
- $\neg LC \in E$.

Consequently, E is either a witness of:

- $\Gamma; \Delta_0 \Rightarrow \Sigma; \Theta, A$;
- $\Gamma; \Delta_0 \Rightarrow \Sigma, \neg B_{i_0}; \Theta$; or
- $\Gamma; \Delta_0 \Rightarrow \Sigma, \neg LC; \Theta$.

Since $|\Delta_0| < |\Delta|$, by induction hypothesis there is thus either:

- a proof α in B_3 of $\Gamma; \Delta_0 \Rightarrow \Sigma; \Theta, A$;
- a proof β in B_3 of $\Gamma; \Delta_0 \Rightarrow \Sigma, \neg B_{i_0}; \Theta$; or
- a proof γ in B_3 of $\Gamma; \Delta_0 \Rightarrow \Sigma, \neg LC; \Theta$.

Therefore, one of the three figures below constitutes a proof of S (again, the respective endsequents of α, β, and γ are explicitly shown):

$$\dfrac{\overset{\alpha}{\Gamma; \Delta_0 \Rightarrow \Sigma; \Theta, A}}{\Gamma; \Delta \Rightarrow \Sigma; \Theta}\,(d_1) \qquad \dfrac{\overset{\beta}{\Gamma; \Delta_0 \Rightarrow \Sigma, \neg B_{i_0}; \Theta}}{\Gamma; \Delta \Rightarrow \Sigma; \Theta}\,(d_2) \qquad \dfrac{\overset{\gamma}{\Gamma; \Delta_0 \Rightarrow \Sigma, \neg LC; \Theta}}{\Gamma; \Delta \Rightarrow \Sigma; \Theta}\,(d_3).$$

□

4. A Sequent Calculus for Disjunctive Default Logic

We now introduce our sequent calculus for brave reasoning for disjunctive default logic which we denote by B_D. Again, the calculus comprises of three kinds of sequents:

(i) sequents for expressing validity in **PL**;
(ii) anti-sequents for expressing non-tautologies; and
(iii) special default inference rules reflecting brave reasoning in DL_D.

As sequents for propositional logic, we use standard two-sided sequents in the sense of Gentzen [23] and a corresponding calculus, LK, which is a slight simplification of the one originally introduced by Gentzen. As a calculus for anti-sequents, we use the one due to Bonatti [37] which he introduced in connection to his calculus for brave reasoning for standard default logic [7,55]; we will denote this calculus by LK^r (note that, independently from Bonatti [37], Goranko [38] developed a similar calculus as part of his refutation systems for different modal logics).

4.1. Postulates of the Calculus

We start with defining the sequent calculus LK for classical sequents.

4.1.1. The Sequent Calculus LK

Definition 9. *A (classical) sequent is an ordered pair of the form $\Gamma \to \Sigma$, where Γ and Σ are finite sets of formulas. Γ is the antecedent and Σ is the succedent of the sequent.*

For a two-valued interpretation I, a sequent $\Gamma \to \Delta$ is true under I if, whenever all formulas in Γ are true under I, then at least one formula in Δ is true under I. Furthermore, a sequent is valid if it is true under each interpretation.

Following customs, we write sequents of the form $\Gamma \cup \{A\} \to \Delta$ simply as "$\Gamma, A \to \Delta$", and if the antecedent or succedent of a sequent is the empty set, then it is omitted from the sequent.

Definition 10. *The postulates of LK are as follows:*

- *axioms of LK are sequents of the form*
  - $\to \top$,
  - $\bot \to$, and
  - $A \to A$, *where A is a formula;*

  and

- *the inference rules of LK are those given in Figure 5.*

Note that the last two rules in Figure 5 are the *weakening rules* of LK. Moreover, from the rules of LK, we can easily obtain derived rules for the defined connectives $\wedge$, $\vee$, and $\equiv$. For instance, the derived rules for $\wedge$ are as follows:

$$\frac{\Gamma, A \to \Sigma}{\Gamma, (A \wedge B) \to \Sigma} \ (\wedge \to)_1 \qquad \frac{\Gamma, B \to \Sigma}{\Gamma, (A \wedge B) \to \Sigma} \ (\wedge \to)_2 \qquad \frac{\Gamma \to \Sigma, A \quad \Lambda \to \Pi, B}{\Gamma, \Lambda \to \Sigma, \Pi, (A \wedge B)} \ (\to \wedge).$$

Soundness and completeness of LK is well known:

$$\frac{\Gamma \to \Sigma, A \quad \Lambda, B \to \Pi}{\Gamma, \Lambda, (A \supset B) \to \Sigma, \Pi} \; (\supset \to) \qquad \frac{\Gamma, A \to \Sigma, B}{\Gamma \to \Sigma, (A \supset B)} \; (\to \supset)$$

$$\frac{\Gamma \to \Sigma, A}{\Gamma, \neg A \to \Sigma} \; (\neg \to) \qquad \frac{\Gamma, A \to \Sigma}{\Gamma \to \Sigma, \neg A} \; (\to \neg)$$

$$\frac{\Gamma \to \Sigma}{\Gamma, A \to \Sigma} \; (wl) \qquad \frac{\Sigma \to \Gamma}{\Sigma \to \Gamma, A} \; (wr)$$

Figure 5. Rules of the sequent calculus LK.

Proposition 6 ([23]). *A sequent $\Gamma \to \Sigma$ is valid iff it is provable in* LK.

In particular, the following relation follows immediately:

Corollary 2. *For every formula A,*

$$\models_2 A \text{ iff the sequent } \to A \text{ is provable in } \mathsf{LK}.$$

4.1.2. The Anti-Sequent Calculus LK^r

Now we introduce our complementary calculus LK^r for axiomatising invalidity in propositional logic, following Bonatti [37] (and Goranko [38]).

Definition 11. *An anti-sequent is an ordered pair of the form $\Gamma \not\to \Theta$, where Γ and Θ are finite sequences of formulas.*

For a two-valued interpretation I, an anti-sequent $\Gamma \not\to \Theta$ is refuted by I, or I refutes $\Gamma \not\to \Theta$, if every formula in Γ is true under I and every formula in Θ is false under I. An anti-sequent $\Gamma \not\to \Theta$ is refutable if there is at least one interpretation that refutes $\Gamma \not\to \Theta$.

Hence, the anti-sequent $\Gamma \not\to \Theta$ is refutable iff the classical sequent $\Gamma \to \Theta$ is invalid. Also, in accordance to the convention for classical sequents, we write "$\not\to \Theta$" and "$\Gamma \not\to$" whenever Γ or Θ is the empty set.

Definition 12. *The postulates of* LK^r *are as follows:*

- *the axioms of* LK^r *are anti-sequents of the form $\Phi \not\to \Psi$, where Φ and Ψ are disjoint finite sets of atomic formulas such that $\bot \notin \Phi$ and $\top \notin \Psi$; and*
- *the inference rules of* LK^r *are those depicted in Figure 6.*

$$\frac{\Gamma \not\to \Theta, A}{\Gamma, (A \supset B) \not\to \Theta} \; (\supset \not\to)_1^r \qquad \frac{\Gamma, B \not\to \Theta}{\Gamma, (A \supset B) \not\to \Theta} \; (\supset \not\to)_2^r$$

$$\frac{\Gamma, A \not\to \Theta, B}{\Gamma \not\to \Theta, (A \supset B)} \; (\not\to \supset)^r$$

$$\frac{\Gamma \not\to \Theta, A}{\Gamma, \neg A \not\to \Theta} \; (\neg \not\to)^r \qquad \frac{\Gamma, A \not\to \Theta}{\Gamma \not\to \Theta, \neg A} \; (\not\to \neg)^r$$

Figure 6. Rules of the anti-sequent calculus LK^r.

Note that, following the general pattern of complementary calculi, the inference rules of LK^r have only single premisses.

We again can obtain corresponding derived rules for the defined connectives. Below we give the ones for $\wedge$:

$$\frac{\Gamma, A, B \not\to \Theta}{\Gamma, (A \wedge B) \not\to \Theta} \ (\wedge \not\to)^r \qquad \frac{\Gamma \not\to \Theta, A}{\Gamma \not\to \Theta, (A \wedge B)} \ (\not\to \wedge)^r_1 \qquad \frac{\Gamma \not\to \Theta, B}{\Gamma \not\to \Theta, (A \wedge B)} \ (\not\to \wedge)^r_2.$$

Soundness and completeness for LK^r was shown by Bonatti [37] (and, independently, by Goranko [38]):

Proposition 7. *An anti-sequent $\Gamma \not\to \Theta$ is refutable iff it is provable in LK^r.*

For formulas, we have then the following immediate corollary:

Corollary 3. *For every formula A,*

$$\not\models_2 A \text{ iff the anti-sequent } \not\to A \text{ is provable in LK}^r.$$

4.1.3. The Default-Sequent Calculus B$_D$

We can now specify our calculus B$_D$ for brave reasoning in disjunctive default logic.

Definition 13. *By a (brave) disjunctive default sequent we understand an ordered quadruple of the form $\Gamma; \Delta \Rightarrow \Sigma; \Theta$, where Γ, Σ, and Θ are finite sets of formulas and Δ is a finite set of disjunctive defaults.*

A disjunctive default sequent $\Gamma; \Delta \Rightarrow \Sigma; \Theta$ is true iff there is an extension E of the disjunctive default theory $\langle \Gamma, \Delta \rangle$ such that $\Sigma \subseteq E$ and $\Theta \cap E = \varnothing$; E is called a witness of $\Gamma; \Delta \Rightarrow \Sigma; \Theta$.

The default sequent calculus B$_D$ consists of sequents, anti-sequents, and disjunctive default sequents. It incorporates the systems LK for sequents and LK^r for anti-sequents, as well as additional axioms and inference rules for disjunctive default sequents, similar to the case of B$_3$.

Definition 14. *The postulates of B$_D$ comprise the following items:*
- *all axioms and inference rules of LK and LK^r;*
- *axioms of the form $\Gamma; \varnothing \Rightarrow \varnothing; \varnothing$, where Γ is a finite set of formulas of PL; and*
- *the inference rules are those depicted in Figure 7.*

$$\frac{\Gamma \to A}{\Gamma; \varnothing \Rightarrow A; \varnothing} \ (l_1)^d \qquad\qquad \frac{\Gamma \not\to A}{\Gamma; \varnothing \Rightarrow \varnothing; A} \ (l_2)^d$$

$$\frac{\Gamma; \varnothing \Rightarrow \Sigma_1; \Theta_1 \quad \Gamma; \varnothing \Rightarrow \Sigma_2; \Theta_2}{\Gamma; \varnothing \Rightarrow \Sigma_1, \Sigma_2; \Theta_1, \Theta_2} \ (mu)^d$$

$$\frac{\Gamma; \Delta \Rightarrow \Sigma; \Theta, A}{\Gamma; \Delta, (A : B_1, \ldots, B_n / C_1 | \cdots | C_m) \Rightarrow \Sigma; \Theta} \ (d_1)^d$$

$$\frac{\Gamma; \Delta \Rightarrow \Sigma, \neg B_i; \Theta}{\Gamma; \Delta, (A : B_1, \ldots, B_i, \ldots, B_n / C_1 | \cdots | C_m) \Rightarrow \Sigma; \Theta} \ (d_2)^d$$

$$\frac{\Gamma; \varnothing \Rightarrow A; \varnothing \quad \Gamma, C_i; \Delta \Rightarrow \Sigma; \Theta, \neg B_1, \ldots, \neg B_n}{\Gamma; \Delta, (A : B_1, \ldots, B_n / C_1 | \cdots | C_i | \cdots | C_m) \Rightarrow \Sigma; \Theta} \ (d_3)^d$$

Figure 7. Additional rules of the calculus B$_D$.

The informal meaning of the nonmonotonic inference rules is similar to the meaning of the rules in B$_3$:

(i) rules $(l_1)^d$ and $(l_2)^d$ combine classical sequents and anti-sequents with disjunctive default sequents, respectively;
(ii) rule $(mu)^d$ again allows the joining of information in case that no default is present; and
(iii) rules $(d_1)^d$, $(d_2)^d$, and $(d_3)^d$ are the default introduction rules, where rules $(d_1)^d$ and $(d_2)^d$ take care of introducing non-active defaults, whilst rule $(d_3)^d$ allows to introduce an active default.

Before we turn to the adequacy of B_D, let us again give an example to illustrate the calculus.

Example 6. *Let us consider the disjunctive default theory from Example 4 dealing with Poole's broken arms scenario [52], which contains the defaults*

$$\frac{: U_l \wedge \neg B_l}{U_l} \quad \text{and} \quad \frac{: U_r \wedge \neg B_l}{U_r},$$

together with the disjunctive default

$$B_l | B_r.$$

This disjunctive default theory has the two extensions:

$$\text{Th}_2(\{B_l, U_r\}) \quad \text{and} \quad \text{Th}_2(\{B_r, U_l\}).$$

Accordingly, the following disjunctive default sequent is true:

$$\emptyset; (: \emptyset / B_l | B_r), (: (U_l \wedge \neg B_l) / U_l), (: (U_r \wedge \neg B_r) / U_r) \Rightarrow B_l, U_r, \neg(U_l \wedge \neg B_l); U_l.$$

A proof, γ, of this sequent in B_D is given below; it uses the two subproofs α and β:

- Proof α:

$$\cfrac{\cfrac{\cfrac{\cfrac{U_r, B_l \not\Rightarrow U_l, B_r}{U_r, B_l, U_r, \neg B_r \not\Rightarrow U_l} (\neg \not\Rightarrow)^r}{U_r, B_l, (U_r \wedge \neg B_r) \not\Rightarrow U_l} (\wedge \not\Rightarrow)^r}{U_r, B_l \not\Rightarrow U_l, \neg(U_r \wedge \neg B_r)} (\not\Rightarrow \neg)^r}{U_r, B_l; \emptyset \Rightarrow \emptyset; U_l, \neg(U_r \wedge \neg B_r)} (l_2)^d$$

- Proof β:

$$\cfrac{\cfrac{\cfrac{B_l \to B_l}{B_l \to U_r, B_l} (wr)}{\cfrac{U_r, B_l \to U_r, B_l}{U_r, B_l; \emptyset \Rightarrow U_r, B_l; \emptyset} (wl)} (l_1)^d \quad \cfrac{\cfrac{\cfrac{\cfrac{\cfrac{B_l \to B_l}{U_r, B_l \to B_l} (wl)}{U_r, B_l, \neg B_l \to \emptyset} (\neg \to)}{U_r, B_l, (U_l \wedge \neg B_l) \to \emptyset} (\wedge \to)_2}{U_r, B_l \to \neg(U_l \wedge \neg B_l)} (\to \wedge)}{U_r, B_l; \emptyset \Rightarrow \neg(U_l \wedge \neg B_l); \emptyset} (l_1)^d \quad \cfrac{\alpha}{U_r, B_l; \emptyset \Rightarrow \neg(U_l \wedge \neg B_l); U_l, \neg(U_r \wedge \neg B_r)} (mu)^d}{U_r, B_l; \emptyset \Rightarrow B_l, U_r, \neg(U_l \wedge \neg B_l); U_l, \neg(U_r \wedge \neg B_r)} (mu)^d$$

- Proof γ:

$$\cfrac{\cfrac{\emptyset \to \top}{\emptyset; \emptyset \Rightarrow \top; \emptyset} (l_1)^d \quad \cfrac{\cfrac{\cfrac{\emptyset \to \top}{U_r \to \top} wl}{U_r; \emptyset \Rightarrow \top; \emptyset} (l_1)^d \quad \beta}{U_r; (: \emptyset / B_l | B_r) \Rightarrow B_l, U_r, \neg(U_l \wedge \neg B_l); \neg(U_r \wedge \neg B_r), U_l} (d_3)^d}{\cfrac{\emptyset; (: \emptyset / B_l | B_r), (: (U_r \wedge \neg B_r) / U_r) \Rightarrow B_l, U_r, \neg(U_l \wedge \neg B_l), \neg(U_l \wedge \neg B_l); U_l}{\emptyset; (: \emptyset / B_l | B_r), (: (U_l \wedge \neg B_l) / U_l), (: (U_r \wedge \neg B_r) / U_r) \Rightarrow B_l, U_r, \neg(U_l \wedge \neg B_l); U_l} (d_2)^d} (d_3)^d$$

4.2. Adequacy of the Calculus

Soundness and completeness of $\mathbf{B_D}$ can be shown by similar arguments as in the case of $\mathbf{B_3}$. We sketch the relevant details.

We again need some preparatory characterisations of extensions, dealing with the introduction of active or non-active defaults.

We start with the notion of a reduct, adapted to the case of $\mathbf{DL_D}$, as introduced by Gelfond, Lifschitz, Przymusinska, and Truszczyński [10].

In what follows, we use the following terminology: By a *disjunctive inference rule*, or simply a *disjunctive rule*, r, we understand an expression of the form

$$\frac{A}{C_1|\cdots|C_m}.$$

We say that a set S of formulas is *closed under* r if, whenever $A \in S$, then $C_i \in S$, for some $i \in \{1,\ldots,m\}$. Moreover, for a set R of disjunctive rules, we say that S is closed under R if S is closed under each $r \in R$.

Definition 15. *Let D be a set of disjunctive defaults and E a set of formulas. The $\mathbf{DL_D}$-reduct of D with respect to E, denoted by D_E^d, is the set consisting of the following disjunctive inference rules:*

$$D_E^d := \left\{ \frac{A}{C_1|\cdots|C_m} \ \middle| \ \frac{A : B_1,\ldots,B_n}{C_1|\cdots|C_m} \in D \text{ and } \{\neg B_1,\ldots,\neg B_n\} \cap E = \emptyset \right\}.$$

A disjunctive rule

$$\frac{A}{C_1|\cdots|C_m}$$

is called $\mathbf{DL_D}$-*residue of a default*

$$\frac{A : B_1,\ldots,B_n}{C_1|\cdots|C_m}.$$

We again allow ourselves to drop the prefix "$\mathbf{DL_D}$-" from "$\mathbf{DL_D}$-reduct" and "$\mathbf{DL_D}$-residue" if no ambiguity can arise.

Towards our characterisation of extensions of disjunctive default theories, we introduce the following notation:

Definition 16. *For a set W of formulas and a set R of disjunctive rules, let $C^R(W)$ be the collection of all sets which*

(i) *contain W,*
(ii) *are closed under propositional consequence, and*
(iii) *are closed under R.*

Furthermore, let $Cn^R(W) = \min(C^R(W))$, i.e., $Cn^R(W)$ contains all minimal sets satisfying (i)–(iii).

Note that, for a disjunctive default theory $T = \langle W, D \rangle$ and a set E of formulas, we obviously have that:

$$Cl_T(E) = C^{D_E^d}(W) \quad \text{and} \quad \Delta_T(E) = Cn^{D_E^d}(W).$$

From this, the following result is immediate:

Theorem 7. *Let $T = \langle W, D \rangle$ be a disjunctive default theory. Then,*

$$E \text{ is an extension of } T \text{ iff } E \in Cn^{D_E^d}(W).$$

Note furthermore that, if $T = \langle W, D \rangle$ is a standard default theory, i.e., if D contains no proper disjunctive defaults, then clearly $Cn^{D_E^d}(W) = \{\text{Th}_2^{D_E^d}(W)\}$ and E is an extension in the standard default logic sense iff $E = \text{Th}_2^{D_E^d}(W)$, where $\text{Th}_2^{D_E^d}(W)$ is the deductive closure operator of classical derivability extended by the (standard) inference rules in D_E^d.

We continue with the following pendant to activeness as defined earlier:

Definition 17. *Let E be a set of formulas. A disjunctive default*

$$\frac{A : B_1, \ldots, B_n}{C_1 | \cdots | C_m}$$

is active in E iff $E \vdash A$ and $\{\neg B_1, \ldots, \neg B_n\} \cap E = \emptyset$.

We again employ our notation $p(d)$ as in case of **DL₃**, but now we define and $j(d)$ for a default d, but now we define

$$j'(d) := \{B_1, \ldots, B_n\} \quad \text{and} \quad c'(d) := \{C_1, \ldots, C_m\},$$

for $d = (A : B_1, \ldots, B_n / C_1 | \cdots | C_m)$.

We obtain the following results corresponding to Lemma 2 and Theorems 2–4, respectively:

Lemma 3. *Let W and E be sets of formulas, R a set of disjunctive inference rules, and*

$$r = \frac{A}{B_1 | \cdots | B_n}$$

a disjunctive inference rule. Then:

1. *If $A \notin E$ and $E \in Cn^R(W)$, then $E \in Cn^{R \cup \{r\}}(W)$.*
2. *If $A \in E$ and $E \in Cn^{R \cup \{r\}}(W)$, then $E \in Cn^R(W \cup \{B\})$, for some formula $B \in \{B_1, \ldots, B_n\}$.*

Theorem 8. *Let $T = \langle W, D \rangle$ be a disjunctive default theory, E a set of formulas, and d a disjunctive default not active in E. Then, E is an extension of $\langle W, D \rangle$ iff E is an extension of $\langle W, D \cup \{d\} \rangle$.*

Theorem 9. *Let E be a set of formulas and d a disjunctive default.*

1. *If E is an extension of $\langle W, D \cup \{d\} \rangle$ and d is active in E, then E is an extension of $\langle W \cup \{C\}, D \rangle$, for some $C \in c'(d)$.*
2. *If E is an extension of the disjunctive default theory $\langle W \cup \{C\}, D \rangle$, for some $C \in c'(d)$, $W \vdash p(d)$, and $\neg j'(d) \cap E = \emptyset$, then E is an extension of $\langle W, D \cup \{d\} \rangle$.*

From this, by similar arguments as in the case of B_3, soundness and completeness of B_D follows.

Theorem 10. *A disjunctive default sequent $\Gamma; \Delta \Rightarrow \Sigma; \Theta$ is provable in B_D iff it is true.*

5. Conclusions

In this paper, we introduced sequent-type calculi for brave reasoning for a three-valued version of default logic [9] and for disjunctive default logic [10], following the method of Bonatti [7]. This form of axiomatisation yielded a particular elegant formulation mainly due to their usage of anti-sequents. In addition, the approach was flexible and could be applied to formalise different versions of nonmonotonic reasoning. Indeed, other variants of default logic besides the versions studied here, including justified default logic [12] and constrained default logic [13,14], have also been axiomatised by this sequent method [60,61].

Related to the sequent approach discussed here are also works employing tableau methods. In particular, Niemelä [62] introduces a tableau calculus for inference under circumscription.

Other tableau approaches, however, do not encode inference directly, rather they characterise models (resp., extensions) associated with a particular nonmonotonic reasoning formalism [63–66].

Variations of our calculi can be obtained by using different calculi for the underlying monotonic logics. As far as the three-valued case is concerned, we opted for the style of calculi as discussed by Rousseau [25] and Zach [26] because they naturally model the underlying semantic conditions of the considered logic. Alternatively, we could have also used two-sided sequent and anti-sequent calculi like the ones described by Avron [24] and Oetsch and Tompits [67], respectively. By employing such two-sided sequents, however, one then deals with calculi having also "non-standard" inference rules introducing two connectives simultaneously. Another prominent proof method for many-valued logics are *hypersequent calculi* [57], which are basically disjunctions of two-sided sequents. However, to the best of our knowledge, no rejection calculus based on hypersequents exist so far and establishing such a system in particular for $Ł_3$ would be worthwhile.

Another topic for future work is to develop calculi for sceptical reasoning for the considered versions of default logic as well as for other variants of default logic discussed in the literature [12–14], similar to the system for sceptical reasoning for standard default logic as introduced by Bonatti and Olivetti [8]. In that work, they also introduced a different version of a calculus for brave default reasoning—extending this calculus to $\mathbf{DL_3}$ and $\mathbf{DL_D}$ would provide an alternative to the calculi discussed here.

Author Contributions: Conceptualization, S.P. and H.T.; formal analysis, S.P. and H.T.; investigation, S.P. and H.T.; writing—original draft preparation, S.P. and H.T.; writing—review and editing, S.P. and H.T.; supervision, H.T. All authors have read and agreed to the published version of the manuscript.

Funding: The first author was supported by the European Master's Program in Computational Logic (EMCL).

Conflicts of Interest: The authors declare no conflict of interest.

References

1. Reiter, R. A Logic for Default Reasoning. *Artif. Intell.* **1980**, *13*, 81–132. [CrossRef]
2. Moore, R.C. Semantical Considerations on Nonmonotonic Logic. *Artif. Intell.* **1985**, *25*, 75–94. [CrossRef]
3. McCarthy, J. Circumscription—A Form of Non-Monotonic Reasoning. *Artif. Intell.* **1980**, *13*, 27–39. [CrossRef]
4. Gelfond, M.; Lifschitz, V. Classical Negation in Logic Programs and Disjunctive Databases. *New Gener. Comput.* **1991**, *9*, 365–385. [CrossRef]
5. Pearce, D. Equilibrium Logic. *Ann. Math. Artif. Intell.* **2006**, *47*, 3–41. [CrossRef]
6. Minsky, M. A Framework for Representing Knowledge. In *Mind Design*; Haugeland, J., Ed.; MIT Press: Cambridge, MA, USA, 1975; pp. 95–128.
7. Bonatti, P.A. Sequent Calculi for Default and Autoepistemic Logic. In Proceedings of the 5th International Workshop on Theorem Proving with Analytic Tableaux and Related Methods (TABLEAUX'96), Lecture Notes in Computer Science, Terrasini, Palermo, Italy, 15–17 May 1996; Migliolli, P., Moscato, U., Mundici, D., Ornaghi, M., Eds.; Springer: Berlin/Heidelberg, Germany, 1996; Volume 1071, pp. 127–142.
8. Bonatti, P.A.; Olivetti, N. Sequent Calculi for Propositional Nonmonotonic Logics. *ACM Trans. Comput. Log.* **2002**, *3*, 226–278. [CrossRef]
9. Radzikowska, A. A Three-Valued Approach to Default Logic. *J. Appl. Non-Class. Logics* **1996**, *6*, 149–190. [CrossRef]
10. Gelfond, M.; Lifschitz, V.; Przymusinska, H.; Truszczyński, M. Disjunctive Defaults. In Proceedings of the 2nd International Conference on Principles of Knowledge Representation and Reasoning (KR'91), Cambridge, MA, USA, 22–25 April 1991; Allen, J.F., Fikes, R., Sandewall, E., Eds.; Morgan Kaufmann: Burlington, MA, USA, 1991; pp. 230–237.
11. Łukasiewicz, J. Philosophische Bemerkungen zu mehrwertigen Systemen des Aussagenkalküls. *Comptes Rendus Des Séances De La Société Des Sciences Et Des Lettres De Varsovie Cl III* **1930**, *23*, 51–77.
12. Łukaszewicz, W. Considerations on Default Logic—An alternative approach. *Comput. Intell.* **1988**, *4*, 1–16. [CrossRef]

13. Schaub, T. *On Constrained Default Theories*; Technical Report AIDA-92-2; FG Intellektik, FB Informatik, TH Darmstadt: Darmstadt, Germany, 1992.
14. Delgrande, J.P.; Schaub, T.; Jackson, W.K. Alternative Approaches to Default Logic. *Artif. Intell.* **1994**, *70*, 167–237. [CrossRef]
15. Mikitiuk, A.; Truszczyński, M. Rational Default Logic and Disjunctive Logic Programming. In Proceedings of the 2nd International Workshop on Logic Programming and Non-Monotonic Reasoning (LPNMR'93), Lisbon, Portugal, 28–30 June 1993; Pereira, L.M., Nerode, A., Eds.; MIT Press: Cambridge, MA, USA, 1993; pp. 283–299.
16. Zhou, Y.; Lin, F.; Zhang, Y. General Default Logic. *Ann. Math. Artif. Intell.* **2009**, *57*, 125–160. [CrossRef]
17. Yue, A.; Ma, Y.; Lin, Z. Four-Valued Semantics for Default Logic. In Proceedings of the 19th Conference of the Canadian Society for Computational Studies of Intelligence (Canadian AI 2006), Lecture Notes in Computer Science, Quebec City, QC, Canada, 7–9 June 2006; Lamontagne, L., Marchand, M., Eds.; Springer: Berlin/Heidelberg, Germany, 2006; Volume 4013, pp. 195–205.
18. Antoniou, G.; Wang, K. Default Logic. In *Handbook of the History of Logic, Volume 8: The Many Valued and Nonmonotonic Turn in Logic*; Gabbay, D., Woods, J., Eds.; North-Holland: Amsterdam, The Netherlands, 2007; pp. 517–632.
19. Beyersdorff, O.; Meier, A.; Thomas, M.; Vollmer, H. The Complexity of Reasoning for Fragments of Default Logic. *J. Log. Comput.* **2012**, *22*, 587–604. [CrossRef]
20. Łukasiewicz, J. O sylogistyce Arystotelesa. *Sprawozdania Z Czynności I Posiedzeń Polskiej Akademii Umiejętności* **1939**, *44*, 220–226.
21. Béziau, J.Y. A Sequent Calculus for Lukasiewicz's Three-Valued Logic Based on Suszko's Bivalent Semantics. *Bull. Sect. Log.* **1999**, *28*, 89–97.
22. Avron, A. Classical Gentzen-Type Methods in Propositional Many-Valued Logics. In *Beyond Two: Theory and Applications in Multiple-Valued Logics*; Fitting, M., Orłowska, E., Eds.; Springer: Berlin/Heidelberg, Germany, 2003; pp. 117–155.
23. Gentzen, G. Untersuchungen über das logische Schließen I. *Math. Z.* **1935**, *39*, 176–210. [CrossRef]
24. Avron, A. Natural 3-Valued Logics—Characterization and Proof Theory. *J. Symb. Log.* **1991**, *56*, 276–294. [CrossRef]
25. Rousseau, G.S. Sequents in Many Valued Logic I. *Fundam. Math.* **1967**, *60*, 23–33. [CrossRef]
26. Zach, R. Proof Theory of Finite-Valued Logics. Master's Thesis, Technische Universität Wien, Institut für Computersprachen, Wien, Austria, 1993.
27. Bogojeski, M.; Tompits, H. On Sequent-Type Rejection Calculi for Many-Valued Logics. In *Reasoning: Games, Cognition, Logic*; Urbański, M., Skura, T., Łupkowski, P., Eds.; College Publications: London, UK, 2020; pp. 193–207.
28. Łukasiewicz, J. Logika dwuwartościowa. *Przegląd Filozoficzny* **1921**, *23*, 189–205.
29. Łukasiewicz, J. *Aristotle's Syllogistic from the Standpoint of Modern Formal Logic*, 2nd ed.; Clarendon Press: Oxford, UK, 1957.
30. Słupecki, J. *Z Badań Nad Sylogistyką Arystotelesa*; Wrocławskie Towarzystwo Naukowe: Wrocław, Poland, 1948; Volume 6, pp. 1–30.
31. Słupecki, J. Funkcja Łukasiewieza. *Zesz. Nauk. Uniw. Wrocławskiego Ser. A* **1959**, *3*, 33–40.
32. Wybraniec-Skardowska, U. Teoria zdań odrzuconych. *Zesz. Nauk. Wyższej Szkoły Pedagog. W Opolu Ser. B Stud. I Monogr.* **1969**, *22*, 5–131.
33. Bryll, G. Kilka uzupelnień teorii zdań odrzuconych. *Zesz. Nauk. Wyższej Szkoły Pedagog. W Opolu Ser. B Stud. I Monogr.* **1969**, *22*, 133–154.
34. Słupecki, J.; Bryll, G.; Wybraniec-Skardowska, U. Theory of Rejected Propositions I. *Stud. Log.* **1971**, *29*, 75–115. [CrossRef]
35. Słupecki, J.; Bryll, G.; Wybraniec-Skardowska, U. Theory of Rejected Propositions II. *Stud. Log.* **1972**, *30*, 97–139. [CrossRef]
36. Tiomkin, M.L. Proving Unprovability. In Proceedings of the Third Annual Symposium on Logic in Computer Science (LICS'88), Scotland, UK, 5–8 July 1988; IEEE Computer Society: Washington, DC, USA, 1988; pp. 22–26.
37. Bonatti, P.A. *A Gentzen System for Non-Theorems*; Technical Report CD-TR 93/52; Christian Doppler Labor für Expertensysteme, Technische Universität Wien: Vienna, Austria, 1993.

38. Goranko, V. Refutation Systems in Modal Logic. *Stud. Log.* **1994**, *53*, 299–324. [CrossRef]
39. Skura, T. A Complete Syntactic Characterization of the Intuitionistic Logic. *Rep. Math. Log.* **1989**, *23*, 75–80.
40. Dutkiewicz, R. The Method of Axiomatic Rejection for the Intuitionistic Propositional Logic. *Stud. Log.* **1989**, *48*, 449–459. [CrossRef]
41. Pinto, L.; Dyckhoff, R. Loop-Free Construction of Counter-Models for Intuitionistic Propositional Logic. In *Symposia Gaussiana, Proceedings of the 2nd Gauss Symposium, Conference A: Mathematics and Theoretical Physics, Munich, Germany, 2–7 August 1993*; Behara, M., Fritsch, R., Lintz, R.G., Eds.; Walter de Gruyter: Berlin, Germany, 1995; pp. 225–232.
42. Skura, T. Aspects of Refutation Procedures in the Intuitionistic Logic and Related Modal Systems. *Acta Univ. Wratislav. Log.* **1999**, *20*, 1–84.
43. Skura, T. Refutation Systems in Propositional Logic. In *Handbook of Philosophical Logic*, 2nd ed.; Gabbay, D., Guenthner, D., Eds.; Springer: Berlin/Heidelberg, Germany, 2011; Volume 16, pp. 115–157.
44. Skura, T. *Refutation Methods in Modal Propositional Logic*; Semper: Warszawa, Poland, 2013.
45. Berger, G.; Tompits, H. On Axiomatic Rejection for the Description Logic $\mathcal{ALC}$. In *Declarative Programming and Knowledge Management–Declarative Programming Days (KDPD 2013)*; Revised Selected Papers; Lecture Notes in Computer Science; Springer: Cham, Switzerland, 2014; Volume 8439, pp. 65–82.
46. Wybraniec-Skardowska, U. On the Notion and Function of the Rejection of Propositions. *Acta Univ. Wratislav. Log.* **2005**, *23*, 179–202.
47. Rosser, J.B.; Turquette, A.R. *Many-Valued Logics*; North-Holland: Amsterdam, The Netherlands, 1952.
48. Malinowski, G. Many-Valued Logic and its Philosophy. In *Handbook of the History of Logic, Volume 8: The Many Valued and Nonmonotonic Turn in Logic*; Gabbay, D., Woods, J., Eds.; North-Holland: Amsterdam, The Netherlands, 2007; pp. 13–94.
49. Wajsberg, M. Aksjomatyzacja trójwartościowego rachunku zdań. *Comptes Rendus Des Séances De La Société Des Sciences Et Des Lettres De Varsovie Cl III* **1931**, *24*, 136–148.
50. Łukaszewicz, W. *Non-Monotonic Reasoning: Formalization of Commonsense Reasoning*; Ellis Horwood: Chichester, UK, 1990.
51. Pearl, J. *Probabilistic Reasoning in Intelligent Systems: Networks of Plausible Inference*; Morgan Kaufmann Publishers: Burlington, MA, USA, 1988.
52. Poole, D. What the Lottery Paradox Tells us about Default Reasoning. In Proceedings of the First International Conference on Principles of Knowledge Representation and Reasoning (KR'89), Toronto, ON, Canada, 15–18 May 1989; Brachman, R.J., Levesque, H.J., Reiter, R., Eds.; Morgan Kaufmann: Burlington, MA, USA, 1989; pp. 333–340.
53. Lin, F.; Shoham, Y. Epistemic Semantics for Fixed-Points Non-Monotonic Logics. In Proceedings of the Third Conference on Theoretical Aspects of Reasoning about Knowledge (TARK'90), Pacific Grove, CA, USA, 4–7 March 1990; Parikh, R., Ed.; Morgan Kaufmann: Burlington, MA, USA, 1990; pp. 111–120.
54. Baumgartner, R.; Gottlob, G. Propositional Default Logics Made Easier: Computational Complexity of Model Checking. *Theor. Comput. Sci.* **2002**, *289*, 591–627. [CrossRef]
55. Bonatti, P.A. *Sequent Calculi for Default and Autoepistemic Logic*; Technical Report CD-TR 93/53; Christian Doppler Labor für Expertensysteme, Technische Universität Wien: Wien, Austria, 1993.
56. Carnielli, W.A. On Sequents and Tableaux for Many-Valued Logics. *J. Non-Class. Log.* **1991**, *8*, 59–76.
57. Avron, A. The Method of Hypersequents in the Proof Theory of Propositional Non-Classical Logics. In *Logic: From Foundations to Applications*; Clarendon Press: Oxford, UK, 1996; pp. 1–32.
58. Hähnle, R. Tableaux for Many-Valued Logics. In *Handbook of Tableaux Methods*; D'Agostino, M., Gabbay, D., Hähnle, R., Posegga, J., Eds.; Kluwer: Dordrecht, The Netherlands, 1999; pp. 529–580.
59. Marek, W.; Truszczyński, M. *Nonmonotonic Logic: Context-Dependent Reasoning*; Springer: Berlin/Heidelberg, Germany, 1993.
60. Egly, U.; Tompits, H. A Sequent Calculus for Intuitionistic Default Logic. In Proceedings of the 12th Workshop on Logic Programming (WLP'97), Forschungsbericht PMS-FB-1997-10, Institut für Informatik, Ludwig-Maximilians-Universität München, Munich, Germany, 17–19 September 1997; pp. 69–79.
61. Lupea, M. Axiomatization of Credulous Reasoning in Default Logics using Sequent Calculus. In Proceedings of the 10th International Symposium on Symbolic and Numeric Algorithms for Scientific Computing (SYNASC 2008), Timisoara, Romania, 26–29 September 2008.

62. Niemelä, I. Implementing Circumscription Using a Tableau Method. In Proceedings of the 12th European Conference on Artificial Intelligence (ECAI'96), Budapest, Hungary, 11–16 August 1996; Wahlster, W., Ed.; John Wiley and Sons: New York, NY, USA, 1996; pp. 80–84.
63. Amati, G.; Aiello, L.C.; Gabbay, D.; Pirri, F. A Proof Theoretical Approach to Default Reasoning I: Tableaux for Default Logic. *J. Log. Comput.* **1996**, *6*, 205–231. [CrossRef]
64. Pearce, D.; de Guzmán, I.P.; Valverde, A. A Tableau Calculus for Equilibrium Entailment. In Proceedings of the 9th International Conference on Automated Reasoning with Analytic Tableaux and Related Methods (TABLEAUX 2000), Lecture Notes in Computer Science, St Scotland, UK, 3–7 July 2000; Dyckhoff, R., Ed.; Springer: Berlin/Heidelberg, Germany, 2000; Volume 1847, pp. 352–367.
65. Cabalar, P.; Odintsov, S.P.; Pearce, D.; Valverde, A. Partial Equilibrium Logic. *Ann. Math. Artif. Intell.* **2007**, *50*, 305–331. [CrossRef]
66. Gebser, M.; Schaub, T. Tableau Calculi for Logic Programs under Answer Set Semantics. *ACM Trans. Comput. Log.* **2013**, *14*, 15:1–15:40. [CrossRef]
67. Oetsch, J.; Tompits, H. Gentzen-Type Refutation Systems for Three-Valued Logics with an Application to Disproving Strong Equivalence. In Proceedings of the 11th International Conference on Logic Programming and Nonmonotonic Resoning (LPNMR 2011), Lecture Notes in Computer Science, Vancouver, BC, Canada, 16–19 May 2011; Delgrande, J.P., Faber, W., Eds.; Springer: Berlin/Heidelberg, Germany, 2011; Volume 6645, pp. 254–259.

© 2020 by the authors. Licensee MDPI, Basel, Switzerland. This article is an open access article distributed under the terms and conditions of the Creative Commons Attribution (CC BY) license (http://creativecommons.org/licenses/by/4.0/).

Article

Kripke-Style Models for Logics of Evidence and Truth

Henrique Antunes [1], Walter Carnielli [2,†], Andreas Kapsner [3,†] and Abilio Rodrigues [1,*,†]

1. Department of Philosophy, Federal University of Minas Gerais, Antônio Carlos Avenue 6627, FAFICH Sala 4035, Belo Horizonte 31270-901, Brazil; antunes.henrique@outlook.com
2. Centre for Logic, Epistemology and the History of Science, and Department of Philosophy, University of Campinas, Sérgio Buarque de Holanda Street 251, Campinas 13083-859, Brazil; walterac@unicamp.br
3. Munich Center for Mathematical Philosophy, Ludwig Maximilian University of Munich, Ludwigstraße 31, 80539 München, Germany; Andreas.Kapsner@lrz.uni-muenchen.de
* Correspondence: abilio.rodrigues@gmail.com
† The second and fourth authors acknowledge support from *CNPq* (The Brazilian National Council for Scientific and Technological Development) Research Grants 307376/2018-4 and 311911/2018-8. The research of the third author has been supported by the Deutsche Forschungsgemeinschaft (DFG, German Research Foundation), project 436508789.

Received: 9 July 2020; Accepted: 5 August 2020; Published: 19 August 2020

Abstract: In this paper, we propose Kripke-style models for the logics of evidence and truth LET_J and LET_F. These logics extend, respectively, Nelson's logic $N4$ and the logic of first-degree entailment (FDE) with a classicality operator $\circ$ that recovers classical logic for formulas in its scope. According to the intended interpretation here proposed, these models represent a database that receives information as time passes, and such information can be positive, negative, non-reliable, or reliable, while a formula $\circ A$ means that the information about A, either positive or negative, is reliable. This proposal is in line with the interpretation of $N4$ and FDE as information-based logics, but adds to the four scenarios expressed by them two new scenarios: reliable (or conclusive) information (i) for the truth and (ii) for the falsity of a given proposition.

Keywords: Kripke models; logics of evidence and truth; paraconsistency

1. Introduction

The aim of this paper is to present Kripke-style models for the logics of evidence and truth LET_J and LET_F, introduced in [1,2]. Both are paraconsistent and paracomplete logics that extend respectively Nelson's logic $N4$ and the logic of first-degree entailment (FDE) with a classicality operator $\circ$ that recovers classical logic for formulas in its scope. The motivation for the logics of evidence and truth is to model contexts of reasoning where one deals with positive and negative evidence, which can be conclusive or non-conclusive. (On the notion of evidence, and $N4$ and FDE as evidence-preserving logics, see [1] Section 2, [3] Section 3 and [2] Section 2.2.1.) Conclusive evidence is subjected to classical logic, and non-conclusive to a paraconsistent and paracomplete logic that is $N4$ in the case of LET_J and FDE in the case of LET_F. According to the interpretation in terms of evidence and truth, a pair of contradictory formulas A and $\neg A$ expresses conflicting non-conclusive evidence for A and $\neg A$, and $\circ A$ means that there is conclusive evidence for the truth or the falsity of A. Conclusive evidence is subjected to classical logic, and so when $\circ A$ holds, A is treated as true or false by the formal systems. Thus, while $A, \neg A \nvdash B$, in these logics it holds that $\circ A, A, \neg A \vdash B$, which means that conflicting evidence cannot be conclusive on pain of triviality. Both LET_J and LET_F are logics of formal inconsistency and undeterminedness [4–6]. Sound and complete valuation semantics were presented for LET_J and LET_F in [1,2], respectively.

It is well known that the logics FDE and $N4$ can be interpreted in terms of preservation of information, the latter in the sense of [7,8]. In terms of information, a formula $\circ A$ can be read as meaning that the information about A is reliable, and LET_J and LET_F can be interpreted in terms of positive and negative information, which can be either reliable or unreliable. This idea fits Belnap and Dunn's proposal of interpreting FDE as a logic to be used by a computer that receives information from different sources [9–11]. The semantic values *True*, *False*, *Both* and *None*, of what became known as Belnap–Dunn 4-valued logic, express the circumstances in which the computer receives, respectively, only positive, only negative, conflicting and no information at all, about a proposition A. In addition to these four scenarios, LET_J and LET_F are capable of representing two additional scenarios: when $\circ A$ does not hold, we have the four scenarios above, but when $\circ A$ holds, exactly one among A and $\neg A$ holds, which means that the information about A, positive or negative, is reliable and subjected to classical logic.

The Kripke-style models to be presented here are intended to represent a database that, as time passes, receives information from different sources that may be either reliable or unreliable. Each stage w represents one of the following six scenarios:

When $w \not\Vdash \circ A$:

1. $w \Vdash A, w \not\Vdash \neg A$: at w the database has only the information that A is true;
2. $w \not\Vdash A, w \Vdash \neg A$: at w the database has only the information that A is false;
3. $w \Vdash A, w \Vdash \neg A$: at w the database has conflicting information about A;
4. $w \not\Vdash A, w \not\Vdash \neg A$: at w the database has no information about A.

When $w \Vdash \circ A$:

5. $w \Vdash A$: at w the database has reliable information that A is true;
6. $w \Vdash \neg A$: at w the database has reliable information that A is false.

These six scenarios can be illustrated by the diagram below:

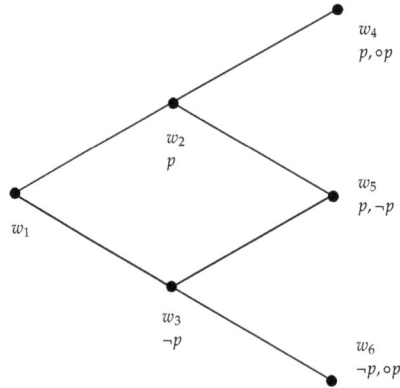

In stage w_1, the database is empty and therefore has no information about p. In w_2 it receives only the information p, which in w_2 is not taken as reliable. From w_2, there are two possibilities: in w_4 the database receives the information that the information about p is reliable, which is expressed by $\circ p$; alternatively, in w_5 the information $\neg p$ is obtained, and so the information about p remains unreliable. Analogous reasoning applies to w_3, which may bifurcate into w_5 or w_6.

In the example above, nothing has been removed from the database. As we will see, LET_J requires persistence for every formula, which means that once some information is inserted in the database it cannot be removed. On the other hand, in the case of LET_F, different persistence clauses may be adopted to express different criteria for revising information.

The remainder of this paper is structured as follows. Section 2 presents the models for LET_J and proves soundness and completeness, and Section 3 does the same regarding LET_F. Section 4 discusses the persistence clauses that can be added to LET_F for revisability of information and gives a proof that the addition of these clauses to the semantics of LET_F does not affect soundness, nor completeness. Section 5 discusses some results related to how the classical behavior propagates across stages in LET_J and LET_F-models, and finally, Section 6, points out some possible further developments.

2. The Logic LET_J

The logic LET_J [1] is an extension of Nelson's paraconsistent logic N4. An interpretation of N4 in terms of positive and negative information can be found in [12]. In [8], a view according to which paraconsistent logics should be interpreted without any ontological or epistemological ingredients in terms of Dunn's notion of information [7] is presented and defended. N4 is FDE plus a semi-intuitionistic implication: Peirce's law does not hold, but the equivalence between $\neg(A \rightarrow B)$ and $A \wedge \neg B$ holds. A Kripke semantics for N4 can be found in [13], p. 164, and it is essentially the local conditions for $\neg, \vee, \wedge$ that mimic the conditions of FDE, the local conditions for $\neg(A \rightarrow B)$ and the intuitionistic global clause for $\rightarrow$.

The language $\mathcal{L}_J$ of LET_J is composed of denumerably many sentential letters $p_1, p_2, \ldots$, the unary connectives $\circ$ and $\neg$, the binary connectives $\wedge, \vee$ and $\rightarrow$ and parentheses. The set of formulas of $\mathcal{L}_J$, which we will also denote by $\mathcal{L}_J$, is inductively defined in the usual way. Henceforth, Roman capitals $A, B, C, \ldots$ will be used as metavariables for the formulas of $\mathcal{L}_J$, while Greek capitals $\Gamma, \Delta, \Sigma, \ldots$ will be used as metavariables for sets of formulas.

Definition 1. *The logic LET_J is defined over $\mathcal{L}_J$ by the following natural deduction rules:*

$$\frac{A \quad B}{A \wedge B} \wedge I \qquad \frac{A \wedge B}{A} \wedge E \quad \frac{A \wedge B}{B}$$

$$\frac{A}{A \vee B} \vee I \frac{B}{A \vee B} \qquad \frac{A \vee B \quad \overset{[A]}{\underset{C}{\vdots}} \quad \overset{[B]}{\underset{C}{\vdots}}}{C} \vee E$$

$$\frac{\overset{[A]}{\underset{B}{\vdots}}}{A \rightarrow B} \rightarrow I \qquad \frac{A \rightarrow B \quad A}{B} \rightarrow E$$

$$\frac{\neg A}{\neg(A \wedge B)} \neg \wedge I \frac{\neg B}{\neg(A \wedge B)} \qquad \frac{\neg(A \wedge B) \quad \overset{[\neg A]}{\underset{C}{\vdots}} \quad \overset{[\neg B]}{\underset{C}{\vdots}}}{C} \neg \wedge E$$

$$\frac{\neg A \quad \neg B}{\neg(A \vee B)} \neg \vee I \qquad \frac{\neg(A \vee B)}{\neg A} \neg \vee E \frac{\neg(A \vee B)}{\neg B}$$

$$\frac{A \quad \neg B}{\neg(A \rightarrow B)} \neg \rightarrow I \qquad \frac{\neg(A \rightarrow B)}{A} \neg \rightarrow E \frac{\neg(A \rightarrow B)}{\neg B}$$

$$\frac{A}{\neg\neg A}\ DNI \qquad \frac{\neg\neg A}{A}\ DNE$$

$$\frac{\circ A \quad A \quad \neg A}{B}\ EXP^\circ \qquad \frac{\circ A}{A \vee \neg A}\ PEM^\circ$$

As is customary, enclosing a formula A in square brackets indicates that A is a discharged hypothesis. The notion of a *derivation* in LET_J can be inductively defined along the lines of the definition presented in [14] (pp. 35–36). It suffices to say here that a derivation is a tree of labeled formulas whose *nodes* are either a hypothesis or the conclusion of applying one of the rules above to formulas that occur previously in the tree. Given $\Gamma \cup \{A\} \subseteq \mathcal{L}_J$, the notation $\Gamma \vdash_J A$ will be used to express that there is a derivation $\mathcal{D}$ in LET_J such that A is the last formula that occurs in $\mathcal{D}$ (its *conclusion*) and all of $\mathcal{D}$'s undischarged hypotheses belong to Γ. $\vdash_J A$ will be treated as a shorthand for $\varnothing \vdash A$. When there is no risk of confusion, we shall write $\vdash$ instead of $\vdash_J$.

Definition 2. *A Kripke model $\mathcal{M}$ for LET_J is a structure $\langle W, \leq, v \rangle$ such that W is a non-empty set of stages, the accessibility relation $\leq$ is a partial order on W, and $v : \mathcal{L}_J \times W \longrightarrow \{0,1\}$ is a valuation function satisfying the following conditions, for every $w \in W$:*

1. $v(A \wedge B, w) = 1$ iff $v(A, w) = 1$ and $v(B, w) = 1$;
2. $v(A \vee B, w) = 1$ iff $v(A, w) = 1$ or $v(B, w) = 1$;
3. $v(\neg\neg A, w) = 1$ iff $v(A, w) = 1$;
4. $v(\neg(A \wedge B), w) = 1$ iff $v(\neg A, w) = 1$ or $v(\neg B, w) = 1$;
5. $v(\neg(A \vee B), w) = 1$ iff $v(\neg A, w) = 1$ and $v(\neg B, w) = 1$;
6. $v(\circ A, w) = 1$ only if exactly one of the following conditions obtains:
   For every $w' \geq w$, $v(A, w') = 1$ and $v(\neg A, w') = 0$;
   For every $w' \geq w$, $v(A, w') = 0$ and $v(\neg A, w') = 1$;
7. $v(A \rightarrow B, w) = 1$ iff for every $w' \geq w$, if $v(A, w') = 1$, then $v(B, w') = 1$;
8. $v(\neg(A \rightarrow B), w) = 1$ iff $v(A, w) = 1$ and $v(\neg B, w) = 1$;

P1. If $v(A, w) = 1$, then for every $w' \geq w$, $v(A, w') = 1$, for every $A \in \mathcal{L}_J$.

Given a Kripke model $\mathcal{M} = \langle W, \leq, v \rangle$ and a stage $w \in W$, we say that a formula A holds in w ($\mathcal{M}, w \Vdash_J A$) if, and only if, $v(A, w) = 1$.

Definition 3. *Let $\Gamma \cup \{A\} \subseteq \mathcal{L}_J$. We say that A is a semantic consequence of Γ ($\Gamma \vDash_J A$) if, and only if, for every model $\mathcal{M} = \langle W, \leq, v \rangle$ and every $w \in W$, if $\mathcal{M}, w \Vdash_J B$, for every $B \in \Gamma$, then $\mathcal{M}, w \Vdash_J A$. A is said to be logically valid if for every model $\mathcal{M}$ and stage $w \in W$, $\mathcal{M}, w \Vdash A$. As in the case of $\vdash_J$, we shall sometimes write $\Vdash$ and $\vDash$ instead of $\Vdash_J$ and $\vDash_J$, respectively.*

Note that Clause 6 of Definition 2 gives only a necessary condition for $v(\circ A, w) = 1$. This mimics the clause for $\circ A$ of the non-deterministic valuation semantics proposed in [1] (p. 3805) and will be important for the results presented later, specially in Section 4. We will now prove that LET_J is sound and complete with respect to $\vDash_J$.

Soundness and Completeness

Theorem 1. *(Soundness Theorem) Let $\Gamma \cup \{A\} \subseteq \mathcal{L}$. If $\Gamma \vdash A$, then $\Gamma \vDash A$.*

Proof. Suppose that $\Gamma \vdash A$. We shall prove that $\Gamma \vDash A$ by induction on the number n of nodes in a derivation $\mathcal{D}$ of A from Γ in LET_J. If $n = 1$, then A is the only formula that occurs in $\mathcal{D}$ and $A \in \Gamma$. Since $\vDash$ is reflexive, it follows that $\Gamma \vDash A$. Suppose now that $n > 1$ and that the result holds for every derivation with fewer nodes than $\mathcal{D}$. It is straightforward to check that for each rule $\mathcal{R}$ of LET_J, if the premises of $\mathcal{R}$ hold in $w \in W$, then so does its conclusion. Let us consider rule $\rightarrow I$ and leave the remaining cases to the reader: suppose that there is a derivation $\mathcal{D}_1$ of B from $\Gamma \cup \{A\}$ in LET_J, and let $\mathcal{D}$ be the derivation (of $A \to B$ from Γ) obtained from $\mathcal{D}_1$ by applying rule $\to I$. Since $\mathcal{D}_1$ has fewer nodes than $\mathcal{D}$, it follows that $\Gamma, A \vDash B$ (by the induction hypothesis). Let $\mathcal{M} = \langle W, \leq, v \rangle$ and $w \in W$ be such that $\mathcal{M}, w \Vdash C$, for every $C \in \Gamma$, and let $w' \geq w$ be such that $v(A, w') = 1$. Since by (P1) the values of the elements of Γ in w remain the same in w', it follows that $\mathcal{M}, w' \Vdash C$, for every $C \in \Gamma$. Hence, $\mathcal{M}, w' \Vdash B$ (since $\Gamma, A \vDash B$ and $v(A, w') = 1$). Therefore, for every $w' \geq w$, $\mathcal{M}, w' \Vdash A$ only if $\mathcal{M}, w' \Vdash B$, i.e., $\mathcal{M}, w \Vdash A \to B$. □

Definition 4. *(Regular set) Let $\Delta \subseteq \mathcal{L}_J$. Δ is a regular set if it satisfies the following three conditions (A regular set, as defined here, corresponds to what is usually called a nontrivial prime theory. For the sake of convenience, we shall adopt the former terminology throughout this paper.):*

1. *Δ is nontrivial: $\Delta \nvdash A$, for some $A \in \mathcal{L}_J$;*
2. *Δ is closed: if $\Delta \vdash A$, then $A \in \Delta$, for every $A \in \mathcal{L}_J$;*
3. *Δ is disjunctive (or prime): if $\Delta \vdash A \vee B$, then $\Delta \vdash A$ or $\Delta \vdash B$, for every $A, B \in \mathcal{L}_J$.*

Definition 5. *Let $\Delta \cup \{A\} \subseteq \mathcal{L}_J$. Δ is said to be maximal with respect to A if, and only if, (i) $\Delta \nvdash A$ and (ii) $\Delta, B \vdash A$, for every $B \notin \Delta$.*

Lemma 1. *If Δ is maximal w.r.t. A, then Δ is a regular set.*

Proof. In order to prove that Δ is a theory, suppose that $\Delta \vdash B$ and that $B \notin \Delta$. Thus, $\Delta, B \vdash A$. By the transitivity of $\vdash$, it follows that $\Delta \vdash A$, which contradicts the initial hypothesis. To prove that Δ is a disjunctive set, suppose that $\Delta \vdash B \vee C$ and that $\Delta \nvdash B$ and $\Delta \nvdash C$, that is, $B \notin \Delta$ and $C \notin \Delta$. Hence, $\Delta, B \vdash A$ and $\Delta, C \vdash A$. Since $\Delta \vdash B \vee C$, it then follows by rule $\vee E$ that $\Delta \vdash A$, which also contradicts the initial hypothesis. □

Proposition 1. *Let $\Gamma \cup \{A\} \subseteq \mathcal{L}_J$. If $\Gamma \nvdash A$, then there is a set $\Delta \supseteq \Gamma$ that is maximal w.r.t. A.*

Proof. Let $B_1, B_2, \ldots$ be a fixed enumeration of $\mathcal{L}_J$ and let the sequence $\langle \Gamma_n \rangle_{n \in \mathbb{N}}$ be defined by:

1. $\Gamma_0 = \Gamma$
2. $\Gamma_{n+1} = \begin{cases} \Gamma_n & \text{if } \Gamma_n, B_n \vdash A \\ \Gamma_n \cup \{B_n\} & \text{if } \Gamma_n, B_n \nvdash A \end{cases}$

It can then be proven by a straightforward induction on n that $\Gamma_n \nvdash A$, for every $n \in \mathbb{N}$. Let $\Delta = \bigcup_{n \in \mathbb{N}} \Gamma_n$. To prove that $\Delta \nvdash A$, it suffices to notice that if A were derivable from Δ, then, by the compactness of $\vdash$, it would also be derivable from Γ_n, for some $n \in \mathbb{N}$. Now, suppose that $C \notin \Delta$ and let n be such that $C = B_n$. Since $B_n \notin \Gamma_{n+1}$ (for $\Gamma_{n+1} \subseteq \Delta$), it follows by construction that $\Gamma_n, B_n \vdash A$. Therefore, $\Delta, C \vdash A$. □

Lemma 2. *Let $\Delta \subseteq \mathcal{L}_J$ be a regular set and $B, C \in \mathcal{L}_J$. Then:*

1. *$B \wedge C \in \Delta$ iff $B \in \Delta$ and $C \in \Delta$;*
2. *$B \vee C \in \Delta$ iff $B \in \Delta$ or $C \in \Delta$;*
3. *$\neg\neg B \in \Delta$ iff $B \in \Delta$;*
4. *$\neg(B \wedge C) \in \Delta$ iff $\neg B \in \Delta$ or $\neg C \in \Delta$;*
5. *$\neg(B \vee C) \in \Delta$ iff $\neg B \in \Delta$ and $\neg C \in \Delta$;*
6. *If $\circ B \in \Delta$, then one of the following conditions obtains:*
   For every regular set $\Sigma \supseteq \Delta$, $B \in \Sigma$ and $\neg B \notin \Sigma$;
   For every regular set $\Sigma \supseteq \Delta$, $B \notin \Sigma$ and $\neg B \in \Sigma$;
7. *$B \to C \in \Delta$ iff for every regular set $\Sigma \supseteq \Delta$, if $B \in \Sigma$, then $C \in \Sigma$;*

8. $\neg(B \to C) \in \Delta$ iff $B \in \Delta$ and $\neg C \in \Delta$.

Proof. Items (1)–(5) and (8) follow immediately from the definition of a regular set together with the rules of LET_J. As for (6), suppose that $\circ B \in \Delta$. By PEM°, it follows that $\Delta \vdash B \vee \neg B$, and so either $B \in \Delta$ or $\neg B \in \Delta$. Let Σ be a regular set such that $\Delta \subseteq \Sigma$ and suppose that $B \in \Delta$. Since $\Delta \subseteq \Sigma$, it then follows that both $\circ B$ and B belong to Σ. Hence, $\neg B \notin \Sigma$, for otherwise Σ would be trivial (in virtue of rule EXP° and the fact that Σ is a regular set). A similar reasoning suffices to show that if $\neg B \in \Delta$, then $\neg B \in \Sigma$ and $B \notin \Sigma$, for every regular set $\Sigma \supseteq \Delta$. Finally, to prove the left-to-right direction of (7), suppose that $B \to C \in \Delta$ and let $\Sigma \supseteq \Delta$ be a regular set such that $B \in \Sigma$. Since $\Delta \subseteq \Sigma$, it follows that $B \to C \in \Sigma$ and so, $C \in \Sigma$ (by rule $\to E$). As for the right-to-left direction, suppose that $B \to C \notin \Delta$. By rule $\to I$, it follows that $\Delta, B \nvdash C$. By Proposition 1 and Lemma 1, there is a regular set $\Delta' \supseteq \Delta \cup \{B\}$ such that $C \notin \Delta'$. Since $\Delta \cup \{B\} \subseteq \Delta'$, $B \in \Delta'$. Therefore, there is a regular set $\Sigma \supseteq \Delta$ such that $B \in \Sigma$ and $C \notin \Sigma$. □

Proposition 2. *If Δ is a regular set, then there is a model $\mathcal{M} = \langle W, \leq, v \rangle$ and a stage $w \in W$ such that:*

$$\mathcal{M}, w \Vdash B \text{ if, and only if, } B \in \Delta, \text{ for every } B \in \mathcal{L}_J.$$

Proof. Let $\mathcal{M} = \langle W, \leq, w \rangle$ be such that:

1. $W = \{\Sigma : \Sigma \text{ is a regular set}\}$;
2. $\leq = \subseteq_W$;
3. $v : \mathcal{L}_J \times W \longrightarrow \{0, 1\}$ is defined by: $v(B, \Sigma) = 1$ iff $B \in \Sigma$, for every $B \in \mathcal{L}_J$.

Since Δ is a regular set, $\Delta \in W$. It follows from the definition of v that $v(B, \Delta) = 1$ if, and only if, $B \in \Delta$. However, in order to complete the proof we are still required to show that v is a valuation, i.e., that it satisfies all clauses of Definition 2. That $\mathcal{M}$ satisfies clauses (1)–(8) is an immediate consequence of Lemma 2 above. Note, moreover, that since $\leq$ has been defined as the set inclusion relation over W, $\mathcal{M}$ also satisfies (P1). □

Theorem 2. *(Completeness Theorem)*

$$\text{Let } \Gamma \cup \{A\} \subseteq \mathcal{L}_J. \text{ If } \Gamma \vDash A, \text{ then } \Gamma \vdash A.$$

Proof. Suppose that $\Gamma \nvdash A$. By Proposition 1 and Lemma 1, there is a regular set $\Delta \supseteq \Gamma$ such that $\Delta \nvdash A$. By Proposition 2, there is a model $\mathcal{M}$ and a stage $w \in W$ such that for every $B \in \mathcal{L}_J$, $\mathcal{M}, w \Vdash B$ if, and only if, $B \in \Delta$. Therefore, $\mathcal{M}, w \Vdash C$, for every $C \in \Gamma$ (since $\Gamma \subseteq \Delta$), and $\mathcal{M}, w \nVdash A$ (for $A \notin \Delta$). □

3. From LET_J to LET_F

The logic LET_F was introduced in [2] as an extension of FDE equipped with both a classicality operator $\circ$ and a non-classicality operator $\bullet$, dual to $\circ$—cf. [2] Section 3.1. (Hilbert and Gentzen-style systems for FDE can be found in [15] Section 2.2.) LET_F can also be obtained from LET_J by dropping the implication and adding $\bullet$, with the respective rules, which say essentially that $\bullet A$ holds if, and only if, $\circ A$ does not hold. As far as we know, classical negation cannot be defined in LET_F, so $\bullet$ had to be introduced as a primitive symbol. In the intended interpretation of the Kripke models presented here, $\bullet A$ means that in the database there is no reliable information about A.

Definition 6. *Let $\mathcal{L}_F$ be the language obtained from $\mathcal{L}_J$ by replacing $\to$ by the unary connective $\bullet$. The logic LET_F results from adding the following rules to the set of LET_J's $\to$-free rules:*

$$\dfrac{\circ A \quad \bullet A}{B} \text{ Cons} \qquad \dfrac{}{\circ A \vee \bullet A} \text{ Comp}$$

We shall use $\vdash_F$ to denote the derivability relation generated by LET_F and abbreviate it to $\vdash$ whenever appropriate.

Definition 7. A Kripke model $\mathcal{M}$ for LET_F is a structure $\langle W, \leq, v \rangle$ as in Definition 2, except that (7), (8) and (P1) are replaced by:

7'. $v(\bullet A, w) = 1$ iff $v(\circ A, w) = 0$

As in the case of LET_J, we say that A holds in w ($\mathcal{M}, w \Vdash_F A$) if, and only if, $v(A, w) = 1$. The definition of LET_F's semantic consequence relation, to be denoted by $\vDash_F$, is like the one for LET_J (Definition 3), with the obvious adjustments. When there is no risk of ambiguity, we write simply $\Vdash$ and $\vDash$ instead of $\Vdash_F$ and $\vDash_F$.

Theorem 3. *(Soundness Theorem)* Let $\Gamma \cup \{A\} \subseteq \mathcal{L}_F$. If $\Gamma \vdash A$, then $\Gamma \vDash A$.

Proof. Suppose that $\Gamma \vdash A$. We shall prove that $\Gamma \vDash A$ by induction on the number n of nodes in a derivation $\mathcal{D}$ of A from Γ in LET_F. If $n = 1$, then $\mathcal{D}$ contains only one formula and so either $A \in \Gamma$ or it is the result of applying rule $Comp$. If $A \in \Gamma$, then $\Gamma \vDash A$, by the reflexivity of $\vDash$. As for the latter case, suppose that A is the formula $\circ B \vee \bullet B$ and let $\mathcal{M} = \langle W, \leq, v \rangle$ and $w \in W$ be arbitrary. By Definition 7(7'), $v(\circ B, w) = 1$ or $v(\bullet B, w) = 1$. It then follows from clause (2) of Definition 7 that $v(\circ B \vee \bullet B) = 1$. Therefore, $\mathcal{M}, w \Vdash A$, and since $\mathcal{M}$ and w were arbitrary, we may conclude that $\Gamma \vDash A$. Suppose now that $n > 1$ and that the result holds for every derivation with fewer nodes than $\mathcal{D}$. It is straightforward to check that for each rule $\mathcal{R}$ of LET_F (other than $Comp$), if the premises of $\mathcal{R}$ hold in $w \in W$, then so does its conclusion. □

The proof of the completeness of LET_F with respect to the class of models characterized in Definition 7 is also similar to the one for LET_J, except for some minor differences. In particular, the definitions of regular and maximal sets (Definitions 4 and 5), and the proofs of Lemma 1 and Proposition 1 will carry over to the case LET_F. Hence, we shall assume those results to hold without presenting their proofs.

Lemma 3. *Let $\Delta \subseteq \mathcal{L}_F$ be a regular set and $B, C \in \mathcal{L}$. Then:*

1. $B \wedge C \in \Delta$ iff $B \in \Delta$ and $C \in \Delta$;
2. $B \vee C \in \Delta$ iff $B \in \Delta$ or $C \in \Delta$;
3. $\neg\neg B \in \Delta$ iff $B \in \Delta$;
4. $\neg(B \wedge C) \in \Delta$ iff $\neg B \in \Delta$ or $\neg C \in \Delta$;
5. $\neg(B \vee C) \in \Delta$ iff $\neg B \in \Delta$ and $\neg C \in \Delta$;
6. If $\circ B \in \Delta$, then one of the following conditions obtains:
   For every regular set $\Sigma \supseteq \Delta$, $B \in \Sigma$ and $\neg B \notin \Sigma$;
   For every regular set $\Sigma \supseteq \Delta$, $B \notin \Sigma$ and $\neg B \in \Sigma$;
7'. $\bullet B \in \Delta$ iff $\circ B \notin \Delta$.

Proof. Items (1)–(5) follow immediately from the definition of a regular set together with the rules of LET_F. As for (6), it can be proven exactly as in the proof of Lemma 2. Finally, to prove (7') it suffices to notice that if $\circ B, \bullet B \in \Delta$, then Δ would be trivial, and that either $\circ B \in \Delta$ or $\bullet B \in \Delta$ (by rule $Comp$ and the assumption that Δ is regular). □

Proposition 3. *If Δ is a regular set, then there is a model $\mathcal{M} = \langle W, \leq, v \rangle$ and a stage $w \in W$ such that:*

$$\mathcal{M}, w \Vdash B \text{ if, and only if, } B \in \Delta, \text{ for every } B \in \mathcal{L}_F.$$

Proof. Let $\mathcal{M} = \langle W, \leq, v \rangle$ be such that:

1. $W = \{\Sigma : \Sigma \text{ is a regular set}\}$;
2. $\leq = \subseteq_W$;
3. $v : \mathcal{L}_F \times W \longrightarrow \{0, 1\}$ is defined by: $v(B, \Sigma) = 1$ iff $B \in \Sigma$, for every $B \in \mathcal{L}_F$.

Since Δ is a regular set, $\Delta \in W$. It then follows from the definition of v that $v(B, \Delta) = 1$ if, and only if, $B \in \Delta$, for every $B \in \mathcal{L}_F$. By Lemma 3 above, Δ satisfies all clauses of Definition 7, and we are done. □

Theorem 4. *(Completeness Theorem)*

$$\text{Let } \Gamma \cup \{A\}. \text{ If } \Gamma \vDash A, \text{ then } \Gamma \vdash A.$$

Proof. Suppose that $\Gamma \nvdash A$. By (the LET_F-analogues of) Proposition 1 and Lemma 1, there is a regular set $\Delta \supseteq \Gamma$ such that $\Delta \nvdash A$ (and so $A \notin \Delta$). By applying Proposition 3, it then follows that there is a model $\mathcal{M} = \langle W, \leq, v \rangle$ and a stage $w \in W$ such that for every $B \in \mathcal{L}_F$, $v(B, w) = 1$ if, and only if, $B \in \Delta$. Therefore, $\mathcal{M}, w \Vdash C$, for every $C \in \Gamma$, but $\mathcal{M}, w \nVdash A$, that is, $\Gamma \nvDash A$. □

Although the persistence clause (P1) of Definition 2 is necessary for proving the soundness of LET_J, it can be completely dispensed with in LET_F. As we shall see in the next section, there are some reasons why supplementing the semantics of LET_F with some weaker versions of (P1) may be desirable. Before we do so, it is worth noting that even in the absence of (P1), for formulas $\circ A$, LET_F already requires the values of A or $\neg A$ to be preserved across stages.

Proposition 4. *Let $\mathcal{M} = \langle W, \leq, v \rangle$ and $A \in \mathcal{L}$. For every $w \in W$, it holds that:*

1. *If $v(\circ A, w) = v(A, w) = 1$, then $v(A, w') = 1$, for every $w' \geq w$;*
2. *If $v(\circ A, w) = v(\neg A, w) = 1$, then $v(\neg A, w') = 1$, for every $w' \geq w$.*

Proof. This is an immediate consequence of clause (6) of Definition 7. □

Thus, in LET_F, whenever $\circ A$ holds in a certain stage w, both A and $\neg A$ will retain their values in every stage w' accessible from w; and since exactly one of A or $\neg A$ holds in w whenever $\circ A$ does, this entails that exactly one of A or $\neg A$ will hold in every such w'.

4. Persistence Clauses and Information Revision

In this section we explore different persistence relations that may hold in a Kripke model for LET_F and indicate how each of those relations may be useful for representing different criteria for revising information.

Recall that, given a model $\mathcal{M}$ and a stage $w \in W$, $v(A, w) = 1$ expresses that positive information A is available at w, while $v(A, w) = 0$ expresses that there is no such information in w. Likewise, $v(\neg A, w) = 1$ indicates the presence at w of negative information $\neg A$, whereas $v(\neg A, w) = 0$ is to be interpreted as the lack of such information. When the information about A is reliable at w, we have $v(\circ A, w) = 1$. For the sake of convenience, we may express the same thing more succinctly by saying that the information conveyed by A (which may assume the form $\neg B$ or $\circ B$) is available at w whenever $v(A, w) = 1$, and that no such information is available at w whenever $v(A, w) = 0$.

Now, how are we to understand the fact that A may assume different values in two $\leq$-related stages? The following definitions may be of some help: given stages $w, w' \in W$ such that $w \leq w'$, we shall say that in w' we have acquired the information conveyed by A whenever $v(A, w) = 0$ and $v(A, w') = 1$; and that we have revised that same information whenever $v(A, w) = 1$ and $v(A, w') = 0$. Using this new terminology, we can then describe the following four scenarios:

1. $v(A, w) = 1$ and $v(A, w') = 1$: the information conveyed by A was available at w and it has not been revised in the process of moving from w to w' (i.e., it remained available);
2. $v(A, w) = 1$ and $v(A, w') = 0$: the information conveyed by A was available at w but it has been revised in the process of moving from w to w';
3. $v(A, w) = 0$ and $v(A, w') = 0$: the information conveyed by A was unavailable at w, nor was it acquired in the process of moving from w to w' (i.e., it remained unavailable);
4. $v(A, w) = 0$ and $v(A, w') = 1$: the information conveyed by A was unavailable at w but it has been acquired in the process of moving from w to w'.

4.1. Persistence Conditions

Having the notions characterized in (1)–(4) at our disposal, we can now categorize the models of LET_F according to the different revisability relations that may or may not hold between formulas and stages. In other words, we can distinguish classes of models in terms of the kinds of information that are allowed to be revised.

Let a literal be a propositional letter or the negation of a propositional letter, and let basic information be the (positive and negative) information conveyed by literals. The models of LET_F can be classified according to whether or not they satisfy one of the following persistence conditions:

P1. *Total non-revisability*

For every $w' \geq w$, if $v(A, w) = 1$, then $v(A, w') = 1$.

P2. *Non-revisability of reliable information*

For every $w' \geq w$, if $v(\circ A, w) = 1$, then $v(\circ A, w') = 1$.

P3. *Non-revisability of reliable information and basic information*

For every $w' \geq w$, if $v(\circ A, w) = 1$, then $v(\circ A, w') = 1$;

For every $w' \geq w$, if $v(p, w) = 1$, then $v(p, w') = 1$;

For every $w' \geq w$, if $v(\neg p, w) = 1$, then $v(\neg p, w') = 1$.

Condition (P1), which was already present in LET_J, amounts to the constraint of total non-revisability: it states that no information whatsoever is allowed to be revised at any stage. In other words, every new piece of information acquired at a certain stage is always passed on to the subsequent stages, leaving no room for data to be removed in the light of new information.

It is to be noted, however, that (P1) does not quite fit in the intended interpretation of LET_F. This is because if $\bullet A$ were to always persist across stages, we would be prevented from acquiring reliable information about A whenever that information had been previously deemed unreliable. On the other hand, in the semantics for LET_J, total non-revisability was required because of the intuitionistic clause for implication. Since $\bullet$ is absent from LET_J, this does not represent a problem there, although the presence of (P1) does prevent revising information in LET_J-models.

(P2) corresponds to the constraint that information already marked as reliable cannot be revised. Thus, once $\circ A$ holds in a stage w, it cannot be removed at any stage $w' \geq w$. Recall that in Proposition 4 we have proved, even in the absence of (P2), that the fact that $\circ A$ holds in a certain stage w is already sufficient for the non-revisability of either A or $\neg A$ (which depends on which of A and $\neg A$ actually holds in w). However, this did not prevent the revisability of $\circ A$ itself; that is, it did not rule out such models as:

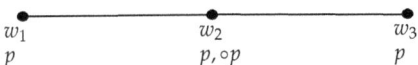

Requiring models to satisfy (P2) will, however, prevent situations in which A ($\neg A$) is non-revisable in virtue of $\circ A$ holding in a certain stage, even though $\circ A$ itself is allowed to be revised at any further stage.

Another important aspect of (P2) is that it entails (actually, is equivalent to):

P2′. If $v(\bullet A, w') = 1$, then for every $w \leq w'$, $v(\bullet A, w) = 1$.

(P2′) says that if some information is unreliable at a stage w', it must have been unreliable in every stage w that precedes w' (To prove that (P2) entails (P2′), suppose that $v(\bullet A, w') = 1$ and let $w \leq w'$. Suppose further that $v(\circ A, w) = 1$. By (P2), it then follows that $v(\circ A, w') = 1$, which contradicts clause

(7′) of Definition 7. Therefore, $v(\bullet A, w) = 1$. Suppose now that $\mathcal{M}$ satisfies (P2′) and let $w \in W$ be such that $v(\circ A, w) = 1$. For any arbitrary $w' \geq w$, $v(\circ A, w') = 1$ or $v(\bullet A, w') = 1$. If $v(\bullet A, w') = 1$, then $v(\bullet A, w) = 1$ (by (P2′)). This result contradicts clause (7′) of Definition 7. Therefore, $v(\circ A, w') = 1$.).

Finally, (P3) adds to (P2) the requirement of non-revisability of basic information. This makes sense if we think of a database in which only literals and formulas of the form $\circ p$ can be inserted. Given a model $\mathcal{M}$ that satisfies (P3), it can be easily proved that any formula A in which neither $\circ$ nor $\bullet$ occur will be preserved across $\leq$-related stages.

Proposition 5. *Let* $\mathcal{M} = \langle W, \leq, v \rangle$ *and* $w \in W$. *Let* $A \in \mathcal{L}_F$ *be such that neither* $\circ$ *nor* $\bullet$ *occur in* A. *If* $\mathcal{M}$ *satisfies (P3) and* $v(A, w) = 1$, *then for every* $w' \geq w$, $v(A, w') = 1$.

Proof. The result can be proved by a straightforward induction on the complexity of A; this proof is left to the reader. □

Concerning the result above, it is to be noted that the condition (P3) added to LET_F does not collapse into (P1) because (P3) does not apply to formulas like $\bullet A$, nor to formulas in which $\circ$ and $\bullet$ appear in the scope of $\neg$.

The conditions (P1), (P2) and (P3) are not exhaustive. The idea of these models as representations of information revision can be developed in different ways, even allowing revisability of reliable information. We can think of different revisability conditions as different levels of access to the database. Total non-revisability (P1) would be a level of access that can insert information but cannot remove anything from the database. Non-revisability of reliable information (P2) fits the idea of two different levels of access: a level-1 access that can only insert basic information but cannot remove nor mark anything as reliable (i.e., cannot insert $\circ A$), and a level-2 access that can remove any information not marked as reliable and mark information as reliable (i.e., can insert and remove literals and insert $\circ A$), but still cannot remove or change reliable information, which is marked with $\circ$. This does not mean, however, that in both cases reliable information cannot be revised once and for all, but only that the model is not able to represent, so to speak, a sort of higher level access to the database.

4.2. Adding Persistence to LET_F

The reader may ask at this point what would be the result of adding the persistence clauses above to the semantics of LET_F. After all, it seems that modifying Definition 7 would restrict the class of models originally characterized in the previous section and, as a result, we should expect LET_F to retain soundness but not completeness with respect to the new, more restricted, classes of models. As we shall now see, though, this is not really the case, for no matter which persistence clause we choose to supplement Definition 7 with, LET_F will continue to be sound and complete with respect to the new class of models. Let us first prove this fact and then explain why none of the persistence clauses (P1)–(P3) interfere with the completeness of LET_F.

Soundness and Completeness with Persistence

To prove that soundness and completeness will continue to hold with respect to the classes of models corresponding to each of the persistence clauses (P1)–(P3), it will suffice to consider the class generated by the most restrictive condition, (P1). In order to establish this result it will be convenient to first introduce some preliminary notation. We shall use the symbol $\mathcal{C}$ to denote the class of models originally characterized in Definition 7—i.e., models with no persistence constraints, except for those stated in Proposition 4—and $\mathcal{C}_i$ ($1 \leq i \leq 3$) to denote the class that results from adding (Pi) to that definition. Note that, for every i, $\mathcal{C}_i$ is properly included in $\mathcal{C}$, $\mathcal{C}_1$ is properly included in both $\mathcal{C}_2$ and $\mathcal{C}_3$, and $\mathcal{C}_3$ in $\mathcal{C}_2$, since every model that satisfies (P1) also satisfies (P2) and (P3), and every model that satisfies (P3) satisfies (P2). Finally, $\vDash_i$ denotes the semantic consequence relation generated by the models in $\mathcal{C}_i$ (We shall continue to use $\vdash$ and $\vDash$ as abbreviations for respectively $\vdash_F$ and $\vDash_F$ throughout this section.). We can then prove that:

Lemma 4. *Let $\mathcal{M} = \langle W, \leq, v \rangle$ be a member of $\mathcal{C}$ and let $w \in W$. Then there is a model $\mathcal{M}_1 = \langle W_1, \leq_1, v_1 \rangle$ in $\mathcal{C}_1$ and a stage $w_1 \in W_1$ such that $\mathcal{M}, w \Vdash B$ if, and only if, $\mathcal{M}_1, w_1 \Vdash B$, for every $B \in \mathcal{L}$.*

Proof. Let $\mathcal{M}_1$ be defined by:

1. $W_1 = \{w\}$;
2. $\leq_1 = \{\langle w, w\rangle\}$; and
3. $v_1 : \mathcal{L}_F \times W_1 \to \{0, 1\}$ is a total function such that for every $B \in \mathcal{L}_F$:

$$v_1(B, w) = 1 \text{ iff } v(B, w) = 1$$

Given that $\mathcal{M}_1$ has only one stage, v_1 (vacuously) satisfies (P1). Hence, all we need to do in order to complete the proof is to show that v_1 satisfies all clauses of Definition 7. Since clauses (1)–(5) and (7′) are all locally formulated, they follow immediately from the definition of v_1. Concerning clause (6), which is the only global clause among (1)-(7′), we may proceed as follows. Suppose that $v_1(\circ C, w) = 1$. Hence, $v(\circ C, w) = 1$, and so exactly one of (I) and (II) below obtains:

(I) For every $w' \in W$ such that $w' \geq w$, $v(C, w') = 1$ and $v(\neg C, w') = 0$;
(II) For every $w' \in W$ such that $w' \geq w$, $v(C, w') = 0$ and $v(\neg C, w') = 1$.

Suppose that (I) holds. Thus, $v(C, w) = 1$ and $v(\neg C, w) = 0$, and, so, $v_1(C, w) = 1$ and $v_1(\neg C, w) = 0$. Since w is the only element of W_1, we may conclude that for every $w' \in W_1$ such that $w' \geq w$, $v_1(C, w') = 1$ and $v_1(\neg C, w') = 0$ (and similarly in the case of (II)). When $v_1(\circ C, w) = 0$, it follows that $v(\circ C, w) = 0$, and there is nothing to be proved since clause 6 has just one direction and is vacuous on this condition. □

Theorem 5. *Let $\Gamma \cup \{A\} \subseteq \mathcal{L}$. Then $\Gamma \vDash A$ if, and only if, $\Gamma \vDash_1 A$.*

Proof. Since every model that belongs to $\mathcal{C}_1$ also belongs to $\mathcal{C}$, it follows immediately that if $\Gamma \nvDash_1 A$, then $\Gamma \nvDash A$. As for the other direction, suppose that $\Gamma \nvDash A$. Hence, there is a model $\mathcal{M}$ in $\mathcal{C}$ and a stage $w \in W$ such that $\mathcal{M}, w \Vdash B$, for every $B \in \Gamma$, and $\mathcal{M}, w \nVdash A$. By Lemma 4 above, there is a model $\mathcal{M}_1 = \langle W_1, \leq_1, v_1 \rangle$ in $\mathcal{C}_1$ and $w_1 \in W_1$ such that $\mathcal{M}_1, w_1 \Vdash B$, for every $B \in \Gamma$, and $\mathcal{M}_1, w_1 \nVdash A$. Therefore, $\Gamma \nvDash_1 A$. □

Lemma 4 states that no matter how many stages a given model $\mathcal{M}$ has, for each stage w of $\mathcal{M}$, we can always find a corresponding model with exactly one stage w_1 such that the same formulas hold in both w and w_1. Notice that because W_1 contains only one stage, $\mathcal{M}_1$ (vacuously) satifies each of the persistence clauses (P1)–(P3). This means that Lemma 4 and Theorem 5 would still be provable in exactly the same way if $\mathcal{C}_1$ (and the corresponding consequence relation $\vDash_1$) were replaced by either $\mathcal{C}_2$ or $\mathcal{C}_3$. As a result, all of $\vDash, \vDash_1, \vDash_2$ and $\vDash_3$ turn out to have the same extension which, together with the soundness and completeness of LET_F, yields:

Corollary 1. *Let $\Gamma \cup \{A\} \subseteq \mathcal{L}_F$. Then:*

1. $\Gamma \vdash A$ *iff* $\Gamma \vDash_1 A$;
2. $\Gamma \vdash A$ *iff* $\Gamma \vDash_2 A$;
3. $\Gamma \vdash A$ *iff* $\Gamma \vDash_3 A$.

How can LET_F be sound and complete with respect to all of $\vDash, \vDash_1, \vDash_2$ and $\vDash_3$, in spite of those relations being characterized in terms of different classes of models? As we shall see, the reason has to do with the fact that in the semantics for LET_F there is no clause that states a sufficient condition for $\circ A$ to hold in a given stage. Before we get to that, however, we first need to take a look at the soundness and completeness proofs of LET_F presented in the previous section, in order to make sure that they would still work had we adopted any of those alternative notions of consequence relation.

That the soundness theorem would continue to hold follows immediately from the fact each C_i is included in C, which, in turn, implies that if ⊢ is sound with respect to the models in C, then it is also sound with respect to models in the more restricted class C_i (given that $\Gamma \vDash A$ implies $\Gamma \vDash_i A$). Notice, moreover, that since nowhere in the proof of Theorem 3 was any of (P1)–(P3) appealed to, the proof would work equally well had we adopted any of $\vDash_1, \vDash_2, \vDash_3$ instead of $\vDash$.

Concerning the completeness theorem, we need to consider the modifications (if any) that would be necessary if the proof were being formulated with respect to models satisfying one of (P1)–(P3). As it turns out, there is precisely one place in the whole proof that requires more attention, viz., Proposition 3.

Recall that it was established in Proposition 3 that for any regular set Δ of formulas of $\mathcal{L}_F$, one can find a model $\mathcal{M}$ and a stage w of $\mathcal{M}$ such that a formula holds in w if, and only if, it belongs to Δ. While proving this result, the model $\mathcal{M}$ was defined in such a way that its stages were all the regular sets of $\mathcal{L}_F$, its accessibility relation $\leq$ was taken to be the inclusion relation $\subseteq$ over W, and its valuation function was defined in terms of the characteristic function of each $\Sigma \in W$. Now, had we proved this result with respect to models that satisfy one of (P1)-(P3), we would have to make sure that $\mathcal{M}$ did indeed satisfy the corresponding clause. In the case of (P1), for example, this would require showing that for every regular sets Σ and Σ' such that $\Sigma \leq \Sigma'$, the fact that formula A belongs to Σ implies that it also belongs to Σ' (and similarly for the other clauses). At this point, it becomes clear, however, that this requirement, as well as the ones corresponding to the other clauses, was already satisfied in the original proof of Proposition 3, given the way $\leq$ was defined (i.e., in terms of $\subseteq$). Hence, as in the case of the proof of the soundness of LET_F, the proof of its completeness would also remain unaltered.

Why does the adoption of any of the persistence clauses above bring no changes whatsoever upon the corresponding deductive system? We can reach a better understanding of this fact by taking a closer look at the proof of Lemma 4, for it is precisely because of that result that we are able to prove the equivalence between $\vDash, \vDash_1, \vDash_2$ and $\vDash_3$.

The proof tells us that given any model $\mathcal{M}$ belonging to C and a stage w in this model, one can always extract a model $\mathcal{M}_1$ out of $\mathcal{M}$ such that w is the only stage of $\mathcal{M}_1$ and the same formulas hold in w with respect to either model. That the semantic values of formulas containing no occurrences of either ◦ or • are carried over to the new model is a consequence of the fact that the semantic conditions of formulas formed with ¬, ∧, ∨ are all local, and so they do not depend on the values their subformulas have at stages other than w.

There is no need to take formulas •A into account here because their semantic conditions are stated directly in terms of those for ◦A. So let us consider what happens with formulas of the form ◦A. Assuming that $\mathcal{M}, w \Vdash ◦A$, the only reason why ◦A could fail to hold in w (w.r.t. $\mathcal{M}_1$) is if there were some $w' \geq w$ in $\mathcal{M}_1$ such that $v(A, w') = v(¬A, w')$. However, since w is the only stage in $\mathcal{M}_1$ and since A and ¬A inherit in $\mathcal{M}_1$ the values they had in $\mathcal{M}$, this cannot happen. What if ◦A did not hold in the original $\mathcal{M}$? Could the elimination of all stages in $\mathcal{M}$ except for w also eliminate all the counterexamples to ◦A in $\mathcal{M}$? The answer is 'no', and the reason for this is that the definition of a Kripke model for LET_F (with or without any of (P1)–(P3)) does not state any sufficient condition for ◦A to hold in a stage. If this were the case, then while moving from $\mathcal{M}$ to $\mathcal{M}_1$ we would have no guarantee that the (sufficient) condition for ◦A to hold in $\mathcal{M}$ would not become satisfied in virtue of there being fewer stages in $\mathcal{M}_1$ than in $\mathcal{M}$—and so ◦A would hold in $\mathcal{M}_1$, even though it failed to hold in $\mathcal{M}$. This situation is thus very different from what takes place in intuitionistic logic. For imagine what would happen if we attempted to prove an analogue of Lemma 4 for intuitionistic logic. Although every formula that holds in w in the original model $\mathcal{M}$ would continue to hold in w in the new model $\mathcal{M}_1$, it could well happen that a formula $A \to B$ that did not hold in w (w.r.t. $\mathcal{M}$) would nonetheless hold in w w.r.t. $\mathcal{M}_1$. This is because all the counter-examples to $A \to B$ could end up being eliminated in $\mathcal{M}_1$. Notice that this phenomenon depends essentially on the fact that in order for $A \to B$ to hold in a stage w in a Kripke model for intuitionistic logic, there can be no stage $w' \geq w$ such that A holds in

w' and B does not hold in w', which amounts to a sufficient condition for $A \to B$ to hold in w. And it is precisely one such condition that is missing in the case of LET_F's $\circ$ operator.

It is worth noting that, as a matter of fact, a semantics for LET_F does not need a global clause for $\circ$, which means that from the strictly technical point of view, Kripke-style models for LET_F collapse into standard models. Nevertheless, the conceptual idea of Kripke models for intuitionistic logic, in which propositions are proved as time passes, has an analogy with the idea of a database that receives information as time passes. Moreover, if we change the 'only if' of the semantic clause for $\circ$ (Definition 2 item 6) to an 'if and only if', we obtain an appealing sufficient condition for $\circ A$: if at a stage w we 'look to the future' and either across all stages A holds or across all stages $\neg A$ holds, then $\circ A$ holds in w (we return to this point in Section 6 below). Therefore, although strictly speaking we have here 'Kripke-style' models rather than Kripke models, from the conceptual point of view our proposal here seems to be quite justified.

Remark 1. *In Omori and Sano [16], p. 162 we find Kripke models for the logic cBS4, which is an extension of LET_J with the following axioms:*

A3. $A \to \circ A \equiv \neg A \to \circ A$,
A8. $\neg \circ A \equiv (A \equiv \neg A)$.

The semantics is given by Kripke models for N4 plus clauses tantamount to the following:

i. $w \Vdash \circ A$ iff $\forall w' \geq w, (w' \Vdash A$ and $w' \nVdash \neg A)$ or $(w' \nVdash A$ and $w' \Vdash \neg A)$;
ii. $w \Vdash \neg \circ A$ iff $\forall w' \geq w, w' \Vdash A$ iff $w' \Vdash \neg A$.

Omori and Sano adopt a Dunn-style relational semantics, with two relations $\Vdash^+$ and $\Vdash^-$, but the result is the same, since $w \Vdash^- A$ is equivalent to $w \Vdash^+ \neg A$. The logics cBS4, BD$\circ$ and BS4 discussed in [16] are indeed related, respectively, to the logics of evidence and truth LET_J, LET_F, and LET_K (the latter is LET_J plus Peirce Law, see [17], pp. 82–83). A more detailed analysis of the similarities and differences between these logics will be done elsewhere.

Although the 'only if' direction of the semantic clause (i) is equivalent to the clause for $\circ$ in LET_J (and in LET_F if persistence for $\circ A$ holds, see Section 4.2), the behavior of the classicality operator $\circ$ in cBS4 is quite different from its behavior in LET_J and LET_F. A formula $\neg \circ A$ in cBS4 has some analogy to a formula $\bullet A$ in LET_F, since in the former $\vdash \circ A \vee \neg \circ A$ and $\circ A, \neg \circ A \vdash B$ hold. However, whether or not $\circ A$ holds in LET_J and LET_F is left undetermined even in those circumstances in which exactly one between A and $\neg A$ holds. The rationale for this is that the information that only A (or $\neg A$) holds may be reliable, and in this case $\circ A$ holds, or unreliable, and so $\circ A$ does not hold. In LET_F this can be expressed by the formulas $A \wedge \circ A$ and $A \wedge \bullet A$. This feature of LET_J and LET_F is essential for the intended interpretation in terms of positive and negative, reliable or unreliable, information.

Modal interpretations for variants of the consistency operator have been proposed before. The first one appears in [18] where $\circ A$ is defined as $A \to \Box A$, obtaining a conceptualization of $\circ$ that preserves all the essential properties of a consistency connective (under a specific negation). In view of its definition, the semantic interpretation of $\circ$ depends naturally on a modal reading. This does not exactly signify assigning a possible-world interpretation to $\circ$, but rather defining a modal formula that behaves like $\circ$. Later on, a modal approach for consistency combined with modal negations was proposed in [19].

5. Some Properties of LET_J and LET_F

The following properties clarify some aspects of LET_J and LET_F that bear directly on their intended interpretations:

Proposition 6. *In LET_J and LET_F the following inferences do not hold:*

1. $\circ A \vdash \circ \neg A$;
2. $\circ A, \circ B \vdash \circ (A * B)$ $(* \in \{\vee, \wedge\})$;

3. $\circ A, \circ B \vdash \circ(A \to B)$ (in LET_J only);
4. $\bullet(A * B) \vdash \bullet A \vee \bullet B$ ($* \in \{\vee, \wedge\}$, in LET_F only).

Proof. Left to the reader. □

It is easy to find counterexamples for all the inferences above. The semantic values of the conclusions are left undetermined by the premises because there is no sufficient condition for $w \Vdash \circ A$. As a consequence, in both LET_J and LET_F propagation rules over $\{\neg, \vee, \wedge, \to\}$ do not hold. On the other hand, let us say that a formula A behaves classically in LET_J or LET_F if $\vdash A \vee \neg A$ and $A, \neg A \vdash B$ hold; so in both LET_J and LET_F, although they do not have propagation rules, the classical behavior propagates over $\{\neg, \vee, \wedge, \to\}$. More precisely:

Proposition 7. *Suppose $\circ \neg^{n_1} A_1, \ldots, \circ \neg^{n_m} A_m$ hold for $n_i \geq 0$ (where $\neg^{n_i}$, $n_i \geq 0$, represents n_i occurrences of negations before the formula A_i).*
Then:

1. *Any LET_F-formula formed with $A_1, \ldots, A_m$ over $\{\wedge, \vee, \neg\}$ behaves classically;*
2. *Any LET_J-formula formed with $A_1, \ldots, A_m$ over $\{\wedge, \vee, \neg, \to\}$ behaves classically.*

Proof. Item (1) has been proved in [2], Fact 31. To prove (2), given that for any $n \geq 0$, $\circ \neg^n A \vdash A \vee \neg A$ and $\circ \neg^n A, A, \neg A \vdash B$, it remains to be proved that: (i) $\circ \neg^n A, \circ \neg^m B \vdash (A \to B) \vee \neg(A \to B)$ and (ii) $\circ \neg^n A, \circ \neg^m B, (A \to B), \neg(A \to B) \vdash C$. The proofs of (i) and (ii) are left to the reader. □

This result establishes that even though, say, $\circ p$ and $\circ q$ do not entail $\circ(p \vee q)$, $\circ(p \wedge q)$, etc., they do entail that every formula formed with p and q over $\{\neg, \vee, \wedge\}$ has a classical behavior. Hence, if formulas of the form $\circ A$ are required to persist across stages in LET_F (i.e., if models are required to satisfy (P2)), this behavior is also transmitted across $\leq$-related stages:

Proposition 8.

1. *In LET_J, if $w \Vdash \circ \neg^{n_1} A_1, \ldots, w \Vdash \circ \neg^{n_m} A_m$, then for any formula B formed with $A_1, \ldots, A_m$ over $\{\wedge, \vee, \neg, \to\}$, and for any $w' \geq w$, B behaves classically in w';*
2. *In LET_F, assuming persistence for formulas $\circ A$, if $w \Vdash \circ \neg^{n_1} A_1, \ldots, w \Vdash \circ \neg^{n_m} A_m$, then for any formula B formed with $A_1, \ldots, A_m$ over $\{\wedge, \vee, \neg\}$, and for any $w' \geq w$, B behaves classically in w'.*

Proof. Item (1) follows from Proposition 7 item 2 above and the fact that persistence holds for every formula in LET_J. Item (2) follows from Proposition 7 item 1 above and the persistence of every formula $\circ A$. □

6. Final Remarks and Further Research

In this paper we proposed Kripke-style models for the logics LET_J and LET_F introduced respectively in [1,2]. The intended interpretation of these models is in terms of a database that receives positive and negative information, that can be either unreliable or reliable, the reliable information being subjected to classical logic. We claim that the semantics is sound with respect to this intended interpretation.

A remarkable feature of these models is that there is no sufficient condition for $\circ A$. This mimics the fact that in the valuation semantics for LET_J and LET_F different values for A and $\neg A$ do not imply $\circ A$, and there is a rationale for this. The information that exactly one of either A or $\neg A$ holds is not enough, for we still need the information that such information is reliable. Note that this is what distinguishes the scenarios 1 and 2 respectively from 5 and 6 mentioned in the Introduction.

There are no introduction rules for $\circ$ in LET_J and LET_F. The idea that the reliability of a formula comes from outside the formal system is appealing, but it could be made more precise. The reliability and conclusiveness of p and $\neg p$ are expressed by logics of evidence and truth as the classicality of p.

Although it is reasonable that no rule concludes $\circ p$, and propositions 7 and 8 show that classical behavior propagates over the standard propositional connectives, it could be an advantage to have propagation rules for $\circ$. This can be obtained simply by changing item 6 of Definition 2, and the corresponding definition for LET_F, putting an 'if and only if' in the place of 'only if'. More precisely, if we make the necessary condition for $w \Vdash \circ A$ also a sufficient condition, the consequent of the result expressed by Proposition 8 becomes stronger: for any formula B formed with $A_1, \ldots, A_m$ over $\{\wedge, \vee, \neg\}$, and for any $w' \geq w$, $w' \Vdash \circ B$. Investigating the consequences of such a change in the semantics presented here, however, will be done elsewhere.

An algebraic semantics for $N4$ was proposed in [20] by means of the $N4$-lattices. In a similar vein, it was proved in [21] Section 9.3 that the logic LET_J is sound and complete with respect to Fidel-structures. As LET_F can be defined from LET_J by dropping the implication and adding the operator • and the rules Cons and Comp, a natural conjecture is that both LET_F and LET_J would be algebraizable (or at least count with an algebraic semantics) by way of the non-deterministic algebraization methods of [22]. This of course has still to be proved.

Author Contributions: Conceptualization: A.R., A.K., and W.C.; formal analysis: H.A. and A.R.; investigation: A.R., A.K., W.C., and H.A.: writing—review and editing, H.A. and A.R. All authors have read and agreed to the published version of the manuscript.

Funding: This research received no external funding.

Conflicts of Interest: The authors declare no conflict of interest.

References

1. Carnielli, W.; Rodrigues, A. An epistemic approach to paraconsistency: A logic of evidence and truth. *Synthese* **2017**, *196*, 3789–3813. [CrossRef]
2. Rodrigues, A.; Bueno-Soler, J.; Carnielli, W. Measuring evidence: A probabilistic approach to an extension of Belnap-Dunn Logic. *Synthese* **2020**. [CrossRef]
3. Carnielli, W.; Rodrigues, A. On epistemic and ontological interpretations of intuitionistic and paraconsistent paradigms. *Log. J. IGPL* **2020**. [CrossRef]
4. Carnielli, W.; Coniglio, M.E. *Paraconsistent Logic: Consistency, Contradiction and Negation*; Springer: Berlin, Germany, 2016.
5. Carnielli, W.; Coniglio, M.E.; Marcos, J. Logics of Formal Inconsistency. In *Handbook of Philosophical Logic*; Gabbay, D.M., Guenthner, F., Eds.; Springer: Berlin, Germany, 2007; Volume 14.
6. Carnielli, W.; Coniglio, M.E.; Rodrigues, A. Recovery operators, paraconsistency and duality. *Log. J. IGPL* **2019**. [CrossRef]
7. Dunn, J.M. Information in computer science. In *Philosophy of Information*; Volume 8 of Handbook of the Philosophy of Science; Adriaans, P., van Benthem, J., Eds.; Elsevier: Amsterdam, The Netherlands, 2008; pp. 581–608.
8. Odintsov, S.; Wansing, H. On the Methodology of Paraconsistent Logic. In *Logical Studies of Paraconsistent Reasoning in Science and Mathematics*; Andreas, H., Verdée, P., Eds.; Springer: Berlin, Germany, 2016.
9. Belnap, N.D. How a computer should think. In *Contemporary Aspects of Philosophy*; Ryle, G., Ed.; Oriel Press: Charleville, UK, 1977.
10. Belnap, N.D. A useful four-valued logic. In *Modern Uses of Multiple Valued Logics*; Epstein, G., Dunn, J.M., Eds.; D. Reidel: Dordrecht, The Netherlands, 1977.
11. Dunn, J.M. Intuitive semantics for first-degree entailments and 'coupled trees'. *Philos. Stud.* **1976**, *29*, 149–168. [CrossRef]
12. Wansing, H. *The Logic of Information Structures*; Springer: Berlin, Germany, 1993.
13. Kapsner, A. *Logics and Falsifications*; Springer: Berlin, Germany, 2014.
14. Van Dalen, D. *Logic and Structure*, 4th ed.; Springer: Berlin, Germany, 2008.
15. Omori, H.; Wansing, H. 40 years of FDE: An Introductory Overview. In *Studia Logica*; Springer: Berlin, Germany, 2017; pp. 1021–1049.
16. Omori, H.; Sano, K. da Costa Meets Belnap and Nelson. In *Recent Trends in Philosophical Logic*; Ciuni, R., Wansing, H., Willkommen, C., Eds.; Springer: Berlin, Germany, 2014.

17. Carnielli, W.; Rodrigues, A. On the philosophy and mathematics of the Logics of Formal Inconsistency. In *New Directions in Paraconsistent Logic*; Springer: Berlin, Germany, 2016.
18. Marcos, J. Nearly every normal modal logic is paranormal. *Log. Anal.* **2005**, *48*, 279–300.
19. Dodó, A.; Marcos, J. Negative modalities, consistency and determinedness. *Electron. Notes Theor. Comput. Sci.* **2014**, *300*, 21–45. [CrossRef]
20. Odintsov, S. Algebraic semantics for paraconsistent Nelson's logic. *J. Log. Comput.* **2003**, *4*, 453–468. [CrossRef]
21. Carnielli, W.; Coniglio, M.E.; Rodrigues, A. On formal aspects of the epistemic approach to paraconsistency. In *Logic and Philosophy of Logic: Recent Trends in Latin America and Spain*; Freund, M., de Castro, M., Ruffino, M., Eds.; College Publications: London, UK, 2018.
22. Coniglio, M.E.; Figallo-Orellano, A.; Golzio, A.C. Non-deterministic algebraization of logics by swap structures. *Log. J. IGPL* forthcoming.

© 2020 by the authors. Licensee MDPI, Basel, Switzerland. This article is an open access article distributed under the terms and conditions of the Creative Commons Attribution (CC BY) license (http://creativecommons.org/licenses/by/4.0/).

Article

Deontic Logics as Axiomatic Extensions of First-Order Predicate Logic: An Approach Inspired by Wolniewicz's Formal Ontology of Situations

Andrzej Malec

Foundation for Computer Science, Logic and Formalized Mathematics, 00-864 Warsaw, Poland; a.malec@amdp.strefa.pl

Received: 14 August 2019; Accepted: 1 October 2019; Published: 6 October 2019

Abstract: The aim of this article is to present a method of creating deontic logics as axiomatic theories built on first-order predicate logic with identity. In the article, these theories are constructed as theories of legal events or as theories of acts. Legal events are understood as sequences (strings) of elementary situations in Wolniewicz's sense. On the other hand, acts are understood as two-element legal events: the first element of a sequence is a choice situation (a situation that will be changed by an act), and the second element of this sequence is a chosen situation (a situation that arises as a result of that act). In this approach, legal rules (i.e., orders, bans, permits) are treated as sets of legal events. The article presents four deontic systems for legal events: AEP, AEPF, AEPOF, AEPOFI. In the first system, all legal events are permitted; in the second, they are permitted or forbidden; in the third, they are permitted, ordered or forbidden; and in the fourth, they are permitted, ordered, forbidden or irrelevant. Then, we present a deontic logic for acts (AAPOF), in which every act is permitted, ordered or forbidden. The theorems of this logic reflect deontic relations between acts as well as between acts and their parts. The direct inspiration to develop the approach presented in the article was the book *Ontology of Situations* by Boguslaw Wolniewicz, and indirectly, Wittgenstein's *Tractatus Logico-Philosophicus*.

Keywords: deontic logic; ontology of situations; semantics of law; formal theory of law; Wittgenstein; Wolniewicz

---

1. Introduction

Boguslaw Wolniewicz in [1] created a formal ontology of situations. Based on his theory, in [2] I proposed a certain semantics of norms. The approach presented in [2] allows a new understanding of deontic logics, which I would like to present below.

Deontic logics formalize the concepts of obligation, prohibition and permission.

Deontic propositional logics use deontic operators whose arguments are sentences, including compound sentences. Usually, deontic operators are defined similarly to modal (aletic) operators. Following the modal logics, iterations of deontic operators are allowed.

Such an approach seems, at least sometimes, not to be intuitive.

Firstly, it does not seem reasonable to apply deontic operators to any sentences. What is the meaning of the sentence "it is mandatory that Mount Everest is the highest mountain in the world", or "it is forbidden that 2 + 2 = 4"?

Secondly, it is not clear what intuitions regarding obligation, prohibition and permission correspond to compound sentences preceded by deontic operators; for example,

$$O\,((p \wedge q) \rightarrow q) \rightarrow (O\,(p \wedge q) \rightarrow O\,q),$$

$$O(p \to (p \vee q)) \to (O\,p \to O\,(p \vee q))?$$

Likewise, what would be the meaning of the sentence "it is mandatory (permitted, forbidden) that if Mount Everest is the highest mountain in the world and 2 + 2 = 4 then 2 + 2 = 4"?

Thirdly, it is not clear what intuitions regarding obligation, prohibition and permission correspond to iterated deontic operators; e.g., O P p, O O p. What does the phrase "it is obligatory that it is permitted" or "it is obligatory that it is obligatory" mean?

Fourthly, such an approach is not free of paradoxical consequences; e.g., the widely discussed

$$O\,p \to O\,(p \vee q)$$

(If it is obligatory to save a drowning person, then it is obligatory to save a drowning person or drink coffee), or

$$F\,p \to O\,(p \to q)$$

(If it is forbidden to kill a man, then it is obligatory to rob this man after killing him).

One can find more information on propositional deontic logics and their paradoxical consequences in [3–5].

In turn, deontic logics other than the propositional are often only partially formalized. On the other hand, sometimes non-standard formal means are used. Operators' arguments are people, norms, acts, states of affairs, and sometimes combinations of the aforementioned. Interesting examples of deontic systems built on a modal calculus of names can be found in [4], where sentences such as "x at the moment t can be y", "x at the moment t is obliged to be y", "x at the moment t is allowed to be y", "x at the moment t is forbidden to be y" are considered. The advantage of such deontic systems is that they usually capture more specific properties of deontic modalities than propositional deontic logics allow, although the downside of such deontic systems, in addition to the formalization issues mentioned above, is the lack of a clear concept as to what domain deontic sentences describe.

However, there are also deontic systems based on a previous in-depth analysis of the domain to which deontic modalities relate. For example, in [6], permission, prohibition and obligation are defined in terms of an action system. The author of [6] aptly assumes that deontic modalities do not relate to states of affairs but to actions. It is actions that are prescribed, prohibited or allowed. Deontic logic should therefore be based on an action system. Having a clear concept of the domain of deontic modalities, the author provides his deontic logics with a strong semantic basis. This approach avoids the paradoxes of propositional deontic logics. When assessing this direction of research as the most promising, two points should be noted. First, actions (acts) are not the only events to which deontic modalities relate. A good example is the so-called "consequence crimes": the law does not prohibit an act itself, but prohibits it if it produces certain effects. In this case, a sequence of situations is prohibited, in which an act is only its initial fragment (and "legal causality" occurs between the act and subsequent elements of the sequence). Deontic logic should therefore include a more general concept than the concept of an act, namely the concept of a deontic event (I use the term "legal event" in this sense). Secondly, it should be noted that an act can be a complex act not only as a sequence of simple acts: situations constituting an act may themselves be complex. The act of saving two out of three drowning people consists of rescuing two people and sacrificing the third. It would be good if deontic logic could also describe such relations; that is, the relations between an act and its parts.

Bearing in mind the above, one may be tempted to create deontic logics which achieve the following:

(1) they shall correspond to the intuitions associated with concepts of obligation, prohibition and permission more accurately than propositional deontic logics do;
(2) they shall use only standard logical tools and shall be based on a clear concept of the domain of deontic modalities;
(3) they shall treat acts as a special case of deontic events;
(4) they shall describe the relations between acts and their parts.

We intend to do this in the following part of this article. At the same time, we want to do this using standard logical means; i.e., means of the first-order predicate logic.

2. Methods

In accordance with Stanislaw Lesniewski's intuitive formalism, formal theories, including logical ones, should have an intuitive interpretation. Of course, the clearer and more intuitive this interpretation is, the better. Therefore, it is worth preceding the selection of axioms and rules of the theory with a careful determination of the domain to be described by this theory.

Such an intuitive interpretation for deontic theories is provided by Boguslaw Wolniewicz's ontology of situations, which is a successful attempt to formally develop the ontology of situations contained in Ludwig Wittgenstein's *Tractatus Logico-Philosophicus*.

In [1], Wolniewicz formally described the logical space from Wittgenstein's *Tractatus*. Wolniewicz creates a mathematical model, namely the structure $<SE, \leq>$, where SE is a non-empty set of objects whose elements he calls "situations" (or "elementary situations"), and $\leq$ is a partial order. Wolniewicz distinguishes in the SE set a set of elementary proper situations (SE") and two inappropriate situations, namely an empty situation (o) and an impossible situation (λ). Wolniewicz assumes that the structure $<SE, \leq>$ is a complete lattice, so each subset of SE has, due to the relation $\leq$, its upper and lower limits. In this lattice, o is the smallest element and λ is the largest; i.e., each elementary situation is contained between the empty situation and the impossible situation. In Wolniewicz's model, every proper elementary situation is an atom or is made of atoms; i.e., elementary situations that cover only the empty situation. The opposites of atoms are possible worlds; i.e., elementary situations that are covered only by the impossible situation. Among the possible worlds, one is singled out as the real world (w_0). A set of elementary situations that are fragments of the real world is the set of real situations, or facts. Other elementary situations are imaginary.

The Wolniewicz's structure above corresponds to a static logical space: reality and alternative worlds at some point of time. Meanwhile, the law orders, bans or permits a situation to be replaced by another; one event shall be followed by another. To reflect these dynamics, in [2], each point of time has a Wolniewicz's structure assigned to it. This way, reality and alternative worlds are represented not by individual possible worlds, but by sequences of possible worlds. Thus, while Wolniewicz's original structure can be compared to a picture of reality and alternative worlds, the elaborate structure resembles a film tape.

This dynamic structure has logical events as its elements. A logical event is a non-empty, finite sequence (string) of proper situations, such that each element of the sequence belongs to a different Wolniewicz's structure. Logical events that meet specific conditions are natural events.

Among natural events, one can distinguish legal events; i.e., natural events subject to legal assessments. In [2], four types of legal events are distinguished:

(1) acts;
(2) multiacts;
(3) indirect acts;
(4) causal events.

The most important of these four types of legal events are acts. They are specific two-element sequences of situations in Wolniewicz's sense: the first element of the sequence is the choice situation (the situation that will be changed by the act), and the second element is the chosen situation (the situation that arises as a result of the act). Multiacts, indirect acts and causal acts are understood in the following way:

$$\textbf{MULTIACT} = \{<x_n, x_{n+1}, x_{n+2}, \ldots, x_{n+m}>: \text{for any i from n to } n+m-1 <x_i, x_{i+1}> \in \textbf{ACT}\},$$

$$\textbf{INDIRECT\_ACT} = \{<x_n, x_{n+1}, x_{n+2}, \ldots, x_{n+m}>: \text{for any i from n to } n+m-1 <x_i, x_{i+1}> \in \textbf{ACT}$$
$$\text{or } <x_i, x_{i+1}> \in \textbf{DET}\},$$

$$\mathbf{CAUSAL\_EVENT} = \{<x_n, x_{n+1}, x_{n+2}, \ldots, x_{n+m}>: <x_n, x_{n+1}> \in \mathbf{ACT} \text{ and for any i from } n+1 \text{ to}$$
$$n+m-1 <x_i, x_{i+1}> \in \mathbf{LEG}\},$$

where **ACT** is the set of acts, **DET** is the set of deterministic changes, and **LEG** is the set of changes governed by so called "legal causal relations" (see [7]).

The above can be symbolically represented in the following way:

$$\text{ACTS} \subset \text{LEGAL EVENTS} \subset \text{NATURAL EVENTS} \subset \text{LOGICAL EVENTS}.$$

In this approach, the deontic concepts of obligation, prohibition and permission obtain a clear interpretation in terms of sets: orders, bans and permits are simply sets of legal events. Orders, bans and permits are called "legal rules". To determine any deontic theory, it is sufficient to determine relations between sets of legal events.

3. Results

3.1. First-Order Predicate Logic as the Basis of Deontic Theories

Deontic logics will be constructed below as theories built upon the classical first-order predicate logic with identity.

As a result, the language and grammar of these deontic theories is the language and grammar of classical first-order predicate logic with identity. No additional symbols or grammar rules will be used.

Non-specific axioms and rules of these deontic theories are as follows:

(1) axioms of classical predicate calculus (including substitutions of axioms of classical propositional calculus);
(2) axioms for the identity predicate:

 (a) $\forall x \, (x = x)$,
 (b) $\forall x \, y \, (x = y \leftrightarrow y = x)$,
 (c) $\forall x \, y \, z \, (x = y \land y = z \rightarrow x = z)$;

(3) rules of classical predicate calculus (including substitutions of rules of classical propositional calculus).

The deontic theories which we will construct below can be divided into:

(1) theories of legal events;
(2) theories of simple acts;
(3) theories of compound acts.

3.2. Theories of Legal Events

The domain of theories of legal events is the set of events as understood in accordance with Section 2. Thus, all propositions of these theories are propositions about events.

We distinguish five unary predicates:

LEV (x)—read "x is a legal event";
PER (x)—read "x is a permitted event";
FOR (x)—read "x is a forbidden event";
OBL (x)—read "x is an ordered event";
IRR (x)—read "x is an irrelevant event".

The specific axioms of these theories are selected in such a way that they determine the relations between sets of ordered, forbidden and permitted events.

3.2.1. Theory 1: All Legal Events are Permitted (AEP)

Adding one specific axiom to non-specific axioms,
A1. $\forall x (LEV(x) \leftrightarrow PER(x))$,
We will get a simple deontic theory: AEP.
This corresponds to the following Venn diagram:

```
+-------------------------------------------+
|                                           |
|     PERMITTED EVENTS = LEGAL EVENTS       |
|                                           |
+-------------------------------------------+
```

AEP does not seem interesting from the point of view of logic.

3.2.2. Theory 2: All Legal Events are Either Permitted or Forbidden (AEPF)

By adding two specific axioms to non-specific axioms,
A1. $\forall x (LEV(x) \leftrightarrow (PER(x) \vee FOR(x)))$,
A2. $\neg \exists x (PER(x) \wedge FOR(x))$,
We will get a deontic theory: AEPF.
This corresponds to the following Venn diagram:

```
+---------------------+-------------------+
|                     |                   |
|  PERMITTED EVENTS   | FORBIDDEN EVENTS  |
|                     |                   |
+---------------------+-------------------+
```

3.2.3. Theory 3: All Legal Events are Either Permitted or Ordered or Forbidden (AEPOF)

By adding three specific axioms to non-specific axioms,
A1. $\forall x (LEV(x) \leftrightarrow (OBL(x) \vee PER(x) \vee FOR(x)))$,
A2. $\neg \exists x (PER(x) \wedge FOR(x))$,
A3. $\forall x (OBL(x) \rightarrow PER(x))$,
We will get a deontic theory: AEPOF.
This corresponds to the following Venn diagram:

```
+-----------------------------+-------------+
|    PERMITTED EVENTS         |             |
|                             |  FORBIDDEN  |
|   +---------------------+   |   EVENTS    |
|   |   ORDERED EVENTS    |   |             |
|   +---------------------+   |             |
+-----------------------------+-------------+
```

3.2.4. Theory 4: All Legal Events are Either Permitted or Ordered or Forbidden or Irrelevant (AEPOFI)

By adding five specific axioms to non-specific axioms,
A1. $\forall x (LEV(x) \leftrightarrow (OBL(x) \vee PER(x) \vee FOR(x) \vee IRR(x)))$,
A2. $\neg \exists x (PER(x) \wedge FOR(x))$,
A3. $\neg \exists x (IRR(x) \wedge FOR(x))$,

A4. ⌐∃x (PER (x) ∧ IRR (x)),
A5. ∀ x (OBL (x) → PER (x)),
We will get a deontic theory: AEPOFI.
This corresponds to the following Venn diagram:

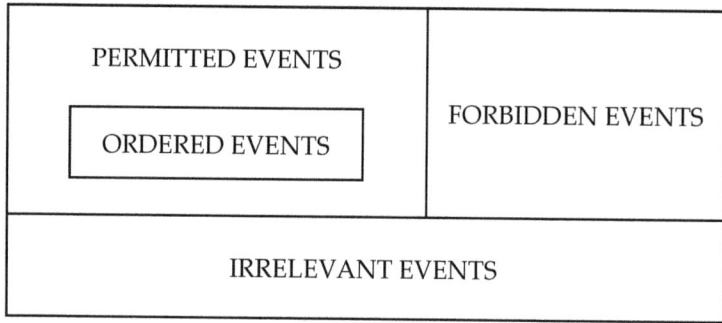

3.2.5. Existence of Legal Events

In the deontic theories set out above, we do not prejudge whether there are legal events. To determine this, a specific axiom should be added to each of these systems:
A0. ∃ x LEV (x).

3.2.6. Selected Theorems of Legal Event Theories

Selected theorems of the theories of legal events are presented below. We omit proofs, because they are quite simple and intuitive.
AEPF, AEPOF, AEPOFI include, in particular, the following theorems:
T1. ∀ x ⌐(PER (x) ∧ FOR (x));
T2. ∀ x (⌐PER (x) ∨ ⌐FOR (x));
T3. ∀ x (PER (x) → ⌐FOR (x));
T4. ∀ x (FOR (x) → ⌐PER (x)).
Of course, we also have in AEPOF and AEPOFI the following theorems:
T5. ∀ x (OBL (x) → ⌐FOR (x));
T6. ∀ x (FOR (x) → ⌐OBL (x));
T7. ∀ x (⌐PER (x) → ⌐OBL (x)).
Theorems T1–T7 have close equivalents in deontic propositional logics.
On the other hand, in AEPF and AEPOF, we have
T8. ∀ x (LEV (x) → (PER (x) ∨ FOR (x)));
T9. ∀ x (LEV (x) → (⌐PER (x) → FOR (x)));
T10. ∀ x (LEV (x) → (⌐FOR (x) → PER (x)));
And consequently, we also have
T11. ∀ x (LEV (x) → (PER (x) ↔ ⌐FOR (x))) which follows from T3, T10;
T12. ∀ x (LEV (x) → (FOR (x) ↔ ⌐PER (x))) which follows from T4, T9.
Theorems T8–T12 have equivalents in deontic propositional logics. The predecessor of these theorems indicates, however, that the relations described by the successor occur only for legal events and not just for any events.

3.3. Theories of Simple Acts

The domain of the theories of acts is the set of situations as understood in accordance with Section 2 above. Thus, all propositions of these theories are propositions about situations.
We distinguish four binary predicates:

ACT (x, y)—read "replacement x by y is an act";
PER (x, y)—read "replacement x by y is permitted";
FOR (x, y)—read "replacement x by y is forbidden";
OBL (x, y)—read "replacement x by y is ordered".

The specific axioms of these theories are selected in such a way that they determine the relations between sets of ordered, forbidden and permitted acts.

We consider only one such theory below, which is an extension of AEPOF.

3.3.1. Theory: All Acts are Either Permitted or Obligatory or Forbidden (AAPOF)

Every act is a legal event. Thus, the first three AAPOF-specific axioms are the exact counterparts of the AEPOF-specific axioms:

A1. ∀ x y (ACT (x, y) ↔ (OBL (x, y) ∨ PER (x, y) ∨ FOR (x, y)));
A2. ¬ ∃ x y (PER (x, y) ∧ FOR (x, y));
A3. ∀ x y (OBL (x, y) → PER (x, y)).

These three axioms determine the relations between any situations x and y, forming one legal event (i.e., forming a sequence of situations < x, y >).

The next three AAPOF-specific axioms define relations involving three situations, x, y, z, forming two legal events (i.e., forming two sequences of situations: < x, y > and < x, z >).

Axiom A4 states that every act is a choice:

A4. ∀ x y (ACT (x, y) → ∃ z (ACT (x, z) ∧ y ≠ z))

(In each choice situation, there are at least two options).

Axiom A5 confirms that the orders are consistent:

A5. ∀ x y z (OBL (x, y) → (y ≠ z → FOR (x, z))

(If in a choice situation x, an option y is ordered, then all other options are prohibited in x).

On the other hand, the axiom A6 states that not everything is forbidden:

A6. ∀ x y (FOR (x, y) → ∃ z (ACT (x, z) ∧ y ≠ z ∧ ¬ FOR (x, z)))

(If in a choice situation x, an option y is forbidden, then some other option is not forbidden in x).

As in the case of the theories of legal events, we do not prejudge whether acts exist. To determine this, it would be necessary to add the specific axiom A0 to AAPOF:

A0. ∃ x y ACT (x, y).

(There are choice situations).

3.3.2. Selected Theorems of AAPOF that are Equivalent to Theorems of AEPOF

In AAPOF, we have exact equivalents of theorems T1–T12 of AEPOF:

T1. ∀ x y ¬ (PER (x, y) ∧ FOR (x, y));
T2. ∀ x y (¬ PER (x, y) ∨ ¬ FOR (x, y));
T3. ∀ x y (PER (x, y) → ¬ FOR (x, y));
T4. ∀ x y (FOR (x, y) → ¬ PER (x, y));
T5. ∀ x y (OBL (x, y) → ¬ FOR (x, y));
T6. ∀ x y (FOR (x, y) → ¬ OBL (x, y));
T7. ∀ x y (¬ PER (x, y) → ¬ OBL (x, y));
T8. ∀ x y (ACT (x, y) → (PER (x, y) ∨ FOR (x, y)));
T9. ∀ x y (ACT (x, y) → (¬ PER (x, y) → FOR (x, y)));
T10. ∀ x y (ACT (x, y) → (¬ FOR (x, y) → PER (x, y)));
T11. ∀ x y (ACT (x, y) → (PER (x, y) ↔ ¬ FOR (x, y)));
T12. ∀ x y (ACT (x, y) → (FOR (x, y) ↔ ¬ PER (x, y))).

3.3.3. Selected AAPOF Theorems Specific to Acts

In AAPOF, we also have theorems that do not have their exact counterparts in AEPOF, which are the consequences of adding specific axioms A4–A6 to the system:

T13. ∀ x y z (OBL (x, y) → (y ≠ z → ⌐ PER (x, z))
(If an option y is ordered in a choice situation x, then no other option is permitted in x);
T14. ∀ x y z (OBL (x, y) → (y ≠ z → ⌐ OBL (x, z))
(If an option y is ordered in a choice situation x, then no other option is ordered in x);
T15. ∀ x y z (OBL (x, y) ∧ OBL (x, z) → y = z)
(If, in a choice situation, two options are ordered, they are identical);
T16. ∀ x y z (y ≠ z → ⌐ (OBL (x, y) ∧ OBL (x, z)))
(In any choice situation, different options cannot be ordered together);
T17. ∀ x y (FOR (x, y) → ∃ z (y ≠ z ∧ PER (x, z)))
(If an option y is forbidden in a choice situation x, then some other option z is permitted in x);
T18. ∀ x y (OBL (x, y) → ∃ z (y ≠ z ∧ FOR (x, z)))
(If an option y is ordered in a choice situation x, then some other option z is forbidden in x);
T19. ∀ x y z (y ≠ z → (OBL (x, y) → ⌐ PER (x, z)))
(If an option y is ordered in a choice situation x, then no other option is permitted in x);
T20. ∀ x y z (ACT (x, y) ∧ ACT (x, z) ∧ y ≠ z ∧ ∀ w (ACT (x, w) → (w = y ∨ w = z)) → ⌐ (FOR (x, y) ∧ FOR (x, z)))
(If there are exactly two options in a choice situation, both cannot be forbidden);
T21. ∀ x y z (ACT (x, y) ∧ ACT (x, z) ∧ y ≠ z ∧ ∀ w (ACT (x, w) → (w = y ∨ w = z)) → (FOR (x, y) → PER (x, z)))
(If, in a choice situation, there are exactly two options, then if one of them is forbidden, the other is permitted);
T22. ∀ x y z (ACT (x, y) ∧ ACT (x, z) ∧ y ≠ z ∧ ∀ w (ACT (x, w) → (w = y ∨ w = z)) → (PER (x, y) ∨ PER (x, z)))
(If, in a choice situation, there are exactly two options, then at least one of them is permitted);
T23. ∀ x y z (ACT (x, y) ∧ ACT (x, z) ∧ y ≠ z ∧ ∀ w (ACT (x, w) → (w = y ∨ w = z)) → (⌐ PER (x, y) → PER (x, z)))
(If, in a choice situation, there are exactly two options, then if one of them is not permitted, the other is permitted);
T24. ∀ x y z w (FOR (x, y) ∧ ∀ z (FOR (x, z) → y = z) → (ACT (x, w) ∧ w ≠ y → PER (x, w)))
(If, in a choice situation, exactly one option is prohibited, then any other option is permitted).

3.4. Theories of Compound Acts

In deontic propositional logics, deontic operators apply to conjunction or alternative of propositions; for example,

$$O (p \wedge q) \rightarrow O p \wedge O q,$$

$$O p \wedge O q \rightarrow O (p \wedge q),$$

$$O p \rightarrow O (p \vee q).$$

Such sentences are intended to formalize the intuition that an obligation, prohibition or permission may relate to situations where one is part of the other.

This intuition can be expressed more precisely by developing AAPOF into the theory of compound acts. We do this by adding axioms defining relations between situations, some of which are parts of the others.

To do so, we need to distinguish further one unary predicate "AT (x)", one binary predicate "ε (x, y)" and one ternary predicate "= + (x, y, z)":

AT (x)—read "x is an atomic situation";
ε (x, y)—read "x is a part of y";
= + (x, y, z)—read "x is the sum (composition) of y and z".

Below, we will write "x ε y" instead of "ε (x, y)" and "x = y + z" instead of "= + (x, y, z)".

3.4.1. AAPOF for Compound Acts

First, we will list axioms that will determine when a situation is a part of another situation, when a situation is the sum (composition) of other situations, and when a situation is an atomic situation.

We use Wolniewicz's approach to define the relation of "being a part of":

A7. $\forall x\ x \varepsilon x$;
A8. $\forall x y z (x \varepsilon y \land y \varepsilon z \to x \varepsilon z)$;
A9. $\forall x y (x \varepsilon y \land y \varepsilon x \to x = y)$.

We also add the A10 axiom for atomic situations:

A10. $\forall x (AT(x) \leftrightarrow \forall y (y \varepsilon x \to y = x))$
(Every atom is a situation that has no proper parts).

Then, we introduce the sum (composition) of situations:

A11. $x = y + z \leftrightarrow y \varepsilon x \land z \varepsilon x \land \forall w (AT(w) \to (w \varepsilon x \to (w \varepsilon y \lor w \varepsilon z)))$
(A situation x is the sum (composition) of situations y and z, when they are parts of it, and each atom of the situation x is a part of the situation y or a part of the situation z).

Using the concept of a part of situation, we can express the intuition that a part of a situation has the same deontic modality as this situation:

A12. $\forall x x_1 y y_1 (x_1 \varepsilon x \land y_1 \varepsilon y \to (OBL(x, y) \to (ACT(x_1, y_1) \to OBL(x_1, y_1))))$;
A13. $\forall x x_1 y y_1 (x_1 \varepsilon x \land y_1 \varepsilon y \to (PER(x, y) \to (ACT(x_1, y_1) \to PER(x_1, y_1))))$;
A14. $\forall x x_1 y y_1 (x_1 \varepsilon x \land y_1 \varepsilon y \to (FOR(x, y) \to (ACT(x_1, y_1) \to FOR(x_1, y_1))))$.

In turn, using the concept of the sum (composition) of situations, we can express intuition, according to which any situation has the same deontic modality as its parts:

A15. $\forall x x_1 x_2 y y_1 y_2 (x = x_1 + x_2 \land y = y_1 + y_2 \to (OBL(x_1, y_1) \land OBL(x_2, y_2) \to OBL(x, y)))$;
A16. $\forall x x_1 x_2 y y_1 y_2 (x = x_1 + x_2 \land y = y_1 + y_2 \to (PER(x_1, y_1) \land PER(x_2, y_2) \to PER(x, y)))$;
A17. $\forall x x_1 x_2 y y_1 y_2 (x = x_1 + x_2 \land y = y_1 + y_2 \to (FOR(x_1, y_1) \land FOR(x_2, y_2) \to FOR(x, y)))$.

3.4.2. Selected AAPOF Theorems Specific to Compound Acts

The consequences of adopting additional specific axioms A7–A17 include, but are not limited to, the following examples:

T25. $\forall x x_1 x_2 y y_1 y_2 (x = x_1 + x_2 \land y = y_1 + y_2 \to (OBL(x, y) \to (ACT(x_1, y_1) \land ACT(x_2, y_2) \to \neg (OBL(x_1, y_1) \land FOR(x_2, y_2))))$
(If an act is ordered, it is not that one part of it is ordered and the other is forbidden);

T26. $\forall x x_1 x_2 y y_1 y_2 (x = x_1 + x_2 \land y = y_1 + y_2 \to (OBL(x, y) \to (ACT(x_1, y_1) \land ACT(x_2, y_2) \to \neg (PER(x_1, y_1) \land FOR(x_2, y_2))))$
(If an act is ordered, it is not that one part of it is permitted and the other is forbidden);

T27. $\forall x x_1 x_2 y y_1 y_2 (x = x_1 + x_2 \land y = y_1 + y_2 \to (OBL(x, y) \to (ACT(x_1, y_1) \land ACT(x_2, y_2) \to \neg (FOR(x_1, y_1) \lor FOR(x_2, y_2))))$
(If an act is ordered, it is not that any part of it is forbidden);

T28. $\forall x x_1 x_2 y y_1 y_2 (x = x_1 + x_2 \land y = y_1 + y_2 \to (OBL(x_1, y_1) \land OBL(x_2, y_2) \to PER(x, y)))$
(If acts are ordered, their composition is permitted);

T29. $\forall x x_1 x_2 y y_1 y_2 (x = x_1 + x_2 \land y = y_1 + y_2 \to (PER(x_1, y_1) \land PER(x_2, y_2) \to \neg FOR(x, y)))$
(If acts are permitted, their composition is not forbidden);

T30. $\forall x x_1 x_2 y y_1 y_2 (x = x_1 + x_2 \land y = y_1 + y_2 \to (FOR(x_1, y_1) \land FOR(x_2, y_2) \to \neg PER(x, y)))$
(If acts are forbidden, their composition is not permitted).

The above relations are useful for reconstructing legal reasoning *a maiori ad minus* and *a minori ad maius*, as well as for reconstructing other similar reasonings.

4. Discussion

A comparison of axioms and theorems of considered deontic theories with axioms and theorems of deontic propositional logics indicates that a number of properties of obligation, prohibition and permission are similarly defined in both approaches.

In particular, the basic theorems of legal event theories, i.e., T1–T7, have close equivalents in deontic propositional logics.

In turn, although theorems T8–T12 have equivalents in deontic propositional logics, their predecessor indicates that successive relations occur only for legal events, not for any events.

For example, T12

$$\forall x\, (\text{LEV}(x) \to (\text{FOR}(x) \leftrightarrow \neg \text{PER}(x)))$$

is a counterpart to the definition of prohibition in propositional logics:

$$F\, p \equiv_{def} \neg P\, p.$$

Interestingly, in none of the four theories of legal events under consideration have we a counterpart of the definition of obligation in propositional logics,

$$O\, p \equiv_{def} \neg P\, \neg p,$$

which is based on a definition from modal (aletic) logics:

$$\Box\, p \equiv_{def} \neg \Diamond\, \neg p.$$

This is because the expression "$P\, \neg p$" has no equivalent in any of these theories. However, this is not the case in theory of acts, where T19

$$\forall x\, y\, z\, (y \neq z \to (\text{OBL}(x, y) \to \neg \text{PER}(x, z)))$$

is a counterpart of the aforementioned definition of obligation in propositional logics:

$$O\, p \equiv_{def} \neg P\, \neg p.$$

Although, of course, the following proposition is not an AAPOF's theorem:

$$\forall x\, y\, z\, (y \neq z \to (\text{OBL}(x, y) \leftrightarrow \neg \text{PER}(x, z))).$$

Further, T16

$$\forall x\, y\, z\, (y \neq z \to \neg(\text{OBL}(x, y) \land \text{OBL}(x, z)))$$

is a counterpart of the theorem

$$\neg(O\, p \land O\, \neg p),$$

while T20

$$\forall x\, y\, z\, (\text{ACT}(x, y) \land \text{ACT}(x, z) \land y \neq z \land \forall w\, (\text{ACT}(x, w) \to (w = y \lor w = z)) \to \\ \neg(\text{FOR}(x, y) \land \text{FOR}(x, z)))$$

is a counterpart of the theorem

$$\neg(F\, p \land F\, \neg p).$$

In turn, T22

$$\forall x\, y\, z\, (\text{ACT}(x, y) \land \text{ACT}(x, z) \land y \neq z \land \forall w\, (\text{ACT}(x, w) \to (w = y \lor w = z)) \to \\ (\text{PER}(x, y) \lor \text{PER}(x, z)))$$

is a counterpart of the theorem

$$P p \vee P \neg p.$$

On the other hand, the A12 axiom

$$\forall x\, x_1\, y\, y_1\, (x_1 \,\varepsilon\, x \wedge y_1 \,\varepsilon\, y \rightarrow (OBL(x, y) \rightarrow (ACT(x_1, y_1) \rightarrow OBL(x_1, y_1))))$$

is a distant counterpart of the theorem

$$O(p \wedge q) \rightarrow O p.$$

In turn, axiom A15

$$\forall x\, x_1\, x_2\, y\, y_1\, y_2\, (x = x_1 + x_2 \wedge y = y_1 + y_2 \rightarrow (OBL(x_1, y_1) \wedge OBL(x_2, y_2) \rightarrow OBL(x, y)))$$

is a distant counterpart of the theorem

$$O p \wedge O q \rightarrow O(p \wedge q).$$

Similarly, axiom A13

$$\forall x\, x_1\, y\, y_1\, (x_1 \,\varepsilon\, x \wedge y_1 \,\varepsilon\, y \rightarrow (PER(x, y) \rightarrow (ACT(x_1, y_1) \rightarrow PER(x_1, y_1))))$$

is a distant counterpart of the theorem

$$P(p \wedge q) \rightarrow P p.$$

While the axiom A16

$$\forall x\, x_1\, x_2\, y\, y_1\, y_2\, (x = x_1 + x_2 \wedge y = y_1 + y_2 \rightarrow (PER(x_1, y_1) \wedge PER(x_2, y_2) \rightarrow PER(x, y)))$$

is a distant counterpart of the theorem

$$P p \wedge P q \rightarrow P(p \wedge q).$$

Similarly, the A14 axiom

$$\forall x\, x_1\, y\, y_1\, (x_1 \,\varepsilon\, x \wedge y_1 \,\varepsilon\, y \rightarrow (FOR(x, y) \rightarrow (ACT(x_1, y_1) \rightarrow FOR(x_1, y_1))))$$

is a distant counterpart of the proposition

$$F(p \wedge q) \rightarrow F p.$$

While the axiom A17

$$\forall x\, x_1\, x_2\, y\, y_1\, y_2\, (x = x_1 + x_2 \wedge y = y_1 + y_2 \rightarrow (FOR(x_1, y_1) \wedge FOR(x_2, y_2) \rightarrow FOR(x, y)))$$

is a distant counterpart of the proposition

$$F p \wedge F q \rightarrow F(p \wedge q).$$

As can be seen, the axioms and theorems of the deontic theories constructed above are usually not the exact equivalents of theorems of deontic propositional logics. They reflect additional restrictions that are necessary for expressing obligation, prohibition and permission in accordance with intuition, but which are inexpressible in propositional logics.

5. Conclusions

Due to the discussed restrictions, the presented systems avoid the non-intuitive properties of propositional deontic logics.

Firstly, deontic sentences do not apply to all domains. They are sentences about legal events, and in particular about acts.

Secondly, in the presented systems we have no equivalents of many non-intuitive sentences of propositional deontic logics, such as those considered in the introduction:

$$O\,((p \wedge q) \to q) \to (O\,(p \wedge q) \to O\,q),$$

$$O\,(p \to (p \vee q)) \to (O\,p \to O\,(p \vee q)).$$

It is a consequence of the accepted limitation that, in the presented systems, deontic sentences are sentences about legal events, and not sentences about any states of affairs.

Thirdly, it is also noteworthy that—for obvious reasons—in the deontic theories presented above, not even far counterparts of propositions that would include iterations of deontic operators exist.

Fourthly, the presented systems have no equivalents to the paradoxical statements of propositional deontic logics such as those considered in the introduction:

$$O\,p \to O\,(p \vee q),$$

$$F\,p \to O\,(p \to q).$$

Once again, it is a consequence of the accepted limitation that, in the presented systems, deontic sentences are sentences about legal events, and not sentences about any states of affairs.

In addition, some axioms and theorems of the deontic theories presented above do not have counterparts in propositional logics at all, and at the same time reflect important intuitions related to deontic modalities. Examples include the A4, A5 and A6 axioms and some theorems obtained with the help of these axioms.

Furthermore, thanks to Wolniewicz's situation ontology, the presented systems are based on a clear concept of deontic modalities: orders, bans and permits are simply sets of legal events.

In the presented approach, a distinction is also made between the deontic properties of any legal events and the deontic properties of acts. The former are described in AEPF, AEPOF, AEPOFI. The latter are expressed, e.g., by axioms A4 to A6 and A7 to A17 of the AAPOF system. Axioms such as A7 to A17 of the AAPOF system also show that it is possible to formally consider the relations between an act and its parts, which is important for the legal applications of deontic logics.

All this leads to the conclusion that deontic theories built on the first-order predicate logic and inspired by Wolniewicz's situation ontology are worthy of attention and development.

Funding: This research received no external funding.

Acknowledgments: I would like to thank Kazimierz Trzęsicki for encouraging me to write this article. I would also thank all the appointed reviewers of the article for their valuable remarks and suggestions, and my son Jakub Malec for the first review of the article and his suggestions.

Conflicts of Interest: The author declares no conflict of interest.

References

1. Wolniewicz, B. *Ontologia Sytuacji (in Polish: Ontology of Situations)*; Państwowe Wydawnictwo Naukowe: Warsaw, Poland, 1985; See also: Wolniewicz, B. *A Formal Ontology of Situations*; Studia Logica 41, 1982; pp. 381–413.
2. Malec, A. Andrzej Malec, Wprowadzenie do semantyki prawa (in Polish: Introduction to semantics of law). Bialystok, Poland, 2018.

3. Ziemba, Z. *Deontic Logic*, in: Witold Marciszewski (Ed.), *Dictionary of Logic as Applied in the Study of Language. Concepts, Methods, Theories*; Martinus Nijhoff Publishers: The Hague, The Netherlands; Martinus Nijhoff Publishers: Boston, Poland; Martinus Nijhoff Publishers: London, UK, 1981; pp. 97–104.
4. Ziemba, Z. *Analityczna Teoria Obowiązku. Studium z Logiki Deontycznej (in Polish: Analytical Theory of Duties. A Study in Deontic Logic)*; Państwowe Wydawnictwo Naukowe: Warsaw, Poland, 1985.
5. Gumański, L. *Istnienie i Logika. Studia z Filozofii (in Polish: Existence and Logic. Studies in Philosophy)*; Wydawnictwo Uniwersytetu Mikolaja Kopernika: Torun, Poland, 2006; pp. 389–446.
6. Czelakowski, J. *Freedom and Enforcement in Action*; Trends in Logic (Studia Logica Library); Springer: Dordrecht, The Netherlands, 2015; Volume 42.
7. Hart, H.L.A. Tony Honore. In *Causation in the Law*, 2nd ed.; Clarendon Press: Oxford, UK, 2002.

© 2019 by the author. Licensee MDPI, Basel, Switzerland. This article is an open access article distributed under the terms and conditions of the Creative Commons Attribution (CC BY) license (http://creativecommons.org/licenses/by/4.0/).

Article

Synthetic Tableaux with Unrestricted Cut for First-Order Theories

Dorota Leszczyńska-Jasion * and Szymon Chlebowski *

Department of Logic and Cognitive Science, Faculty of Psychology and Cognitive Science, Adam Mickiewicz University, ul. Szamarzewskiego 89a, 60-568 Poznań, Poland
* Correspondence: Dorota.Leszczynska@amu.edu.pl (D.L.-J.); Szymon.Chlebowski@amu.edu.pl (S.C.)

Received: 14 August 2019; Accepted: 15 November 2019; Published: 29 November 2019

Abstract: The method of synthetic tableaux is a cut-based tableau system with synthesizing rules introducing complex formulas. In this paper, we present the method of synthetic tableaux for Classical First-Order Logic, and we propose a strategy of extending the system to first-order theories axiomatized by *universal axioms*. The strategy was inspired by the works of Negri and von Plato. We illustrate the strategy with two examples: synthetic tableaux systems for identity and for partial order.

Keywords: synthetic tableaux; principle of bivalence; cut; first-order theory; universal axiom

Cut? Don't eliminate, introduce!

Gentzen's *Hauptsatz* is rightly considered to be a milestone in the development of structural proof theory. For decades, it was thought that cut-elimination, yielding analyticity of the system, is a goal per se. However, today, it is well-known that eliminating cuts frequently increases the size and length of proofs. One can find examples showing that, in the worst case, cut-elimination produces non-elementarily larger and longer proofs [1–3]. For this reason, techniques of cut-introduction are being studied in order to shorten proofs (see [4,5]). In [4], it is shown that the technique of atomic cut-introduction is able to provide an exponential compression in the length of proofs. In [5], the authors studied the introduction of non-atomic formulas by the rule of cut.

Minimizing proofs is not the only reason to study proof systems with the rule of cut. Cut-formulas represent lemmatas, introduced to a proof in order to improve its structure or to bring in a new concept. It makes the proof more legible for a human. Instead of introducing cut to cut-free proofs (as is done in the above-mentioned papers), it may be useful, or perhaps more natural, to study proofs constructed within a cut-based system, i.e., a system in which the rule of cut cannot be eliminated. This approach is present in [6,7], where the authors introduced sequent calculi which are cut-based. The calculi are actually sequent-variants of tableau system **KE**.

What is the method of synthetic tableaux?

This paper presents the system of synthetic tableaux for First-Order Logic, which is an extension of the method for Classical Propositional Logic presented by Urbański [8] and Urbański [9], but the inspiration for the first-order version comes from D'Agostino [10] and Mondadori [11]. The method was explored for some cases of propositional logics by Urbański [12] and Urbański [13]; however, an extension to the first-order level substituted a challenging task. On the propositional level, the closest proof-theoretical relative of the method of synthetic tableaux seems to be the calculus **KI**, which is an "inversion" of **KE** (see [10,11] for **KI** and [14,15] for **KE**). However, the calculus for First-Order Logic presented here differs substantially from the version of **KI** for First-Order Logic.

The system of synthetic tableaux presented here is equipped with the so-called rule of the *Principle of Bivalence*, which is a form of cut. This rule is not eliminable from the system. In the case of system **KE**, non-eliminable cut is the remedy the authors propose for the computational collapse of analytic tableau system (in fact, of any proof system deprived of a representation of the principle of bivalence, see [14,15]). The situation is exactly analogous in the case of the system of synthetic tableaux for the propositional level. However, as may be expected, the situation complicates on the level of first-order; we demonstrate by examples the problems with restricting the applications of cut to analytic applications. Possible solutions to this problem will be examined in the future.

In Section 1, we describe the method of synthetic tableaux for the propositional case. In Section 2, we make a reference to system **KI**, which is an inversion of the more famous **KE**. The completeness proof presented in Section 3 was inspired by the completeness proofs of **KI** and **KE** with respect to an axiomatic system. In Section 4, we sketch some results concerning relative complexity of proof systems. The results motivate our research on proof systems like synthetic tableaux. In Sections 5–7, we describe the system of synthetic tableaux for the First-Order Logic, together with the proofs of soundness and completeness. Finally, Section 8 presents our strategy of extending the synthetic tableaux to first-order theories.

1. The Method of Synthetic Tableaux for CPL

Below, we present the synthetic tableaux system (ST-system, for short) for Classical Propositional Logic (CPL, for short). We describe the rules of tableau construction and define the notion of proof. In the case of CPL, there is exactly one binary branching rule and a collection of linear rules called *synthetic* or *synthesizing* since they build complex formulas from their subformulas or from their negations. As the reader shall see, there is a clear analogy between the synthesizing rules and natural-deduction rules introducing a connective, or sequent-calculus rules.

We use $\mathcal{L}_{\mathsf{CPL}}$ for the language of CPL with $\neg, \rightarrow, \vee, \wedge$. By A, B, C, F, we refer to formulas of $\mathcal{L}_{\mathsf{CPL}}$. $\mathcal{VAR}$ is for the set of propositional variables, $Sub(A)$ for the set of all subformulas of formula A, understood in the usual manner. By $\neg Sub(A)$, we mean the set of negations of the subformulas of A, that is, $\neg Sub(A) = \{\neg F : F \in Sub(A)\}$.

The number of linear rules depends on the number of logical connectives in the language; in the account presented in [16], which we follow here, there is 10 such rules (displayed below in Table 1).

Table 1. Linear synthesizing rules of the ST-system for CPL.

$$\frac{\neg B}{B \rightarrow C}\ \mathbf{r}^1_\rightarrow \qquad \frac{C}{B \rightarrow C}\ \mathbf{r}^2_\rightarrow \qquad \frac{B,\ \neg C}{\neg(B \rightarrow C)}\ \mathbf{r}^3_\rightarrow \qquad \frac{B}{B \vee C}\ \mathbf{r}^1_\vee \qquad \frac{C}{B \vee C}\ \mathbf{r}^2_\vee \qquad \frac{\neg B,\ \neg C}{\neg(B \vee C)}\ \mathbf{r}^3_\vee$$

$$\frac{\neg B}{\neg(B \wedge C)}\ \mathbf{r}^1_\wedge \qquad \frac{\neg C}{\neg(B \wedge C)}\ \mathbf{r}^2_\wedge \qquad \frac{B,\ C}{B \wedge C}\ \mathbf{r}^3_\wedge \qquad \frac{B}{\neg\neg B}\ \mathbf{r}_\neg$$

The premises of $\mathbf{r}^3_\rightarrow$, $\mathbf{r}^3_\vee$, and $\mathbf{r}^3_\wedge$ in Table 1 may occur in any order. If one wonders where, for example, this "C" in $\mathbf{r}^1_\rightarrow$ comes from, the following proviso comes to the rescue. A linear rule may be applied in the construction of a synthetic tableau for formula A provided that each premise *and conclusion* of the rule belongs to the set $Sub(A) \cup \neg Sub(A)$. Thus, in the case of rule $\mathbf{r}^1_\rightarrow$, any C such that $B \rightarrow C$ is in $Sub(A) \cup \neg Sub(A)$ is fine.

The branching rule is simply the rule of cut on literals. Following the insight of D'Agostino (see [14,15]) we call it the "PB-rule", or simply "PB", from the *Principle of Bivalence*.

$$p_i \quad \neg p_i \quad \text{PB}$$

Further, we say that the branching is *performed on p_i* or that the PB-rule was applied *with respect to p_i*. As in the case of the synthetic rules, applications of the PB-rule are subject to a restriction: the PB-rule may be applied with respect to p_i in the construction of a synthetic tableau for formula A provided that $p_i \in Sub(A)$. As a matter of fact, the restriction is built in the definition of a synthetic tableau (see Definition 1). With this restriction, the branching rule permits only analytic atomic cuts (for the notion of analytic cut, see [17–19]).

In the original account (see [8,16]), Mariusz Urbański defined the notion of *synthetic inference*, which is a sequence of formulas regulated by the above rules in a well-defined manner. Then, the notion of proof comes as a family of such interconnected sequences. The family-of-sequence account was motivated by the close relationship between synthetic tableaux and the so-called erotetic search scenarios, which were also defined as families of sequences, the so-called *erotetic derivations*. (*Erotetic search scenario* is a concept defined on the grounds of the logic of questions called *Inferential Erotetic Logic*. The reader can find more information about the relationship in [16]. The books by Urbański [9] and Urbański [20] contain broad exposition of the matter but are written in Polish. For erotetic search scenarios, see [21], and, for Inferential Erotetic Logic, the best recommendation is [22].) However, this is out of proof-theoretical tradition, where proofs are usually trees or sequences, and nowadays erotetic search scenarios may also be defined as trees (see [22,23]) so there is no need to stand by the sequence format. Below, we adopt the more common account of trees.

Trees and tableaux. We assume the proof-theoretic account of trees as partially ordered sets. *Branch* of a tree is its subset, which is a chain maximal with respect to inclusion. Below, in the proof of Lemma 2, we use the notion of *size of a finite tree*, which is the number of nodes of the tree.

As is practiced in structural proof theory, tableaux (derivations, proofs) are defined as trees labeled with formulas (or sequents, see ([24], p. 8), compare also [25]). Thus a (synthetic) tableau comes as a labeled tree $\langle X, R, \ell \rangle$, where ℓ is a function assigning formulas to nodes of $\langle X, R \rangle$. If $\mathcal{T} = \langle X, R, \ell \rangle$ is a labeled tree, then by $\mathbf{r}_\mathcal{T}$ we mean the *root of* $\mathcal{T}$. The notion of proof in the ST-system for CPL is given by the following definitions.

Definition 1. *A synthetic tableau for formula A is a finite labeled tree*

$$\mathcal{T} = \langle X, R, \ell \rangle$$

generated by the rules: $\mathbf{r}^1_\rightarrow, \mathbf{r}^2_\rightarrow, \mathbf{r}^3_\rightarrow, \mathbf{r}^1_\vee, \mathbf{r}^2_\vee, \mathbf{r}^3_\vee, \mathbf{r}^1_\wedge, \mathbf{r}^2_\wedge, \mathbf{r}^3_\wedge, \mathbf{r}_\neg,$ PB, *and such that*

$$\ell : X \setminus \{\mathbf{r}_\mathcal{T}\} \longrightarrow Sub(A) \cup \neg Sub(A)$$

and each leaf is labeled either with A or with $\neg A$.

Definition 2. *A proof of A in the ST-system is a synthetic tableau $\mathcal{T}$ for A such that each leaf of $\mathcal{T}$ is labeled with A.*

Here are two examples of synthetic tableaux. The first one is a synthetic tableau for formula $(p \rightarrow q) \rightarrow (\neg q \rightarrow \neg p)$ and is a proof of the formula in the ST-system. If a formula was obtained by a linear rule, we indicate the rule to the right.

Example 1. Let $F = (p \to q) \to (\neg q \to \neg p)$.

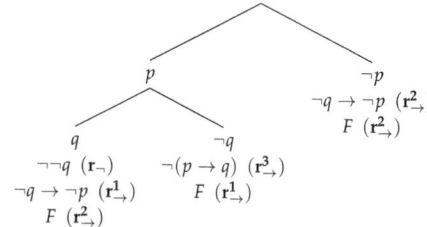

The second example is a synthetic tableau for $(p \to q) \to (\neg p \to \neg q)$ and is not its proof, as the third (from the left) branch of the tableau ends with the negation of this formula.

Example 2. Let $F = (p \to q) \to (\neg p \to \neg q)$.

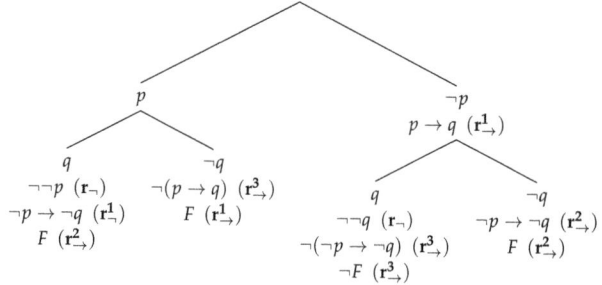

The above synthetic tableaux start with an application of the PB-rule. This is a general feature of the system—since PB is the only no-premise rule, it must be the one starting the construction of a synthetic tableau. Consequently, each synthetic tableau has more than one branch.

The system is sound and complete with respect to standard semantics for CPL; the most detailed proofs of these facts may be found in [9] (in Polish). Soundness is proved indirectly by the use of the concept of *minimal error point*. The idea is that, if A is not valid although it has a proof, then there must be an element in the tableau, which "introduces error" into the structure. We search for the element trying to establish the highest point in the structure of the tableau where the error appears. Then, a contradiction is derived, as it occurs that every error must have some erroneous predecessor. Completeness is proved by establishing a procedure of construction of a special kind of synthetic tableau, called *canonical synthetic tableau*. Canonical synthetic tableau for formula A starts with successive branchings on all propositional variables of A. If the number of distinct variables of A is k, then the number of branches of the canonical synthetic tableau for A is 2^k. In the margin, this result shows that the ST-system for CPL is a *standard proof system* in the sense of D'Agostino and Mondadori (see, e.g., [14]), which is not at all surprising, as synthetic tableaux for CPL in the canonical version constitute a formal representation of the familiar truth-tables method.

In the case of ST-system for CPL, soundness and completeness may be proved by quite simple techniques using, e.g., Hintikka sets and Hintikka's Lemma. However, since we aim at FOL, we use some more general tools. Completeness is proved with respect to axiomatic account, i.e., by simulating *Modus Ponens*—this idea is taken from the works by D'Agostino and Mondadori, which we briefly refer in the next section. The proof of soundness is inspired by the use of *semantic trees* in the proofs for resolution systems. Here, we rely on the version of this technique presented in [26].

There is one important difference between the version presented here and the original one by Urbański; however, the difference does not influence the metalogical properties mentioned above. In the case of the original system, the definition of synthetic inference warrants that, if p_i is its term, then $\neg p_i$ is not. On the level of a synthetic tableau, it means that the branching rule (our PB-rule) is never applied on the same branch more than once with respect to the same propositional variable. This condition warrants consistency of each synthetic inference. In the account presented here, this condition is neglected as it proved to be a hindrance in designing the ST-system for FOL. There has been an attempt to generalize the ST-system to the first-order case in a way which saves the property of consistent branch. The outcomes are presented in the research report [27]; however, the basic metalogical results—soundness and completeness—are missing.

2. System KI for CPL

Before synthetic tableaux were independently designed by Urbański, a system similar in spirit, called **KI**, was considered by D'Agostino and Mondadori. Interestingly, both synthetic tableaux and **KI** were to some extent inspired by Kalmár's work (see [28]). Even the notion of *synthetic* rule occurs both in [10] and in Urbański's work.

KI is a system which satisfies Prawitz' *inversion principle* with respect to a much better known system **KE**. The latter was developed by Marco Mondadori in the late 1980s, and analyzed carefully by D'Agostino and Mondadori [14,15]. Information about system **KI** can be found in Section 3.7 of [10] and in [11].

System **KI** for CPL is expressed in a language with truth signs ([10], the unsigned version is also considered.) It contains introduction rules and the following version of the PB-rule:

$$t(A) \quad f(A)$$

which is *not* restricted to propositional variables. As mentioned above, "PB" is for the *Principle of Bivalence*, as the rule clearly embodies the idea that A is either true or false. When one accepts *arbitrary* formulas to be introduced by the PB-rule, one must also accept inconsistencies on branches. This is the price to be paid for the unrestricted use of cut. One of the foundational ideas of the method of synthetic tableaux by Urbański was that they formalize reasoning in which the final conclusion is derived from all the possible consistent sets of atoms that build it (this is the Kalmár's inspiration). Hence, the restriction of the PB-rule to syntactical atoms gains an additional justification, irrespective of efficiency of this kind of system.

On the other hand, if inconsistent branches are the price to be paid for unrestricted (and possibly more efficient) use of cut, it is the price we can bear, especially when we realize that *the price for consistent branches* is extremely high. The problem is that it is highly improbable to describe in the framework of a restricted ST-system any logic which is not both propositional and extensional. Apparently, any finitely valued logic that may be characterized by finite matrices, *and only such logics*, may be successfully described in this framework. It explains why the only successful attempts to describe ST-systems where the cases of CPL, three-valued extensional Ł3 (see [12]) and paraconsistent CLuN, which may be fully characterized by semi-valuations (see [13]). Consistency of every branch is rather the property of these logics than a desired property of ST-systems.

Going back to **KI**, since the system permits inconsistent branches, it needs the notions of an open and a closed branch. A branch is called *closed* if, for some formula F, it contains both "$t(F)$" and "$f(F)$", otherwise it is called *open*.

The introduction rules of **KI** for CPL are displayed in Table 2. We quote the names of the rules after [10].

Table 2. Introduction rules of **KI**.

$$\frac{f(B)}{t(B \to C)} \text{ It} \to 1 \qquad \frac{t(B)}{t(B \vee C)} \text{ It} \vee 1 \qquad \frac{f(B)}{f(B \wedge C)} \text{ If} \wedge 1 \qquad \frac{t(B)}{f(\neg B)} \text{ If} \neg$$

$$\frac{t(C)}{t(B \to C)} \text{ It} \to 2 \qquad \frac{t(C)}{t(B \vee C)} \text{ It} \vee 2 \qquad \frac{f(C)}{f(B \wedge C)} \text{ If} \wedge 2 \qquad \frac{f(B)}{t(\neg B)} \text{ It} \neg$$

$$\frac{\begin{array}{c} t(B) \\ f(C) \end{array}}{f(B \to C)} \text{ If} \to \qquad \frac{\begin{array}{c} f(B) \\ f(C) \end{array}}{f(B \vee C)} \text{ If} \vee \qquad \frac{\begin{array}{c} t(B) \\ t(C) \end{array}}{t(B \wedge C)} \text{ It} \wedge$$

As the reader can see in Table 2, the introduction rules of **KI** for CPL are signed versions of the linear (synthetic) rules of ST-system for CPL. In [11], the author considered system called *canonical restriction* of **KI**, where the use of the PB-rule is restricted to atoms and a tree for A is built from its subformulas. Thus, the canonical restriction of **KI** fully corresponds to ST-system by Urbański, modulo truth signs. (More specifically, it is easy to see that the two systems for CPL, namely the canonical version of **KI** and the ST-system of Urbański, polynomially simulate each other. We say more about this issue in Section 4. For the notion of *p*-simulation, see [10].)

The notion of proof is introduced as follows. Let S be a finite set of signed formulas. S can be empty. A **KI**-*tree for S* is an expansion tree regulated by the rules of **KI**, starting from the elements of S. When S is empty, the origin of the tree is labeled with $\emptyset$.

Now, let Γ be a set of formulas (without the truth signs) and let A be a formula (with no truth sign). A **KI**-*proof of A* from Γ is a **KI**-tree for $\{t(B) : B \in \Gamma\}$, such that $t(A)$ occurs in every open branch. Finally, A is a **KI**-*theorem*, symbolically $\vdash_{\mathbf{KI}} A$, if A is provable from the empty set of formulas.

Completeness of this system has been proved by the authors with respect to the axiomatic system for CPL. This has inspired us to use the same technique in proving completeness of the first-order version. It is also worth mentioning that completeness of both **KI** and **KE** may be proved exactly by the same argument.

3. Completeness Proof with Respect to the Axiomatic Account of CPL

Here, we present how the completeness-proof strategy works for the case of **KI** and CPL in order to use the same pattern in the completeness proof of the ST-system for FOL (see the next section).

The axiom schemes and the rule of *Modus Ponens* (MP, for short) presented in Table 3 constitute the axiomatic proof system $\mathcal{F}$ for CPL. In the presentation, we rely on the conventions introduced in [29].

Table 3. Axiom schemes of $\mathcal{F}$.

1. $A \to (B \to A)$
2. $(A \to B) \to ((A \to (B \to C)) \to (A \to C))$
3. $A \to A \vee B$
4. $B \to A \vee B$
5. $(A \to C) \to ((B \to C) \to (A \vee B \to C))$
6. $(A \to B) \to ((A \to \neg B) \to \neg A)$
7. $\neg \neg A \to A$
8. $A \wedge B \to A$
9. $A \wedge B \to B$
10. $A \to (B \to A \wedge B)$

$$\frac{A \to B \quad A}{B} \text{ MP}$$

The proof of completeness requires a demonstration that every axiom scheme has its scheme of proof. As an example, here is a scheme of **KI**-proof of Axiom 2. For conciseness, let:

$$F = (A \to B) \to ((A \to (B \to C)) \to (A \to C))$$

$$Cons = (A \to (B \to C)) \to (A \to C)$$

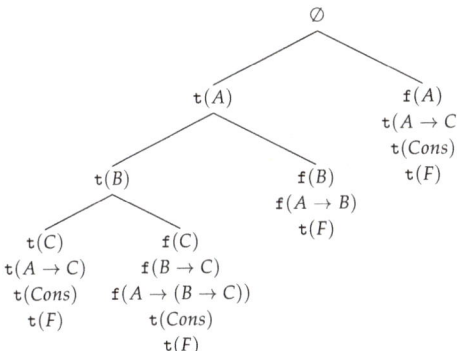

All the other axioms of $\mathcal{F}$ may be easily proved in **KI**. The next theorem states that **KI** simulates the only rule of $\mathcal{F}$.

Theorem 1. *If* $\vdash_{KI} A$ *and* $\vdash_{KI} A \to B$, *then* $\vdash_{KI} B$.

Proof. (After [10,11].) Suppose that $\vdash_{KI} A$ and $\vdash_{KI} A \to B$, and let $\mathcal{T}_1$ and $\mathcal{T}_2$ stand for the proofs of, respectively, formulas A and $A \to B$. Then, the following tree:

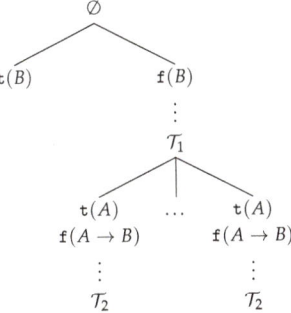

is a **KI**-proof of B, since each branch, except for the leftmost one, goes through $f(A \to B)$ and ends with $t(A \to B)$, thus the leftmost branch is the only open branch of the tree. □

As mentioned above, this strategy of completeness proof requires unrestricted usage of the PB-rule in the system. That would be a drawback of the system obviously, but we truly adore the Gentzen-like spirit of the proof. We believe that the proof reveals important connections between axiomatic systems, Gentzen sequent systems with cut, and **KI** or ST-systems but with unrestricted cut. We elaborate on this topic in the next section.

4. Synthetic Tableaux and Other Deductive Systems for CPL: A Note on Relative Complexity

There are another reasons to consider axiomatic systems, Gentzen sequent systems with cut, **KI**, and ST-system with unrestricted cut, as belonging to one family of deductive systems. Below, we cite the results established in [10,11], concerning the relative complexity of deductive systems for CPL. However, it is not the aim of this paper to describe the formal tools and report all the details needed to establish the quoted results. All this may be found in [30,31] (for the background), [10,11]. Our aim here is to indicate the context in which the method of synthetic tableaux appears especially attractive.

That a proof system **S'** *linearly/polynomially simulates* another proof system **S** means, informally, that there is a function computable in linear/polynomial time which maps every proof of a formula A in **S** to a proof of the same formula in **S'**.

Systems **KE** and **KI** (for CPL) can linearly simulate each other. We also know (see [11]) that the canonical restriction of **KI** (that is, **KI** with the use of the PB-rule restricted to atoms) can linearly simulate truth-tables, but not vice versa. Hence, canonical **KI**, as well as ST-system for CPL, are actually *improvements* on truth-tables in terms of systems complexity. On the other hand, analytic restriction of **KI**, where the PB-rule is restricted to subformulas of the formula to be proved, but is not restricted to atoms, can polynomially simulate both truth-tables and the analytic tableau system (see [32]) but, again, not vice versa. Finally, canonical **KI**, and thus also ST-system for CPL, cannot polynomially simulate the analytic tableau system. Here is what we lose when cut is restricted to atoms.

Again, let **S** and **S'** stand for propositional proof systems. Following D'Agostino [10], we write: **S'** $\leq_p$ **S** to mean that **S** polynomially simulates (p-simulates) **S'**. The inscription **S'** $\equiv_p$ **S** means that the relation of p-simulation holds in both directions, and **S'** $<_p$ **S** means that **S** p-simulates **S'** but not vice versa. The following holds (let us recall that the systems are considered as pertaining to CPL):

1. Gentzen system with cut $\equiv_p$ Natural Deduction $\equiv_p$ Frege systems
2. Resolution $<_p$ any system from (1)
3. Cut-free Gentzen system $<_p$ any system from (1)

It can also be shown that **KE**, and thus also **KI**, can linearly simulate Natural Deduction.

To sum up, it is clear that the presence of a version of cut increases efficiency of proof systems. All the above pertains to systems for CPL; however, we believe that the relation between ST-systems, especially with cut unrestricted, and axiomatic systems is worthy of further research. The use of unrestricted cut can be beneficial if the system was used to support automated deduction with FOL. Clearly, the issue needs further research.

5. The First-Order Case

The presentation of syntax, semantics, and axiomatic account of FOL is based on the conventions introduced in [29].

FOL is expressed in language called $\mathcal{L}_{FOL}$, containing:

- propositional connectives: $\neg, \wedge, \vee, \rightarrow$;
- infinite set of variable symbols; we use $x, y, z, \ldots$ as metasymbols for variables;
- quantifiers $(\exists x), (\forall x)$;
- function symbols of arbitrary arities; $f, g, h, \ldots$ are used as metasymbols, function symbols of arity 0 are called *constant symbols*; and
- relation symbols of arbitrary arities; $P, Q, R, \ldots$ are used as metasymbols.

The notions of *term*, *atomic formula*, and *formula* are defined in the usual manner, similarly for *free* and *bound occurrence of variable*, and *sentence*. $FORM_{\mathcal{L}_{FOL}}$ is for the set of all formulas of the language.

We use $t, s, \ldots$ as metasymbols for terms. The result of *substitution of term t for x in A* is denoted by $A(t/x)$, and defined in the usual manner; *int.al.*, whenever $A(t/x)$ is considered, it is assumed that term t is free for x in A. After [29], we adopt the simplifying conventions for denoting substitutions, according to which:

- If "$A(x)$" and "$A(t)$" are written in the same context, this means that $A = A(x)$ is a formula and that $A(t)$ is $A(t/x)$; using "$A(x)$" neither presupposes that x occurs free in A nor that it occurs in A at all.
- If "$A(t)$" and "$A(s)$" are written in the same context, then this is to mean that A is a formula, x is a variable and $A(t) = A(t/x), A(s) = A(s/x)$.

For semantics of FOL, we use:

- $\mathcal{M} = \langle M, f^{\mathcal{M}} \rangle$ for an interpretation of $\mathcal{L}_{\text{FOL}}$, where M is the domain of $\mathcal{M}$ and $f^{\mathcal{M}}$ is the interpreting function; and
- σ, σ^* for object assignments, that is, mappings from the set of variables to the domain M of $\mathcal{M}$.

We write "$\mathcal{M} \vDash A [\sigma]$" for "formula A is satisfied in $\mathcal{M}$ under σ", and "$\mathcal{M} \vDash A$" for "A is true in $\mathcal{M}$".

5.1. Axiomatic System

Axiomatic proof system $\mathcal{F}_{\text{FO}}$ for FOL contains the 10 axiom schemes listed in Table 3, and, additionally, the following two axiom schemes, in which A is any formula and t is any term free for x in A.

11. $A(t) \to (\exists x) A(x)$; and
12. $(\forall x) A(x) \to A(t)$.

The set of rules of $\mathcal{F}_{\text{FO}}$ contains MP and the following two quantifier rules of inference, where x does not appear freely in C; "GC" is for the generalization over consequent and "GA" for the generalization over antecedent.

$$\frac{C \to A(x)}{C \to (\forall x) A(x)} \, GC \qquad \frac{A(x) \to C}{(\exists x) A(x) \to C} \, GA$$

$\mathcal{F}_{\text{FO}}$ is sound and complete with respect to the model semantics referred to above (see [29]).

5.2. Synthetic-Tableaux System

The ST-system for FOL consists of the linear (synthesizing) rules, the PB-rule and the notion of proof. In this case, the PB-rule is not a subject to any restrictions, thus:

$$F \quad \neg F$$

may be applied at any time, in a tableau constructed for a formula A, and F is arbitrary. This unrestricted form of cut (PB-rule) is necessary, as we have seen, to prove completeness of **KI** with respect to CPL in the "Gentzen-way". In the ST-system for FOL, the rule is left unrestricted for the same reason. However, one can think of restrictions for practical applications. For example, it seems that there are no obstacles to restrict F to be an element of $Sub(A)$, but we do not consider this restriction here. Needless to say, no other counterpart of cut elimination, except for possible restrictions of applicability of the rule, is possible in the ST-system.

The linear rules of the system are those for propositional connectives, listed in Table 1, and the following two rules for existential quantification, where t is a term free for x in A:

$$\frac{A(t)}{(\exists x) A(x)} \, r_\exists \qquad \frac{\neg A(t)}{\neg (\forall x) A(x)} \, r_{\neg \forall}$$

As in the propositional case, the applications of linear rules *could* be restricted to the set of suitably defined subformulas of the formula to be proved. The problem, however, lies in the notion of *subformula*. First, it must take into account substitutions, to the effect that, e.g., $A(t)$ is a subformula of $(\exists x)A(x)$. What is more, however, the problem gets more complicated in the case of the global rules introduced below. For this reason, we resign from the subformula restrictions in this version of the system.

For the next step, we need the notion of a subtableau of a tableau. Let $R|_Y$ stand for the restriction of relation R to set Y. If $\mathcal{T} = \langle X, R \rangle$ is a tree, then a *subtree* of $\mathcal{T}$ is any $\mathcal{S} = \langle Y, R|_Y \rangle$ such that: (i) $Y \subseteq X$; (ii) if w is in Y, then each R-successor of w is in Y; and (iii) $\mathcal{S}$ has a root, that is, $\mathcal{S}$ is itself a tree. If $\mathcal{T}$, together with a labeling function, is a synthetic tableau, then its subtrees, together with suitably restricted labeling functions, may be considered as *subtableaux* of $\mathcal{T}$; however, we want the subtableaux to start with a branching, just as tableaux do. Moreover, a restriction of the labeling function must leave the root empty. The following example illustrates this idea.

Example 3. *The second tree, $\mathcal{T}_2$, is a subtree of tree $\mathcal{T}_1$, and a subtableau of tableau $\mathcal{T}_1$. The third, $\mathcal{T}_3$, is not a subtree of $\mathcal{T}_1$ since the node labeled with "f", an R-successor of the node labeled with "d", is missing. Finally, $\mathcal{T}_4$ is a subtree of tree $\mathcal{T}_1$, but is not a subtableau of tableau $\mathcal{T}_1$, as it does not start with branching.*

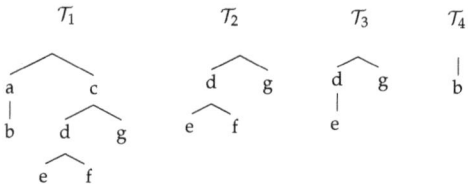

As in **KI**, we also need the following notion:

Definition 3. *If for certain formula F, a branch of a labeled tree carries both F and $\neg F$, then we say that the branch is* closed, *otherwise we say that the branch is* open.

Every linear rule of ST-system for FOL considered so far is *local* in that it acts on a *single* branch extending it with one node carrying the conclusion of the rule. The following rules for universal generalization are *global*, since they act on *each* open branch of a subtableau which, in particular, may be the whole tableau. Global rules are not necessary at the propositional level, but here we need it to incorporate universal generalization. For clarity, we present the rules in a descriptive form.

UG1 If $\mathcal{T}^*$ is a subtableau of $\mathcal{T}$ such that every open branch of $\mathcal{T}^*$ ends with formula $C \to A(x)$, where x does not occur freely in C, **and no formula in $\mathcal{T}^*$ has been synthesized with the use of a premise which is not on $\mathcal{T}^*$**, then $\mathcal{T}$ may be extended by adding $C \to (\forall x)A(x)$ to each open branch of $\mathcal{T}^*$.

UG2 If $\mathcal{T}^*$ is a subtableau of $\mathcal{T}$ such that every open branch of $\mathcal{T}^*$ ends with formula $A(x) \to C$, where x does not occur freely in C, **and no formula in $\mathcal{T}^*$ has been synthesized with the use of a premise which is not on $\mathcal{T}^*$**, then $\mathcal{T}$ may be extended by adding $(\exists x)A(x) \to C$ to each open branch of $\mathcal{T}^*$.

For preciseness, let us state the proviso written in bold in a more detailed way (we call it "the bold proviso", although it is rather quite moderate):

the bold proviso if a formula lying on a branch of $\mathcal{T}^*$ gets here by a local rule, then the premises necessary to derive it precede it on the same branch of $\mathcal{T}^*$

Naturally, it can happen that a premise is present somewhere above the subtableau $\mathcal{T}^*$, but nevertheless the same formula must be "available" at the appropriate branch of $\mathcal{T}^*$.

Here is an example of a synthetic tableau where the global rules are applied. The first tree, $\mathcal{T}_1$, is obtained by local linear rules ($\mathbf{r}_\exists$, $\mathbf{r}^2_\to$, on the left branch, and $\mathbf{r}_{\neg\forall}$, $\mathbf{r}^1_\to$ on the right branch). Since x is not free in $(\exists x) R(x,y)$, rule **UG2** is applicable. One obtains a tree with formula $(\exists x)(\forall y) R(x,y) \to (\exists x) R(x,y)$ in both leaves, to which **UG1** is applied. In this way, tree $\mathcal{T}_2$ is obtained from $\mathcal{T}_1$ by **UG2** and **UG1**. In both cases, the subtableau $\mathcal{T}^*$ specified in **the bold proviso** is identical to the whole tableau, hence the proviso is satisfied.

Example 4.

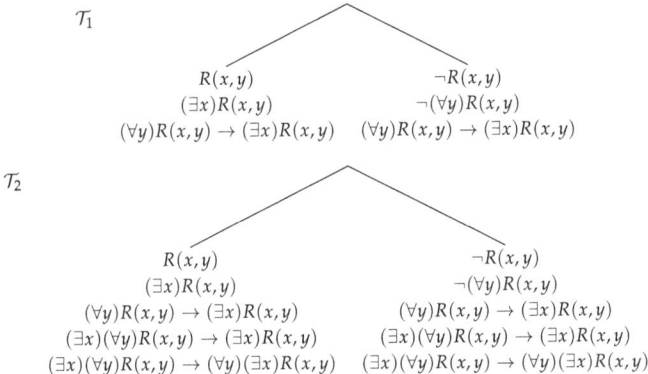

The above tableau illustrates also the difficulty with restricting the applications of the rules of the system to subformulas of the formula to be proved. As mentioned above, the problem is in defining the very notion of subformula in the case of a language with quantifiers. Considering the above case, the definition would have to make formula $(\forall y)R(x,y) \to (\exists x)R(x,y)$ a subformula of $(\exists x)(\forall y)R(x,y) \to (\forall y)(\exists x)R(x,y)$. Summarizing, the following definition abandons the restriction to subformulas. Consequently, it abandons the reference of a synthetic tableau to a particular formula A. However, the reference to a particular formula occurs in the notion of a proof.

Definition 4. *A synthetic tableau in the ST-system for $\mathcal{L}_{\mathsf{FOL}}$ is a finite labeled tree $\mathcal{T} = \langle X, R, \ell \rangle$ generated by: the unrestricted PB-rule, the local linear rules $\mathbf{r}^1_\to$, $\mathbf{r}^2_\to$, $\mathbf{r}^3_\to$, $\mathbf{r}^1_\vee$, $\mathbf{r}^2_\vee$, $\mathbf{r}^3_\vee$, $\mathbf{r}^1_\wedge$, $\mathbf{r}^2_\wedge$, $\mathbf{r}^3_\wedge$, $\mathbf{r}_\neg$, $\mathbf{r}_\exists$, $\mathbf{r}_{\neg\forall}$ and/or the global rules* **UG1**, **UG2**, *and such that:*

$$\ell : X \setminus \{\mathbf{r}_\mathcal{T}\} \longrightarrow FORM_{\mathcal{L}_{\mathsf{FOL}}}$$

Definition 5. *A proof of A in the ST-system for* FOL *is a synthetic tableau $\mathcal{T}$ for A such that each leaf of an open branch of $\mathcal{T}$ is labeled with A.*

The first tree in Example 4 is a proof of formula $(\forall y)R(x,y) \to (\exists x)R(x,y)$ in the ST-system, and the second is a proof of $(\exists x)(\forall y)R(x,y) \to (\forall y)(\exists x)R(x,y)$ in the ST-system.

Finally, here is the announced proof of the completeness of the ST-system for FOL with respect to $\mathcal{F}_{\mathsf{FO}}$.

Theorem 2. *Every axiom of $\mathcal{F}_{\mathsf{FO}}$ is provable in the ST-system for* FOL.

Proof. As to axiom Schemes 1–10, one example is presented in Section 3. The other proofs are easy to find. Let us consider the quantifier cases.

For each axiom of the form $A(t) \to (\exists x)A(x)$, where t is free for x in A, there is the following tree:

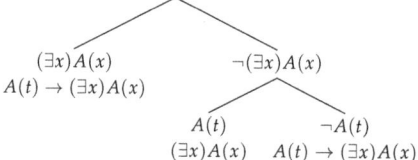

where the application of R$_\exists$ on the second branch is permitted because t is free for x in A. Similarly, the following tree:

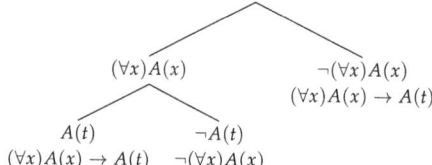

constitutes a proof for an axiom of the form $(\forall x)A(x) \to A(t)$, where t is free for x in A. □

Theorem 3. *The ST-system for FOL can simulate the inference rules of $\mathcal{F}_{FO}$. More specifically:*

1. *If a formula of the form $C \to A(x)$, where x is not free in C, has a proof in the ST-system for FOL, then $C \to (\forall x)A(x)$ has it as well (rule GC).*
2. *If a formula of the form $A(x) \to C$, where x is not free in C, has a proof in the ST-system for FOL, then $(\exists x)A(x) \to C$ has it as well (rule GA).*
3. *If A and $A \to B$ have proofs in the ST-system for FOL, then there is also a proof of B (rule MP).*

Proof. The result is obvious in the case of the rules of Universal Generalization—rules GC and GA are simulated by **UG1** and **UG2**, respectively. Suppose that a formula of the form $C \to A(x)$, where x is not free in C, has a proof $\mathcal{T}$ in the ST-system for FOL. Since $\mathcal{T}$ is its own subtableau, the "bold proviso" is satisfied. Thus, an application of **UG1** results in a proof of $C \to (\forall x)A(x)$. The reasoning is analogous in the case of rule GA.

For the case of MP, the proof of Theorem 1 may be repeated in the first-order setting. Thus, suppose that A and $A \to B$ have proofs $\mathcal{T}_1$ and $\mathcal{T}_2$ (respectively) in the ST-system for FOL. Then, the tree displayed on Figure 1, page 13, is a proof of B in the ST-system for FOL, since the leftmost branch is the only open branch of the tree. □

Now, we can explain why the rules of universal generalization need the "bold proviso". Consider the following simplified versions of **UG1** and **UG2**:

UG1^0   If $\mathcal{T}$ is a proof of formula $C \to A(x)$ in ST-system, where x does not occur freely in C, then each open branch of $\mathcal{T}$ may be extended with $C \to (\forall x)A(x)$. The result is a proof of $C \to (\forall x)A(x)$.

UG2^0   If $\mathcal{T}$ is a proof of formula $A(x) \to C$ in ST-system, where x does not occur freely in C, then each open branch of $\mathcal{T}$ may be extended with $(\exists x)A(x) \to C$. The result is a proof of $(\exists x)A(x) \to C$.

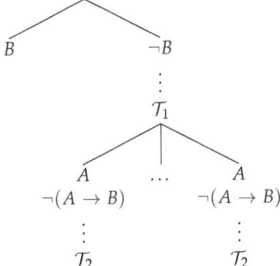

Figure 1. A simulation of *MP*.

The problem shows up when one tries to conduct the proof of Theorem 3 for *MP*. We cannot simply "glue" trees $\mathcal{T}_1$, $\mathcal{T}_2$ with the branch starting with $\neg B$, because the potential applications of **UG1⁰**, **UG2⁰** in $\mathcal{T}_1$, $\mathcal{T}_2$ would miss their justification: with the left branch (with formula B), the relevant tableaux would no longer be proofs of the respective premises $C \to A(x)/A(x) \to C$. Thus, the restriction concerning subtableaux is necessary to save our strategy of proving completeness via axiomatic system. It remains an open question whether the ST-system with **UG1⁰**, **UG2⁰** instead of **UG1**, **UG2** is complete.

Theorem 4. *Suppose that A is a formula such that $\mathcal{M} \vDash A$ for every interpretation $\mathcal{M}$ of $\mathcal{L}_{\mathsf{FOL}}$. Then, A has a proof in the ST-system for* **FOL**.

Proof. Since A is valid, we know that A has a proof $\mathcal{P}$ in $\mathcal{F}_{\mathsf{FO}}$. By Theorems 2 and 3, and by induction on the length of $\mathcal{P}$, A has a proof in the ST-system. □

5.3. Derivability of Universal Generalization

We have shown that the ST-system for **FOL** is complete with respect to $\mathcal{F}_{\mathsf{FO}}$. However, it may seem more natural to use the more common form of universal generalisation:

$$\text{if } \vdash A(x), \text{ then } \vdash (\forall x)A(x)$$

instead of *GC* and *GA*. Consequently, one may consider a counterpart of universal generalization in the synthetic tableaux framework. Below, we show that such a rule is derivable.

Theorem 5. *Let $A(x)$ be a formula which is provable in the ST-system for* **FOL**. *Then, $(\forall x)A(x)$ is also provable.*

Proof. Let $\mathcal{D}$ be an ST-proof of $A(x)$, and let $d_1, \ldots, d_n$ stand for the open branches of $\mathcal{D}$ (by Definition 5, each open branch ends with the formula $A(x)$). First, by using $\mathbf{r}_\to^2$ with formula $A(x)$ as the premise, we add $(\exists x)A(x) \to A(x)$ to each d_i ($1 \le i \le n$). Since x is not free in the antecedent, we transform the tree into a proof of $(\exists x)A(x) \to (\forall x)A(x)$ by **UG1**. The result is displayed on Figure 2. Finally, we extend each open branch of the tableau by means of the application of the following sequence of rules: PB-rule peformed on $(\forall x)A(x)$, then on the right branch, $(\exists x)A(x)$ is derived from $A(x)$ by $\mathbf{r}_\exists$, with x as term t (x is free for x in A). Then, the last formula is derived by $\mathbf{r}_\to^3$. Naturally, the branch is closed. The derivation is displayed on Figure 3.

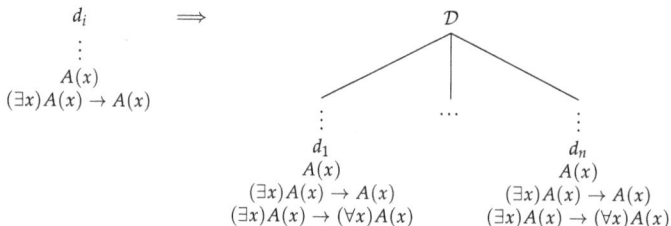

Figure 2. Universal Generalization, part 1.

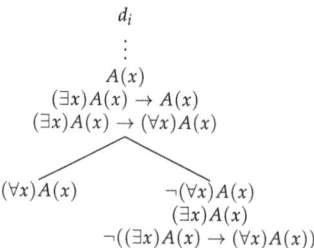

Figure 3. Universal Generalization, part 2.

The only possibly open branches in the modified tree are those going through d_i, $A(x)$ and ending with $(\forall x)A(x)$. Thus, we have obtained an ST-proof of that formula. □

Theorem 5 shows that the following global rule is derivable in the ST-system for FOL:

UG If $\mathcal{T}$ is a proof of formula $A(x)$ in the ST-system for FOL, then each open branch of $\mathcal{T}$ may be extended by adding $(\forall x)A(x)$. The result is a proof of $(\forall x)A(x)$ in the system.

The rule is not a subject to any restriction akin to the bold proviso. Again, this follows from the strategy adopted at the metalevel, that is, from the fact that in the proof of Theorem 5 we act on the whole tableau of $A(x)$. We use this version of generalization in the formalization of first-order theories in Section 8.

*5.4. System **KI** for FOL*

At the end of this section, let us refer briefly to the version of **KI** for FOL. The construction of the system is described in [11]. The author found the system by putting sequences of formulas into nodes of a **KI**-tree, which enables him to formulate restrictions on the rules introducing quantifiers on a branch. The rules presented there are local and they actually deal with quantifiers in a manner characteristic to analytic tableaux. For this reason, it is hard to compare the system of **KI** and the ST-system for FOL. Thus, we conclude with the modest remark that the main difference between the ST-system for FOL and **KI** for FOL lies in the treatment of quantifiers: **KI** treats quantifiers in a way characteristic to analytic tableau systems, whereas the ST-system deals with quantifiers by the use of global rules and unrestricted cut, which makes it more similar to natural-deduction or Gentzen systems.

6. Soundness of the ST System for FOL

Let us start with the following observation:

Lemma 1. *Each local rule of the ST-system for* FOL *preserves the property of being satisfied under an object assignment σ in an interpretation $\mathcal{M}$ of $\mathcal{L}_{FOL}$.*

Proof. We consider $\mathbf{r}^1_\to$ and $R_\exists$. The other cases are analogous.

Suppose that, for some interpretation $\mathcal{M}$ and σ in $\mathcal{M}$, $\mathcal{M} \models \neg B\,[\sigma]$. Then, $\mathcal{M} \not\models B\,[\sigma]$ and thus $\mathcal{M} \models A \to B\,[\sigma]$. This argument shows that if the premise of rule $\mathbf{r}^1_\to$ is satisfied under σ in $\mathcal{M}$, then the conclusion is satisfied as well.

Suppose that t is free for x in A, and that $\mathcal{M} \models A(t)\,[\sigma]$. Let u stand for the interpretation of term t in $\mathcal{M}$ under σ. Let σ^* stand for an object assignment in $\mathcal{M}$ such that $\sigma^*(x) = u$ and $\sigma^*(y) = \sigma(y)$ for every $y \neq x$. Then, $\mathcal{M} \models A(x)\,[\sigma^*]$. This means that $\mathcal{M} \models (\exists x)A(x)\,[\sigma]$. □

The following fact expresses the Principle of Bivalence with respect to the satisfaction relation in model-theoretic semantics.

Fact 1. *Let $\mathcal{M}$ and σ be arbitrary, and let F stand for an arbitrary formula.*

$$\mathcal{M} \models F\,[\sigma] \text{ or } \mathcal{M} \models \neg F\,[\sigma].$$

The idea of the proof of Lemma 2 comes from using semantic trees in the proof of resolution systems (see [26], Section 3.8 for details).

We say that a branch $\mathcal{B}$ of a synthetic tableau $\mathcal{T}$ is *compatible with σ in $\mathcal{M}$* iff for every formula F that labels $\mathcal{B}$, it is the case that $\mathcal{M} \models F\,[\sigma]$. Let us observe that:

Fact 2. *If a branch of a synthetic tableau is compatible with some object assignment σ, then the branch cannot be closed.*

Lemma 2. *Let $\mathcal{T}$ be a synthetic tableau. Let $\mathcal{M}$ be an arbitrary but fixed interpretation and let σ stand for an arbitrary but fixed object assignment in $\mathcal{M}$. There is a branch $\mathcal{B}$ of $\mathcal{T}$ which is compatible with σ in $\mathcal{M}$.*

Proof. The proof is by induction on the size of $\mathcal{T}$, that is, the number of its nodes.

<u>Base case.</u> The smallest possible synthetic tableau is an empty root. (By Definition 4, a synthetic tableau is a finite tree, and a tree must have a root. Hence, it follows that a one-element tree, containing the root only, is the smallest possible synthetic tableau.) Since there is no label, $\mathcal{T}$ is compatible with any assignment of an arbitrary model.

<u>Induction step.</u> Assume the following <u>induction hypothesis</u>: for synthetic tableaux of size up to n, for any σ of an arbitrary model $\mathcal{M}$, there is a branch in the tableau compatible with σ. Suppose that the size of $\mathcal{T}$ is $n + 1$. We need to consider the last rule that acted upon the tableau.

Suppose that it was the PB-rule. Let $\mathcal{T}^*$ stand for the tableau to which the rule was applied, and assume that $\mathcal{B}^*$ is the particular branch that the rule acted upon. $\mathcal{T}^*$ has $n - 1$ nodes, thus, by the induction hypothesis, there is a branch $\mathcal{B}$ in $\mathcal{T}^*$ compatible with σ in $\mathcal{M}$. If the branch is not $\mathcal{B}^*$, then it is present also in the tableau $\mathcal{T}$, and thus it follows that there is a branch in $\mathcal{T}$ compatible with σ in $\mathcal{M}$, as required. If $\mathcal{B} = \mathcal{B}^*$, then, by Fact 1, one of the two branches created from $\mathcal{B}^*$ by the PB-rule is compatible with σ in $\mathcal{M}$.

Now, suppose that the last rule applied was a local linear rule. Then, we consider a tableau $\mathcal{T}^*$ to which the rule was applied, we use the induction hypothesis, and—similarly as in the above case—we conclude that, if the branch of $\mathcal{T}^*$ compatible with σ is the one modified by the local rule, then, by Lemma 1, the resulting branch of $\mathcal{T}$ is compatible with σ in $\mathcal{M}$.

Finally, the difficult case. The last rule applied was one of the general rules; suppose that it was **UG1**, the reasoning is analogous in the case of **UG2**. Let $\mathcal{T}^*$ be the synthetic tableau to which the rule was applied. The size of $\mathcal{T}^*$ is not grater than n, thus, by the induction hypothesis, there is a branch $\mathcal{B}^*$ in $\mathcal{T}^*$

compatible with σ in $\mathcal{M}$. By Fact 2, the branch must be open, so it has a formula of the form $C \to A(x)$ in its leaf, where x does not occur freely in C.

Suppose that $\mathcal{M} \not\models C \to (\forall x)A(x)\,[\sigma]$. Therefore, $\mathcal{M} \models C\,[\sigma]$ and $\mathcal{M} \not\models (\forall x)A(x)\,[\sigma]$. Hence, it follows that there is an object u such that if σ^* is an object assignment in $\mathcal{M}$ such that $\sigma^*(x) = u$ and $\sigma^*(y) = \sigma(y)$ for each $y \neq x$, then (i) $\mathcal{M} \not\models A(x)\,[\sigma^*]$. Since x does not occur freely in C, still (ii) $\mathcal{M} \models C\,[\sigma^*]$. However, since $\mathcal{T}^*$ satisfies the induction hypothesis, it satisfies the hypothesis also with respect to object assignment σ^* in $\mathcal{M}$. Thus, there is an open branch of $\mathcal{T}^*$ compatible with σ^*. Each open branch of $\mathcal{T}^*$ ends with $C \to A(x)$, thus $\mathcal{M} \models C \to A(x)\,[\sigma^*]$. However, this contradicts the previous Arrangements (i) and (ii). Thus, it follows that $\mathcal{M} \models C \to (\forall x)A(x)\,[\sigma]$. Hence, the branch resulting from $\mathcal{B}^*$ is compatible with σ in $\mathcal{M}$. □

Theorem 6. *If A has a proof in the ST-system for* FOL, *then A is valid.*

Proof. Suppose that $\mathcal{T}$ is a proof of A and that A is not valid. Let $\mathcal{M}$ be an arbitrary (but fixed) interpretation of $\mathcal{L}_{\mathsf{FOL}}$, and let σ be an arbitrary (but fixed) object assignment in $\mathcal{M}$ such that $\mathcal{M} \not\models A\,[\sigma]$. By the previous lemma, there is a branch $\mathcal{B}$ of $\mathcal{T}$ compatible with σ. The branch must be open, and since $\mathcal{T}$ is a proof of A, the formula labels the leaf of $\mathcal{B}$. Thus, $\mathcal{M} \models A\,[\sigma]$, which is a contradiction. □

7. Some Further Remarks on Relations between the ST-System and the Axiomatic System

Despite finishing the completeness proof, for some time, the first author was not able to prove anything interesting in the system. Let us illustrate the difficulties with the following formula F, where x does not occur free in A (for simplicity, we omit the xs after A and B):

$$F = (\forall x)(A \to B) \to (A \to (\forall x)B)$$

The first attempt to prove F was the following:

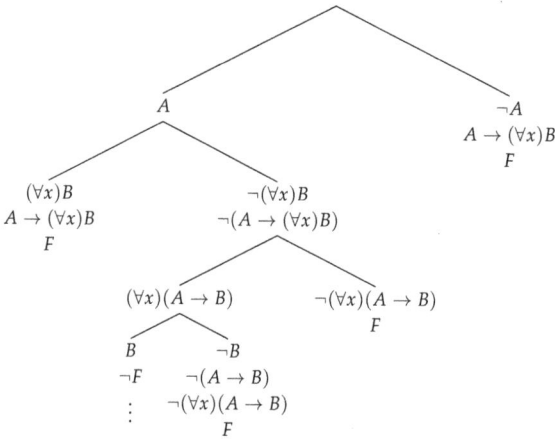

which is not satisfactory; the branch with $\neg F$ should be closed, but it is not clear how to derive a contradiction. In a system of analytic tableaux, one would instantiate on $\neg \forall x B$ introducing some $\neg B(a)$, but in this system $\neg B(a)$ may only come by branching, and the problem then is with closing the left branch with $B(a)$ on it. The first author overcame this difficulty after recalling a proof of the formula in axiomatic system $\mathcal{F}_{FO}$:

1. $(\forall x)(A \to B) \to (A \to B)$ — Axiom 12
2. $((\forall x)(A \to B) \to (A \to B)) \to ((\forall x)(A \to B) \land A \to B)$ — Thesis of FOL
3. $(\forall x)(A \to B) \land A \to B$ — MP:2,1
4. $(\forall x)(A \to B) \land A \to (\forall x)B$ — GC:3
5. $((\forall x)(A \to B) \land A \to (\forall x)B) \to ((\forall x)(A \to B) \to (A \to (\forall x)B))$ — Thesis of FOL
6. $(\forall x)(A \to B) \to (A \to (\forall x)B)$ — MP:5,4

In the second line of the proof, the converse of the exportation law is used in order to generalize on B. This is the point.

The following tree represents a proof of formula $C = (\forall x)(A \to B) \land A \to (\forall x)B$, where x does not occur free in A, in the ST-system. In the subtableau starting with A, the leaves are derived by **UG1**.

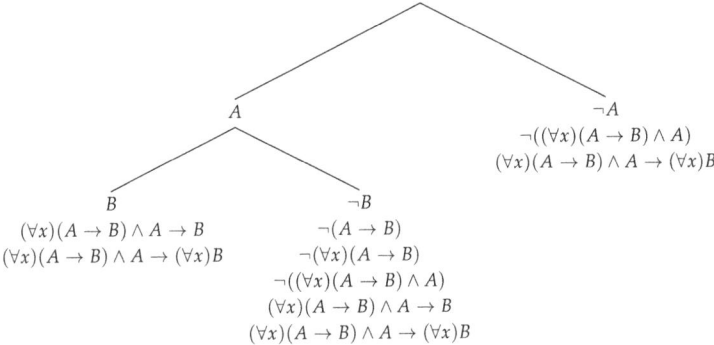

Finally, Figure 4 presents a proof of F where the problematic branch is closed by contradicting formula C. We need to use it, since without "importing" A to the antecedent we cannot generalize on B. Let us also explain that the fourth (from the left) branch contains a kind of a detour: formula $(\forall x)(A \to B) \land A \to B$ is derived here to make **UG1** applicable in the subtableau starting with A. After deriving formula C, we need to extend the branch with F (obtained by $\mathbf{r}^1_{\to}$, which is a local rule), to make the tree a proof of F. The whole tableau is a good example illustrating the fact that the synthetic tableaux system is not a "tableau system" in the common sense of the term.

The interesting reflection is that about the relation between the axiomatic system and that of ST. The latter may seem bizarre, if an axiomatic proof is needed in order to get a hint on how to prove a formula. It appears that heuristics of proving theorems in axiomatic systems may serve also as heuristics of proving theorems in the ST-system. What is the benefit, if any? Does it work also the other way round, that is, is it the case that heuristics suitable for the ST-system can work for axiomatic systems as well?

At the moment, we are not able to answer this questions satisfactorily. However, the example with formula F sheds some light in this darkness. Although we need a hint to prove it, the use of the PB-rule is still restricted to subformulas (as mentioned above, the notion of subformula must take into account all substitutions) of formula F. The more complex formulas are built from the "atoms" introduced by the PB-rule. Albeit this "composition" derives us from the set of subformulas of F, it does so only via the export-import manipulation necessary to generalize on the consequent. This may be a matter of the particular form of rules for universal generalization in this ST-system.

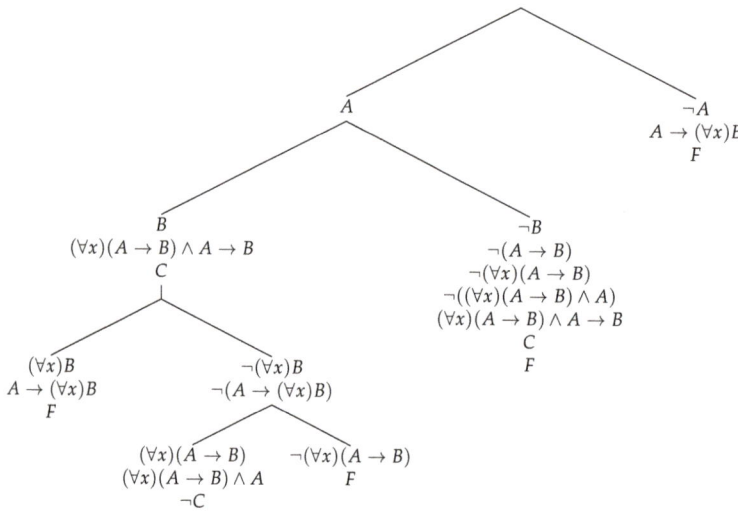

Figure 4. A proof of F.

8. ST-Systems for First-Order Theories

In [33], the authors presented a strategy of transforming axioms of certain forms into sequent calculus rules, in order to obtain a sequent calculus for a given first-order theory. In this section, we show that similar techniques may be applied beyond the domain of structural proof theory.

8.1. Universal Axioms

After Negri and von Plato [33], by *universal axioms* we mean sentences of $\mathcal{L}_{\mathsf{FOL}}$ of the following form:

$$(\forall x_1)\ldots(\forall x_k)(A_1 \wedge \ldots \wedge A_n \to B_1 \vee \ldots \vee B_m) \tag{ax}$$

where A_i and B_j are *atomic formulas*. For simplicity, in this section, we assume that both conjunction and disjunction may have an arbitrary number of arguments. We show how the universal axioms can be converted into ST-rules. When $n, m \geq 1$, the rule has one of the forms indicated below.

$$\dfrac{\begin{array}{c} A_1 \\ \vdots \\ A_n \end{array}}{B_1 \vee \ldots \vee B_m}\,(R_{ax}^1) \qquad \dfrac{\begin{array}{c} A_1 \\ \vdots \\ A_n \end{array}}{B_1 \mid \ldots \mid B_m}\,(R_{ax}^{1*}) \qquad \dfrac{A_1 \wedge \ldots \wedge A_n}{B_1 \vee \ldots \vee B_m}\,(R_{ax}^2) \qquad \dfrac{A_1 \wedge \ldots \wedge A_n}{B_1 \mid \ldots \mid B_m}\,(R_{ax}^{2*})$$

Rules (R_{ax}^{1*}) and (R_{ax}^{2*}) cause branching to m subtrees. If the local rules for $\vee$ and $\wedge$ are present in the system, then each of the four rules may be derived from any of the other three. We show two examples of such derivations below, the others are easy to find. On the left: rule (R_{ax}^1) is assumed, rule (R_{ax}^{1*}) is derived. On the right: rule (R_{ax}^{2*}) is assumed, rule (R_{ax}^1) is derived.

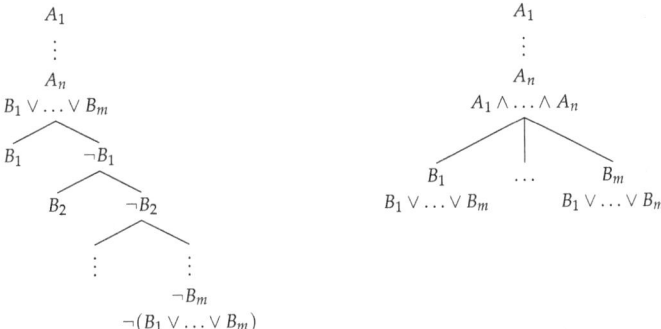

The general scheme of the proof of (ax) in an ST-system, assuming only atomic cuts and rule (R^1_{ax}), looks as follows:

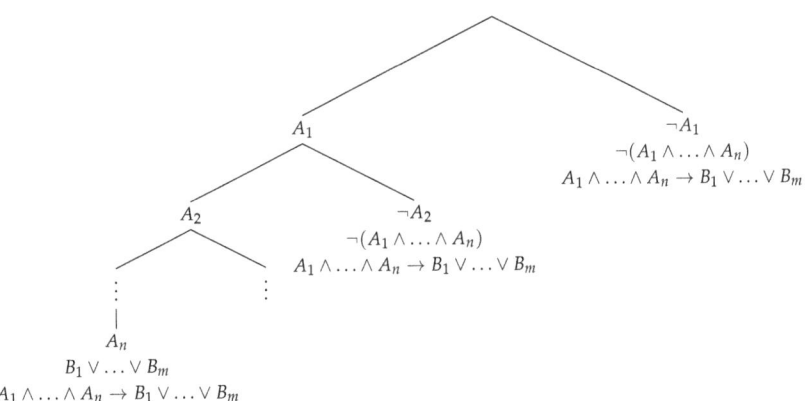

If we allow cut-formulas to be non-atomic, we can use the second rule and the proof can be simplified:

$$
\begin{array}{cc}
A_1 \wedge \ldots \wedge A_n & \neg(A_1 \wedge \ldots \wedge A_n) \\
B_1 \vee \ldots \vee B_m & A_1 \wedge \ldots \wedge A_n \to B_1 \vee \ldots \vee B_m \\
\end{array}
$$
$$A_1 \wedge \ldots \wedge A_n \to B_1 \vee \ldots \vee B_m$$

Further, for universal axioms (ax) with $n = 0$ and $m \geq 1$, we obtain the following no-premises rules:

$$\dfrac{}{B_1 \vee \ldots \vee B_m}\ (R^{\emptyset 1}_{ax}) \qquad \dfrac{}{B_1 \mid \ldots \mid B_m}\ (R^{\emptyset 1*}_{ax})$$

At each stage of a derivation, one can introduce the formula (in the case of $(R^{\emptyset 1}_{ax})$) or extend a tree by means of m subtrees (in the case of $(R^{\emptyset 1*}_{ax})$). As before, each of this rule is derivable from the other.

If $n \geq 1$ and $m = 0$, the rules can be formulated as follows:

$$\frac{\begin{array}{c} A_1 \\ \vdots \\ A_n \end{array}}{C} \, (R_{ax}^{\varnothing 2}) \qquad \frac{A_1 \wedge \ldots \wedge A_n}{C} \, (R_{ax}^{\varnothing 2*})$$

C is an arbitrary formula. If we had *Falsum* ($\bot$) in the language, we could put $C = \bot$. Again, these rules are inter-derivable in the system. Below to the left, we assume ($R_{ax}^{\varnothing 2*}$) and derive ($R_{ax}^{\varnothing 2}$); and vice versa to the right, where F is for $\neg(A_1 \wedge \ldots \wedge A_n)$:

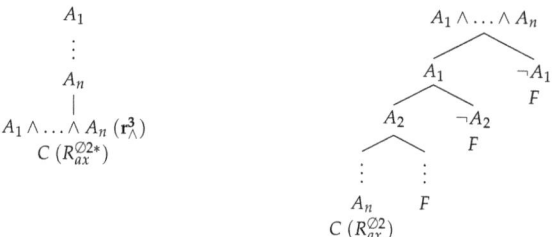

8.2. First Example: Identity

We show how to use the general approach in order to tackle the First-Order Logic with identity. First, we have to extend the language $\mathcal{L}_{\mathsf{FOL}}$ by means of a special predicate symbol, =, representing identity. Axiomatic system for FOL with identity, $\mathcal{F}_{\mathsf{FO}}^=$, can be obtained from $\mathcal{F}_{\mathsf{FO}}$ by the addition of the following two axioms:

$$(\forall x)(x = x) \qquad (ref_=)$$

$$t_i = t_j \wedge A(t_i) \to A(t_j // t_i) \qquad (rep_=)$$

where the notation "$A(t_j // t_i)$" indicates that some occurrences (possibly all of them) of a term t_i in a formula A has been replaced by a term t_j. Similarly as in the case of substitutions, if a formula of the form $A(t_j // t_i)$ is considered, it is assumed that the replacement operation was performed "correctly", that is, the variables of t_i are not bounded in A, and t_j is free for each variable of t_i in A.

According to the introduced strategy, the axioms correspond to the following two rules:

$$\frac{}{t = t} \, (R_{ref=}^{\varnothing 1}) \qquad \frac{\begin{array}{c} t_i = t_j \\ A(t_i) \end{array}}{A(t_j // t_i)} \, (R_{rep=}^1)$$

The assumed semantics is model-theoretic, with the identity predicate interpreted as identity in the domain in every interpretation. The rules have a local character, therefore, for the soundness of the whole ST-system, it is enough to check that they have a property expressed in Lemma 1. We skip the details.

ST-system for FOL with identity results from the system designed for FOL in Section 5 by means of extending the latter with the two rules for identity. Let us call the resulting calculus $\mathsf{ST}_{\mathsf{FOL}}^=$.

Theorem 7 (Completeness). *If a formula is provable in $\mathcal{F}_{\mathsf{FO}}^=$, then it is also provable in $\mathsf{ST}_{\mathsf{FOL}}^=$.*

Proof. We know that all the rules of inference of $\mathcal{F}_{\mathsf{FO}}^=$ can be simulated in $\mathsf{ST}_{\mathsf{FOL}}^=$ (Theorem 3). What we have to show is that both axioms are derivable in our system. This is in fact the case as the following derivations show. The first axiom is derivable by:

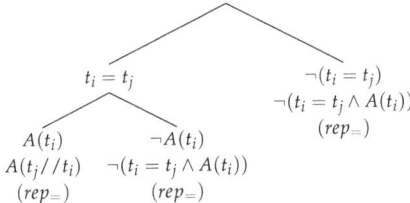

The derivation starts with branching on $x = x$. Then, on the right, we use rule $(R^{\emptyset 1}_{ref=})$, to introduce $x = x$, which makes the branch closed. Now, the only open branch of the tree is the left one, where the rule **UG** is applied (let us recall that it is derivable in the ST-system for FOL).

The second axiom, $(rep_=)$, is derivable by means of the following tree:

$$
\begin{array}{c}
\begin{array}{cc}
t_i = t_j & \neg(t_i = t_j) \\
& \neg(t_i = t_j \wedge A(t_i)) \\
& (rep_=)
\end{array} \\
\begin{array}{cc}
A(t_i) & \neg A(t_i) \\
A(t_j//t_i) & \neg(t_i = t_j \wedge A(t_i)) \\
(rep_=) & (rep_=)
\end{array}
\end{array}
$$

This finishes the proof. □

In some systems for FOL with identity, such as natural deduction system (see [33]), it is possible to use a version of the replacement rule restricted solely to atomic formulas. In our setting, such a rule would have the following form:

$$\dfrac{t_i = t_j \quad P(t_i)}{P(t_j//t_i)} \; (repAt)$$

Naturally, the formulas expressing symmetry and transitivity of identity:

$$(\forall x)(\forall y)(x = y \to y = x) \qquad (sym_=)$$

$$(\forall x)(\forall y)(\forall x)(x = y \wedge y = z \to x = z) \qquad (trans_=)$$

are derivable in $ST^=_{FOL}$, and both are derivable by the use of $(repAt)$. Figure 5 presents a derivation of $(sym_=)$. In Figure 6, $(trans_=)$ is proved, where on the leftmost branch rule $(repAt)$ is used with $x = y$ for $A(t_i), y = z$ for $t_i = t_j$, and $x = z$ for $A(t_j)$.

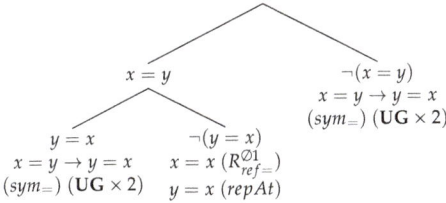

Figure 5. Symmetry.

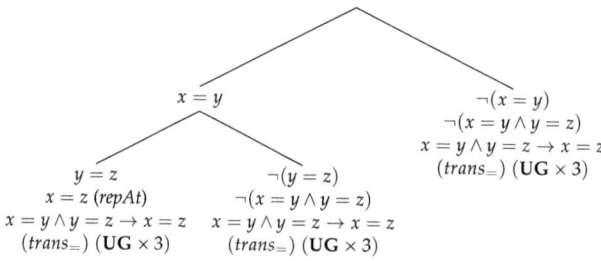

Figure 6. Transitivity.

8.3. Second Example: Partial Order

The second example of a first-order theory is that of a partial order. We add to the language $\mathcal{L}_{\mathsf{FOL}}$ a new predicate symbol, $\leq$, and we build $\mathcal{F}_{\mathsf{FO}}^{\leq}$ by adding two specific axioms to $\mathcal{F}_{\mathsf{FO}}$:

$$(\forall x)(x \leq x) \tag{$ref_\leq$}$$

$$(\forall x)(\forall y)(\forall z)(x \leq y \land y \leq z \to x \leq z) \tag{$trans_\leq$}$$

Following the presented strategy, we transform the first axiom into the following rule:

$$\frac{}{x \leq x}\ (R^{\varnothing 1}_{ref_\leq})$$

which states that at each branch $x \leq x$ can be synthesized without any additional information present.

The second axiom corresponds to the rule:

$$\frac{x \leq y \quad y \leq z}{x \leq z}\ (R^{1}_{trans_\leq})$$

We can synthesize $x \leq z$ on the condition that we have already synthesized $x \leq y$ and $y \leq z$. If we add these two rules to $\mathsf{ST}_{\mathsf{FOL}}$, we obtain a system which we call: $\mathsf{ST}_{\mathsf{FOL}}^{\leq}$.

Theorem 8 (Completeness). *If a formula is provable in $\mathcal{F}_{\mathsf{FO}}^{\leq}$, then it is also provable in $\mathsf{ST}_{\mathsf{FOL}}^{\leq}$.*

Proof. Again, we can derive the rules of inference of $\mathcal{F}_{\mathsf{FO}}^{\leq}$ as in Theorem 3. We show how the axioms can be derived.

The first axiom is derivable in a manner similar to reflexivity axiom in FOL with identity:

$$\frac{x \leq x \quad \neg(x \leq x)}{(\forall x)(x \leq x) \quad x \leq x}$$

The second axiom is derived as displayed on Figure 7. □

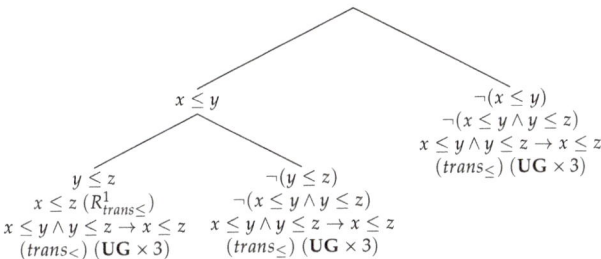

Figure 7. A proof of $(trans_\leq)$.

The tree displayed on Figure 8 represents a proof of formula $F = \neg(x \leq z) \to (\neg(x \leq y) \vee \neg(y \leq z))$. The formula has been synthesized on every open branch. There is one closed branch containing both $x \leq z$ and its negation.

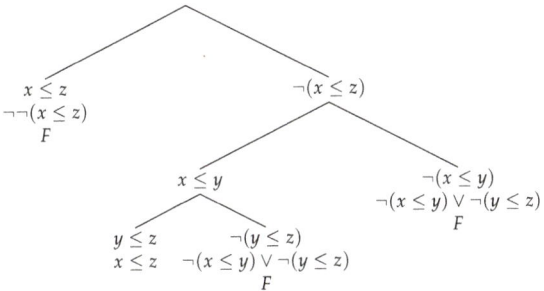

Figure 8. A proof of $F = \neg(x \leq z) \to (\neg(x \leq y) \vee \neg(y \leq z))$.

As in the case of identity, the issue of soundness is settled by verifying that the additional local rules satisfy Lemma 1.

Let us note that the theory of liner order can be obtained from the theory of partial order by means of the addition of the linearity axiom:

$$x \leq y \vee y \leq x \qquad (lin_\leq)$$

which corresponds to one of the following rules:

$$\frac{}{x \leq y \vee y \leq x}\ (R^{\emptyset 1}_{lin_\leq}) \qquad \frac{}{x \leq y \mid y \leq x}\ (R^{\emptyset 1*}_{lin_\leq})$$

One can easily show that the ST-system for liner order is capable of simulating corresponding axiomatic system, which implies completeness.

9. Conclusions

In this paper, we describe a synthetic tableaux system for First-Order Logic and a general strategy for providing proof systems for some of its axiomatic extensions. We show that the resulting systems are complete with respect to the corresponding axiomatic systems. Synthetic tableaux method was developed to formalize direct validity checking and was inspired by Kalmár's completeness proof (see [28]). ST system for First-Order Logic is different from the system for propositional logic, due to the fact that one has to introduce *global rules*, which act on the whole proof tree, while on the propositional level *local* rules are sufficient.

The rule of cut is, in a sense, *essential* to this method—it is the rule which starts every derivation introducing the information needed to synthesize more complex formulas. The presence of the cut rule in a system may be an advantage and a disadvantage at the same time. From the *proof search oriented perspective*, an application of the cut rule is problematic: the rule is unrestricted, so finding the right formula to cut on is hard, even if cut is restricted to subformulas of a goal-formula only. On the other hand, once the right cut formula has been found, the resulting proof may have significantly smaller size than that of proofs constructed by more mechanical methods; we indicate this issue in Section 4.

A step in the same direction has been made before in [34]. The authors proposed a proof procedure for First-Order Logic based on tableaux method with the cut rule. Our work is similar to this proposal in the sense that the cut rule is used while proving theorems and not as the post-processing tool employed to transform already existing derivations to reduce their size. The main difference between the approach in [34] and our approach concerns the proof method used and the form of syntactic structures being processed—in the cited work, the rules act on clauses, whereas our solution does not assume deriving a clausal form.

Finally, we wish to point out some open problems related to the introduced systems. Clearly, there is a connection between Frege systems and ST method, which we think should be further investigated. In particular, a question arises whether proof heuristics working well in one system can be transferred to the other. Another problem pertains to the lack of a subformula property being an effect of unrestricted cut applications. Is there a way to restrict the class of formulas appearing in a proof tree for a given formula?

Author Contributions: The first author is mainly responsible for Sections 1–7, and the second author is mainly responsible for Sections 8 and 9; however, the general issues as well as many details in the whole paper were discussed and corrected jointly. Conceptualization, D.L.-J. and S.C.; Formal analysis, D.L.-J. and S.C.; Investigation, D.L.-J. and S.C.; Methodology, D.L.-J. and S.C.; Supervision, D.L.-J.; Writing—original draft, D.L.-J. and S.C.; and Writing—review and editing, D.L.-J. and S.C.

Funding: The first version of this paper, prepared by Dorota Leszczyńska-Jasion, was part of the project supported by funds of the National Science Centre, Poland, grant No. 2012/04/A/HS1/00715. The work of Szymon Chlebowski was supported financially by the National Science Centre, Poland, grant No. 2017/26/E/HS1/00127.

Conflicts of Interest: The authors declare no conflict of interest. The funders had no role in the design of the study; in the collection, analyses, or interpretation of data; in the writing of the manuscript, or in the decision to publish the results.

References

1. Boolos, G. Don't eliminate cut. *J. Philos. Log.* **1984**, *13*, 373–378. [CrossRef]
2. Orevkov, V.P. Lower bounds for increasing complexity of derivations after cut elimination. *J. Sov. Math.* **1982**, *20*, 2337–2350. [CrossRef]
3. Statman, R. Lower Bounds on Herbrand's Theorem. *Proc. Am. Math. Soc.* **1979**, *75*, 104–107.
4. Woltzenlogel Paleo, B. Atomic cut introduction by resolution: proof structuring and compression. In *Logic for Programming, Artificial Intelligence and Reasoning (LPAR- 16)*; Lecture Notes in Computer Science; Clark, E., Voronkov, A., Eds.; Springer: Berlin, Germany, 2010; Volume 6355, pp. 463–480.
5. Ebner, G.; Hetzl, S.; Leitsch, A.; Reis, G.; Weller, D. On the Generation of Quantified Lemmas. *J. Autom. Reason.* **2019**, *63*, 95–126. [CrossRef]
6. Finger, M.; Gabbay, D. Cut and pay. *J. Log. Lang. Inf.* **2006**, *15*, 195–218. [CrossRef]
7. Finger, M.; Gabbay, D. Equal Rights for the Cut: Computable Non-analytic Cuts in Cut-based Proofs. *Log. J. IGPL* **2007**, *15*, 553–575. [CrossRef]
8. Urbański, M. Remarks on Synthetic Tableaux for Classical Propositional Calculus. *Bull. Sect. Log.* **2001**, *30*, 194–204.

9. Urbański, M. *Tabele Syntetyczne a Logika pytań (Synthetic Tableaux and the Logic of Questions)*; Wydawnictwo UMCS: Lublin, Poland, 2002.
10. D'Agostino, M. *Investigations into the Complexity of Some Propositional Calculi*; Technical Monograph; Oxford University Computing Laboratory, Programming Research Group: Oxford, UK, 1990.
11. Mondadori, M. Efficient Inverse Tableaux. *J. IGPL* **1995**, *3*, 939–953. [CrossRef]
12. Urbański, M. Synthetic Tableaux for Łukasiewicz's Calculus Ł3. *Log. Anal.* **2002**, *177–178*, 155–173.
13. Urbański, M. How to Synthesize a Paraconsistent Negation. The Case of CLuN. *Log. Anal.* **2004**, *185–188*, 319–333.
14. D'Agostino, M.; Mondadori, M. The Taming of the Cut. Classical Refutations with Analytic Cut. *J. Log. Comput.* **1994**, *4*, 285–319. [CrossRef]
15. D'Agostino, M. Are Tableaux an Improvement on Truth-Tables? Cut-Free proofs and Bivalence. *J. Log. Lang. Comput.* **1992**, *1*, 235–252. [CrossRef]
16. Urbański, M. Synthetic Tableaux and Erotetic Search Scenarios: Extension and Extraction. *Log. Anal.* **2001**, *173–175*, 69–91.
17. Smullyan, R.M. Analytic cut. *J. Symb. Log.* **1968**, *33*, 560–564. [CrossRef]
18. D'Agostino, M. Tableau Methods for Classical Propositional Logic. In *Handbook of Tableau Methods*; D'Agostino, M., Gabbay, D.M., Hähnle, R., Posegga, J., Eds.; Kluwer Academic Publishers: Dordrecht, The Netherlands, 1999; pp. 45–123.
19. Indrzejczak, A. *Natural Deduction, Hybrid Systems and Modal Logics*; Springer: Berlin, Germany, 2010.
20. Urbański, M. *Rozumowania abdukcyjne. Modele i Procedury (Abductive Reasoning. Models and Procedures)*; Wydawnictwo Naukowe UAM: Poznań, Poland, 2009.
21. Wiśniewski, A. Erotetic Search Scenarios. *Synthese* **2003**, *134*, 389–427. [CrossRef]
22. Wiśniewski, A. *Questions, Inferences, and Scenarios*; Studies in Logic, Logic and Cognitive Systems; College Publications: London, UK, 2013; Volume 46.
23. Leszczyńska-Jasion, D. *Erotetic Search Scenarios as Families of Sequences and Erotetic Search Scenarios as Trees: Two Different, Yet Equal Accounts*; Technical Report 1(1); Institute of Psychology, Adam Mickiewicz University: Poznań, Poland, 2013.
24. Troelstra, A.S.; Schwichtenberg, H. *Basic Proof Theory*, 2nd ed.; Cambridge University Press: Cambridge, UK, 2000.
25. Negri, S.; von Plato, J. *Structural Proof Theory*; Cambridge University Press: Cambridge, UK, 2001.
26. Fitting, M. *First-Order Logic and Automated Theorem Proving*, 2nd ed.; Springer: Berlin, Germany, 1996.
27. Urbański, M. *First-order Synthetic Tableaux*; Technical Report; Institute of Psychology, Adam Mickiewicz University: Poznań, Poland, 2005.
28. Kalish, D.; Montague, R. *Logic. Techniques of Formal Reasoning*, 2nd ed.; Harcourt Brace Jovanovich College Publishers: Fort Worth, TX, USA, 1980.
29. Buss, S.R. Chapter I. "An Introduction to Proof Theory". In *Handbook of Proof Theory*; Buss, S.R., Ed.; Elsevier: Amsterdam, The Netherlands, 1998; pp. 1–78.
30. Cook, S.A.; Reckhow, R.A. On the lenght of proofs in the propositional calculus. In Proceedings of the 6th Annual Symposium on the Theory of Computing, Seattle, WA, USA, 30 April–2 May 1974; pp. 135–148.
31. Cook, S.A.; Reckhow, R.A. The Relative Efficiency of Propositional Proof Systems. *J. Symb. Log.* **1979**, *44*, 36–50. [CrossRef]
32. Smullyan, R.M. *First-Order Logic*; Springer: Berlin/Heidelberg, Germany; New York, NY, USA, 1968.
33. Negri, S.; von Plato, J. *Proof Analysis: A Contribution to Hilbert's Last Problem*; Cambridge University Press: Cambridge, UK, 2011.
34. Lettmann, M.P.; Peltier, N. A Tableaux Calculus for Reducing Proof Size. In Proceedings of the 9th International Joint Conference on Automated Reasoning, IJCAR 2018, Oxford, UK, 14–17 July 2018; pp. 64–80.

© 2019 by the authors. Licensee MDPI, Basel, Switzerland. This article is an open access article distributed under the terms and conditions of the Creative Commons Attribution (CC BY) license (http://creativecommons.org/licenses/by/4.0/).

Article

On Certain Axiomatizations of Arithmetic of Natural and Integer Numbers

Urszula Wybraniec-Skardowska

Department of Philosophy, Cardinal Stefan Wyszyński University in Warsaw, Wóycickiego 1/3, 01-938 Warsaw, Poland; skardowska@gmail.com; Tel.: +48-22-569-6801

Received: 1 July 2019; Accepted: 1 September 2019; Published: 4 September 2019

Abstract: The systems of arithmetic discussed in this work are non-elementary theories. In this paper, natural numbers are characterized axiomatically in two different ways. We begin by recalling the classical set *P* of axioms of Peano's arithmetic of natural numbers proposed in 1889 (including such primitive notions as: set of natural numbers, zero, successor of natural number) and compare it with the set *W* of axioms of this arithmetic (including the primitive notions like: set of natural numbers and relation of inequality) proposed by Witold Wilkosz, a Polish logician, philosopher and mathematician, in 1932. The axioms *W* are those of ordered sets without largest element, in which every non-empty set has a least element, and every set bounded from above has a greatest element. We show that *P* and *W* are equivalent and also that the systems of arithmetic based on *W* or on *P*, are categorical and consistent. There follows a set of intuitive axioms *PI* of integers arithmetic, modelled on *P* and proposed by B. Iwanuś, as well as a set of axioms *WI* of this arithmetic, modelled on the *W* axioms, *PI* and *WI* being also equivalent, categorical and consistent. We also discuss the problem of independence of sets of axioms, which were dealt with earlier.

Keywords: axiomatizations of arithmetic of natural and integers numbers; second-order theories; Peano's axioms; Wilkosz's axioms; axioms of integer arithmetic modeled on Peano and Wilkosz axioms; equivalent axiomatizations; metalogic; categoricity; independence; consistency

1. Introduction

The notion of natural numbers counts amongst the oldest, being one of the most universal abstract notions. Natural numbers belong to the fundamental subject of study of theoretical arithmetic concerned with defining all kinds of numbers, as well as studying their properties and relations between numbers of the same or different kinds. Theoretical arithmetic deals with examination of different types of numbers and their axiomatization, including that of natural numbers and integers. While defining its notions, we base ourselves on second-order logic and set theory.

In the paper, we will discuss some different axiomatizations of arithmetic of natural numbers **NA** and arithmetic of integer numbers **IA** (the presentation is based on results originally published in Polish by various authors, and which, as a consequence of their being available only in Polish, are not known among the vast majority of mathematical logicians). Presented theories will be non-elementary second-order theories and alphabet of languages which will assume two sorts of variables: individual variables and variables ranging over sets of individuals, i.e., natural numbers or integers, respectively.

We will start with the original axiomatization of **NA** proposed by Giuseppe Peano [1] by the set *P* of axioms on which is based the deductive system *PA* (the axiomatic non-elementary deductive theory; for short: the system *PA*) and will compare it with the little known axiomatization of the arithmetic **NA** by the set *W* of axioms, which was provided by Witold Wilkosz [2], a Polish logician, mathematician and philosopher of Kraków. The deductive system based on Wilkosz's set *W* of axioms will be denoted by *WA*.

Then, we will expand both sets *P* and *W* of axioms to the sets of axioms of arithmetic of integer numbers **IA**, which are modeled on them: the set of axioms *PI* by Iwanuś [3] and mine *WI* [4,5], which will be compared with each other and also with the set *SI* of axioms given by Sierpiński [6].

We will also give several metalogical theorems of the systems of arithmetic, which are presented.

2. Two Simple Axiomatizations of NA

2.1. Peano's Axioms for *PA*

Historically, the first axiomatic system of arithmetic of natural numbers, which is characterized by unique simplicity, was that presented by Italian mathematician Giuseppe Peano in 1889, in his book in Latin [1]. The essential ideas of axiomatization of NA were first published by Dedekind [7] in 1888. Peano's axioms specify the ideas, but there can be no doubt concerning the originality of Peano's work (see [8], p. 101). Peano's original formulation of the axioms assumes the following three primitive notions: *number (positive integer)* N, *unity* 1 and the *successor of a number*; the modern set memberships relation ∈ comes from Peano's relation ε (*is*) that he used in [1] (see [8], chapter VII). The most modern formulations of Peano's axioms use 0 as the "first" natural number instead of 1 and the set of all natural numbers as N. In this work the successor of a number is a unary function defined on natural numbers and denoted by the symbol *. In modern presentations, Peano's axioms are written using the symbolism of mathematical logic and set theory. They are axioms of a non-elementary theory of natural numbers, including set theory. Recollecting them, we use the convention that the individual variables m, n, k, l, ... ranging over the set N, while those of X, Y, Z ... over subsets of the set N. Peano's axioms of the system *PA* are the following:

P1. $0 \in N$,
P2. $n^* \in N$,
P3. $n^* \neq 0$,
P4. $m^* = n^* \Rightarrow m = n$,
P5. $0 \in X \land \forall k \in X (k^* \in X) \Rightarrow N \subseteq X$ (the induction principle).

Axioms P1–P4 are elementary ones, whereas axiom P5, called *induction axiom*, is an axiom of the 2nd order, a non-elementary one. In the first-order Peano arithmetic (elementary arithmetic) which is weaker than *PA* (see, e.g., [9], chapter II, section 7, chapter III, section 5 and [10], chapter 5), it is reformulated by the induction axiom schema. The induction axiom is applied in inductive proofs of theorems of the form T(n), where n denotes a natural number.

If T(n) with the free variable n is an expression of arithmetic **NA** and T(0) is its true expression and from the assumption that T(k) is true for $k \geq 0$ it follows that T(k*)—its truthfulness for number k*—then T(n) is true for any natural number of set N. Such proofs are based on the following schema of the rule of inductive proof of theorem T(n) for all n:

T(0)
T(k) ⇒ T(k*) for any $k \geq 0$
_____
T(n) for any $n \in N$

In proofs of theorems based on the set of axioms *P*, the following generalized theorems on induction are also made use of:

T1. $m \in X \land \forall k \in X (m \leq k \Rightarrow k^* \in X) \Rightarrow \forall n \in N (m \leq n \Rightarrow n \in X)$,
T2. $\forall m (\forall k < m (k \in X) \Rightarrow m \in X) \Rightarrow \forall n \in N (n \in X)$.

In compliance with T1, if any set of natural numbers to which m belongs satisfies the condition that for each number k of the set, which is not smaller than m, its successor k* also belongs to this set, then to this set belong all the natural numbers not smaller than m. With m = 0, T1 = P5. With theorem T1 corresponds the rule of inductive proof based on the following schema:

T(m) for m ∈ N
T(k) ⇒ T(k*) for k ≥ m
_____
T(n) for all n ≥ m, n ∈ N

In compliance with T2, if each natural number m satisfies the condition: if any number smaller than m belongs to a given set of natural numbers, then m also belongs to this set, then each natural number does. With theorem T2 corresponds the rule of inductive proof based on the following schema:

for each m ∈ N
T(k) ⇒ T(m) for any k < m
_____
T(n) for all n ∈ N

The proof of theorem T1 is based on the induction axiom P5 and the elementary theorems of system *PA*, whereas in the proof of theorem T2 the minimum principle and the elementary theorems of system *PA* are made use of. The former follows from the induction principle P5 and requires introducing additional definitions into system *PA*, including the definition of relation of less than, <, and that of non-greater than, ≤.

It can be proved that the induction principle P5, the maximum principle Pmax and the minimum principle Pmin are equivalent to one another on the basis of the elementary theorems of system *PA*, since the following relations of implication hold:

$$P5 \to Pmax \to Pmin \to T2 \to P5$$

where

Pmax. In any non-empty set of natural numbers for which there is an upper bound element, there is the greatest number. Symbolically:

$$\exists k\, (k \in X) \wedge \exists n \forall m \in X\, (m \leq n) \Rightarrow \exists n \in X \forall m \in X\, (m \leq n).$$

Pmin. In any non-empty set of natural numbers, there is the least number. Symbolically:

$$\exists k\, (k \in X) \Rightarrow \exists n \in X \forall m \in X\, (n \leq m).$$

Thus, we obtain the first metalogical theorem:

MT1. The principles P5, Pmax and Pmin are mutually equivalent on the basis of the elementary theorems of system *PA*.

Remark. *Each of these non-elementary expressions could then be the only non-elementary axiom of arithmetic of natural numbers NA if—from it and suitably selected elementary axioms—each elementary theorem follows (cf. Słupecki et al. [11] and Sierpiński [6]).*

The principles Pmax and Pmin are noted in system *PA* by means of relations of less than, <, or non-greater than, ≤, but the former is defined by means of the operation of addition +.

The definitions of the relation < and that of ≤ in system *PA* are the following:

D3. $m < n \Leftrightarrow \exists k \in N \setminus \{0\}\, (m + k = n)$,
D4. $m \leq n \Leftrightarrow m = n \vee m < n$.

The definitions of the operations of addition + and multiplication · are recursive in *PA*:

D1a. $m + 0 = 0$,
b. $m + n^* = (m + n)^*$.

D2a. $m \cdot 0 = 0$,
b. $m \cdot n^* = m \cdot n + m$.

These operations satisfy the well-known properties of a commutative semi-ring with unity ($1 = 0^*$), and it can be proved that the relation < (relation ≤) in *PA* well-orders set N (we differentiate two well-known notions of a relation ordering a set: strict ordering (<) a set and weak ordering (≤) a set).

The structure $< N, +, \cdot, 0, 1, \leq >$ is an ordered commutative semi-ring.

2.2. Wilkosz's Axioms for System *WA*

The primitive notions in Wilkosz's axiomatic system *WA* are: the set of natural numbers N and the relation of less than, <. We write Wilkosz's axioms, accepting that the variables m, n, l, k, ... run over set N, X is a subset of N.

W1. $\exists n\ (n \in N)$—there is a natural number,
W2. $m \neq n \Rightarrow m < n \vee n < m$—trichotomy,
W3. $(m < n \Rightarrow \sim (n < m))$—anti-symmetry of relation <,
W4. $m < n \wedge n < k \Rightarrow m < k$—transitivity of relation <,
W5. $m < n \Rightarrow m, n \in N$—the field of relation < is set N,
W6. $\exists k\ (k \in X) \Rightarrow \exists n \in X \forall m \in X\ (n \leq m)$—the minimum principle,
W7. $\exists k\ (k \in X) \wedge \exists n \forall m \in X\ (m \leq n) \Rightarrow \exists n \in X \forall m \in X\ (m \leq n)$—the maximum principle,
W8. $\sim \exists m \forall n\ (m \neq n \Rightarrow n < m)$—there is not the greatest number in set N.

It is easy to see that in system *WA*, the relation < well-orders set N (in the sense of strict order).

Relation <—a primitive notion in Wilkosz's system *WA*—is a notion defined in Peano's system *PA* (see D3), and the primitive notions of system *PA*, which are not primitive ones in Wilkosz's system *WA*, are defined in it in the following way:

(1) $k = 0 \Leftrightarrow \forall n\ (k \leq n)$—0 is the least natural number,
(2) $k = n^* \Leftrightarrow k \in \{m \in N \mid n < m\} \wedge \forall i \in \{m \in N \mid n < m\}(k \leq i)$—$n^*$ is the least natural number among numbers greater than n.

Relation ≤ less than or equal (not greater) has the following definition:
$m \leq n \Leftrightarrow m = n \vee m < n$.

It can be proved that the definitions (1) and (2) are correct: there is precisely one natural number k satisfying the definiens of definition (1) and there is precisely one number k (the successor of number n), which satisfies the definiens of definition (2). In the proofs the axioms W2–W4 are used.

Relying on, in Wilkosz's system *WA*, the definitions of zero and the successor function of a natural number, we can define the operations of addition + and multiplication, in the same way as in system *PA* (by means of definitions D1a,b and D2a,b).

2.3. Equivalence of the Deductive Systems *PA* and *WA*

It needs reminding that, in accordance with Tarski's inferential definition of two equivalent sets of sentences of a deductive system (see [12]), two sets of sentences are *equivalent* if sets of all their consequences (deduced from them sentences) are equal. Thus, most often, for the equivalence of axiomatic deductive system the following definition (cf. [13,14]) is used:

(*) Two axiomatic deductive systems are *equivalent* if the set of axioms and definitions of one of them is *equivalent* to the set of axioms and definitions of the other system, i.e., if each axiom and definition of one of them is a theorem or definition of the other system and the other way round—each axiom and definition of the other system is a theorem or definition of the first system.

Let us note that

(i1) Axioms of system *WA* are theorems of system *PA* since W1 follows directly from P1; W8 follows from the fact that in system *PA* there holds the theorem that n < n* for any n ∈ N, and n* ∈ N (P2); W3–W5 in *PA* follow from the theorem that N is a set ordered by the relation <; W6 and W7 (principles minimum and maximum, respectively) follow from the induction axiom P5 (see MT1).

(i2) Definitions (1) and (2) of zero and the successor of a natural number in *WA* are theorems in system *PA*.

(i3) Definitions of the operations addition and multiplication in system *WA* are the same as in system *PA*.

(i4) Each axiom and definition of system *WA* is a theorem or definition in system *PA*.

On the other hand

(j1) Axioms of system *PA* are theorems of system *WA*, since from the correctness of definitions (1) and (2) in *WA*, in particular axioms P1 and P2 follow; also P3 is a theorem in *WA*, because if it would be possible that n* = 0, then it would follow from (1) that n* ≤ n and from (2) that n < n*, and hence that 0 ≤ n and n < 0, that is on the basis of W3: ~0 < n and n = 0, that is 0 < 0, which leads to contradiction according to W3; next, P4 follows from (2) and from the property of relation <, as one ordering set N. Axiom P5—the induction principle follows from those of maximum and minimum (W7 and W6; see MT1).

(j2) Definition D3 of relation < in *PA* can be derived from axioms and definitions of system *WA*.

(j3) Each axiom and definition of system *PA* is a theorem of system *WA*.

From (i4) and (j3), in compliance with (*), we obtain the following metatheorem:

MT2. The systems of *PA* and *WA* are equivalent.

This equivalence was sketched in the booklet by Wilkosz [2] under the title *Arytmetyka liczb całkowitych* (The Arithmetic of Integers). Equivalence of Wilkosz's and Peano's systems was the subject of my MA thesis.

2.4. Independence of Axioms in Systems *PA* and *WA*

As is well known from Gödel's first incompleteness theorem given in 1931 [15], no finite set of axioms of natural numbers is complete, or even each infinite, countable set of axioms of arithmetic is incomplete. There arises the problem, however, whether it is possible to reduce the number of axioms of *PA* and *WA* without depleting the set of theorems which can be proved about natural numbers.

It can be shown that

MT3a. The set of axioms of *PA* arithmetic system is independent (Sierpiński [6]).

b. The set of axioms of *WA* system is dependent and can be reduced to the set:

{W1, W3, W5, W6, W7, W8}.

Axiom W2 follows from axiom W6, while axiom W4 follows from axioms W3, W5 and W6.

Hence, it follows that the axioms of the theory of well-ordered sets in regard to relation < can be based on axioms W3, W5 and W6, while Wilkosz's system *WA* can be based on the axioms:

W1'. ∃n (n∈ N)—there is a natural number,
W2.' ∀m∃n (m ≤ n)—in set N there is not the greatest number,
W3.' ∀m∀n (m < n ⇔ ~ (n < m))—asymmetry of relation < in N,
W4.' ∃k (k ∈ X) ⇒ ∃n∈X∀m∈X (n ≤ m)—the minimum principle,
W5.' ∃k (k ∈ X) ∧ ∃n∀m∈X (m ≤ n) ⇒ ∃n∈X∀m∈X (m ≤ n)—the maximum principle.

To prove independence of the axioms one can, as it is known, use the method of interpretation, which consists in finding such an interpretation of primitive terms of the given system that makes all the axioms, apart from one, e.g., Ai, true at the interpretation. If we find it, then the given axiom Ai is independent from the others.

2.5. Categoricity of Arithmetic Systems **PA** and **WA**

Let us recall the definition of the notion of categoricity (see, e.g., [16–18]):

(**) A deductive system is *categorical* if and only if all its models are isomorphic.

As we mentioned, **PA** and **WA** as second-order, non-elementary systems, as well as elementary Peanos arithmetic, are not complete, yet we can show that they are categorical (cf. [7,10,18,19], chapter 8).

A *model* of Peano's arithmetic system **PA** is each triple $<N, 0, S>$ assigned to the triple $<N, 0, *>$ of primitive terms of system **PA**, where N is an infinite set, $0 \in N$, and S: $N \to N$, which satisfies Peano's axioms P1–P5.

A *model* of Wilkosz's arithmetic system **WA** is each tuple $<N, \prec>$ assigned to the tuple $<N, <>$ of primitive terms of **WA** system, where N is an infinite, countable set and $\prec$ a binary relation with the field N, which satisfies Wilkosz's axioms W1–W8 (W1'–W5').

(m1) Two models of **PA**: $P_1 = <N_1, 0_1, S_1>$ and $P_2 = <N_2, 0_2, S_2>$ are isomorphic if and only if there is bijection f: $N_1 \to N_2$ such that f is homomorphism from P_1 to P_2, that is $f(0_1) = 0_2$ and $f(S_1(m)) = S_2(f(m))$ for any $m \in N_1$.

(m2) Two models of **WA**: $W_1 = <N_1, \prec_1>$ and $W_2 = <N_2, \prec_2>$ are isomorphic if and only if there is bijection f: $N_1 \to N_2$ being homomorphism from W_1 to W_2, that is $m \prec_1 n \Rightarrow f(m) \prec_2 f(n)$ for any $m, n \in N_1$. Dedekind already in [7] proved that

(m3) Each two models of arithmetic system **PA** are isomorphic. In the book by Słupecki et al. [11], there is a proof that

(m4) Each two models of arithmetic system **WA** are isomorphic.

Hence, we have the metalogic corollary:

MT4. The deductive systems **PA** and **WA** of natural numbers arithmetic **NA** are categorical; they are in power $\aleph_0$, so they are aleph-null categorical systems.

Thus, Peano's and Wilkosz's second-order systems have only one model, up to isomorphism.

This is not so when we consider the systems of arithmetic of natural numbers as systems (elementary theories) of the first-order. According to the upward Löwenheim–Skolem's theorem, there are non-standard models of Peano's elementary arithmetic system of all infinite cardinality (see e.g., [9], chapter III, section 5, [20], chapter VI).

2.6. Set-Theoretical Models for **PA** and **WA**

Peano's arithmetic possesses a "natural" set-theoretical model deriving from Frege.

Let $\mathbb{N}$ be an infinite set of all cardinal numbers of finite subsets of any (infinite) set U, i.e.,

$$\mathbb{N} = \{card(X) \mid X \in Fin(U)\},$$

where $Fin(U)$ is the smallest family of sets to which the empty set $\emptyset$ belongs and which is closed under the relation S:

$$X \, S \, Y \Leftrightarrow \exists x \in U \setminus X \, (Y = X \cup \{x\}) \text{ for any } X, Y \in Fin(U).$$

The formal definition of the set $Fin(U)$ is the following:

$$Fin(U) = \cap\{A \subseteq P(U) \mid \emptyset \in A \wedge \forall X \in A \, \exists Y \in A \, (X \, S \, Y \Rightarrow Y \in A)\}.$$

(mP) The set-theoretical model for **PA** is the triple $<\mathbb{N}, card(\emptyset), S^*>$,
where for $m = card(X)$ and $X \, S \, Y$, $S^*(m) = m + 1 = card(Y)$, for $X, Y \in Fin(U)$.

(mW) The set-theoretical model for **WA** is the triple $<\mathbb{N}, <>$,
where $<$ is the relation of less than for the cardinal numbers of set $\mathbb{N}$:

$$m < n \Leftrightarrow m \leq n \wedge n \neq m,$$

$m \leq n \Leftrightarrow \exists X, Y \, (\text{card}(X) = m \wedge \text{card}(Y) = n \wedge \exists Z \subseteq Y (\text{card}(Z) = m)).$

From (mP) and (mW) we get two next metalogic corollaries:

MT5. *PA* and *WA* systems are consistent.

(since it follows from the theorem of categoricity (MT4) that all theorems of these systems are true, because they are true in each model of these systems).

MT6. Systems *PA* and *WA* are (treated as) fragments of set theory.

As we know, the theorem MT6 is of great importance to studies on the foundations of mathematics.

3. Simple Axiomatizations of Arithmetic of Integers, Based on Systems *PA* and *WA*

Axiomatic systems for integer arithmetic **IA** are most often based on notions of operations of addition and multiplication defined on the set I of integers. In this part of the work, we will give an axiomatization of integer arithmetic **IA** modelled on the systems *PA* and *WA* respectively for the arithmetic of natural numbers **NA**, extending these systems accordingly and comparing them with Sierpiński's system *SIA* [6], including addition and multiplication as its primitive notions.

3.1. Iwanuś's Axioms for IA, Modelled on the Axioms of System PA

We will give here two systems of axioms proposed by Bolesław Iwanuś [3] for **IA** system. They are interesting due to their intuitive character. The first system based on them will be denoted as P^1IA, and the other one—P^2IA. The primitive notions of P^1IA are: set N* of all non-negative integers, set *N of all non-positive numbers, integer 0 and two unary operations in N*∪*N of successor and predecessor of an integer. The successor and the predecessor of integer i will be denoted as i* and *i, respectively. In the intuitive meaning, $i^* = i + 1$ and $^*i = i - 1$.

We assume that i, j, k, l, ... are variables ranging over the set N*∪*N, while variables A, B, C ... range over the subsets of this set.

3.1.1. Axioms of System P^1IA Are the Symmetric Axioms for Numbers of the Sets N* and *N:

A*1. $0 \in N^*$,   *A1. $0 \in {}^*N$,
A*2. $i \in N^* \Rightarrow i^* \in N^*$,   *A2. $i \in {}^*N \Rightarrow {}^*i \in {}^*N$,
A*3. $i \in N^* \Rightarrow i^* \neq 0$,   *A3. $i \in {}^*N \Rightarrow {}^*i \neq 0$,
A*4. $0 \in A \wedge \forall i \in A \, (i^* \in A) \Rightarrow N^* \subseteq A$,   *A4. $0 \in A \wedge \forall i \in A \, ({}^*i \in A) \Rightarrow {}^*N \subseteq A$.
A5. $i \in N^* \cup {}^*N \Rightarrow {}^*(i^*) = i = ({}^*i)^*$.

Axioms A*1–A*3 and *A1–*A3 are modelled on those of Peano (P1–P3). Axioms A*4 and *A4 correspond to that of induction P5. Axiom A5 establishes relations between the successor and the predecessor operation and does not allow identification of these notions with each other, nor identification of sets N* and *N. The set I of all integers is defined as follows:

D0. $I = N^* \cup {}^*N$.

The counterparts of Peano's axiom P4

$i, j \in N^* \wedge i^* = j^* \Rightarrow i = j$ and $i, j \in {}^*N \wedge {}^*i = {}^*j \Rightarrow i = j$

are direct consequences of A5.

It is easy to notice that with the assumption that the set I of integers is a primitive notion of the system of arithmetic **IA**, the symmetrical axioms of P^1IA can be replaced by weaker ones, deriving from Słupecki in [11]:

A1. $0 \in I$,
A2. $i \in I \Rightarrow i^*, {}^*i \in I$,
A3. $i \in I \Rightarrow i^* \neq i$,
A4. $A \subseteq I \wedge 0 \in A \wedge \forall i \in A \, (i^*, {}^*i \in A) \Rightarrow I = A$,
A5. $i \in I \Rightarrow ({}^*i)^* = {}^*(i^*) = i$,

In this system, there are theorems that, to a certain extent, are similar to Peano's axiom A3, which are in force:

$(i \in I \wedge i \neq {}^*0) \Rightarrow i^* \neq 0$,
$(i \in I \wedge i \neq 0^*) \Rightarrow {}^*i \neq 0$.

In Iwanuś's system P^1IA, there are the following definitions of the operations of: addition +, subtraction − and multiplication:

D^I1 a. $i + 0 = i$,
    b. $i + j^* = (i + j)^*$,
    c. $i + {}^*j = {}^*(i + j)$,

D^I2 a. $i − 0 = i$,
    b. $i − j^* = {}^*(i − j)$,
    c. $i − {}^*j = (i − j)^*$,

D^I3a. $i \cdot 0 = i$,
    b. $i \cdot j^* = i \cdot j + i$,
    c. $i \cdot {}^*j = i \cdot j − i$.

It is assumed that $1 = 0^*$ and it is proved that

I1. ${}^*N = (I − N^*) \cup \{0\}$,
I3. $i^* = i + 1$,

I2. $k = j − i \Leftrightarrow i + k = j$,
I4. ${}^*i = i − 1$.

In proofs of the theorems of system P^1IA the following meta-theorem is made use of:

MT7. If α is an expression of system P^1IA, in which—beside primitive notions—there are exclusively the defined terms + and ·, then α is a theorem of this system if expression α^d, dual with respect to α, is a thesis of this system; expression α^d is *dual* to α, when the terms:

$$N^*, {}^*N, (\ )^*, {}^*(\ ), +, \cdot$$

which occur in it, are substituted in each place of their appearance with the following ones, respectively:

$${}^*N, N^*, {}^*(\ ), (\ )^*, +, \cdot$$

In proofs of theorems on the basis of axioms A*4 and *A4, the following rules of mathematical induction for integers based on the given below schemata are applied:

T(0)
T(k)⇒T(k*) for any k ≥ 0
―――――――――――
T(i) for any i ∈ N*

T(0)
T(k)⇒ T(k*) ∧ T(*k) for any k
―――――――――――
T(i) for any i ∈ I

Remark 1. *On the basis of system P^1IA one can prove all the axioms of the commutative ring.*

The inequality relation less-than, <, in I is determined by the following definition added to P^1IA:

D^I4. $i < j \Leftrightarrow \exists k \in N^*\setminus\{0\}\ (i + k = j)$.

Remark 2. *In system P^1IA, one can prove all the theorems of arithmetic of integers IA relating to relation* <.

3.1.2. The Other System of Arithmetic of Integers Built by Iwanuś [3] and Modelled on *System PA*

The system is denoted by P^2IA and based only on the following three primitive notions: set I of all integers, the function of successor * and number 0.
The following formulas are the axioms of system P^2IA:

(I1) $i \in I \Rightarrow \exists j \in I\ (i = j^*)$,
(I2) $i, j \in I \wedge i^* = j^* \Rightarrow i = j$,
(I3) $\exists A \subseteq I\ (0 \in A \wedge \forall i \in A\ (i^* \in A \wedge i^* \neq 0))$,
(I4) $A \subseteq I \wedge 0 \in A \wedge \forall i \in A\ (i^* \in A \wedge \exists j \in A\ (i = j^*)) \Rightarrow I \subseteq A$.

Axiom I3 assumes the existence of a certain subset of set I, about which—on the base of the above accepted set of axioms—it can be proved that it is isomorphic due to function * to the set of all natural numbers. Axiom I4 is a postulate of induction in the set of integers.

If we introduce into system P^2IA still one more primitive term N (as a name of a subset of set I which is isomorphic to the set of natural numbers), then axiom I3 can be substituted with the following set of axioms:

I3a. $N \subseteq I$,
  b. $0 \in N$,
  c. $i \in N \Rightarrow i^* \in N$,
  d. $i \in N \Rightarrow i^* \neq 0$.

Axiom I3 is weaker than axioms I3a–d, because I3 follows from these axioms, although not all of I3a–d follow from I3.

In P^2IA system the primitive notions of P^1IA system are defined in the following way:

D^I1'. $i, j \in I \Rightarrow (*i = j \Leftrightarrow i = j^*)$,
D^I2'. $i \in N^* \Leftrightarrow \forall A \subseteq I\, (0 \in A \wedge \forall j \in A\, (j^* \in A)) \Rightarrow i \in A$,
D^I3'. $i \in *N \Leftrightarrow \forall A \subseteq I\, (0 \in A \wedge \forall j \in A\, (*j \in A)) \Rightarrow i \in A$.

All the remaining definitions of system P^1IA are the same in system P^2IA.

Iwanuś proves that

MT8. Systems P^1IA and P^2IA are equivalent.

B. Iwanuś also proves in [3] that these systems are equivalent to Sierpiński's system of arithmetic of integers SIA [6], based on primitive notions: the set I, operations of addition + and multiplication ·, zero 0, one 1 and the set N^*, satisfying the axioms of the ring without zero divisors:

R1. $i, j \in I \Rightarrow i + j \in I \wedge i \cdot j \in I$,
R2. $i, j, k \in I \Rightarrow i + j = j + i \wedge i \cdot j = j \cdot i \wedge (i + j) + k = i + (j + k) \wedge (i \cdot j) \cdot k = i \cdot (j \cdot k) \wedge i \cdot (j + k) = i \cdot j + i \cdot k$,
R3. $\forall i, j \in I\, \exists k \in I (i + k = j)$,
R4. $\forall i \in I\, (i + 0 = i) \wedge \forall i \in I\, (i \cdot 1 = i) \wedge 1 \in I$,
R5. $i, j \in I \wedge i \cdot j = 0 \Rightarrow i = 0 \vee j = 0$,
R6. $N^* \subset I$, R7. $0 \in N^*$, R8. $i \in N^* \Rightarrow i + 1 \in N^*$,
R9. $0 \in A \wedge \forall i \in A\, (i^* \in A) \Rightarrow N^* \subseteq A$,
R10. $\forall i \in A \setminus N^*\, \exists j \in N^* (i + j = 0)$.

Definitions of primitive terms of system P^1IA are introduced into system SIA as follows:

D^S1. $*N = (I \setminus N^*) \cup \{0\}$,
D^S2. $k = j - i \Leftrightarrow i + k = j$,
D^S3. $i^* = i + 1$,
D^S4. $*i = i - 1$.

Definition D^I4 of relation < of system P^1IA is the same as in system SIA.

MT9. System P^1IA (P^2IA system), modelled on Peano's system of natural numbers arithmetic PA, and system SIA are equivalent.

3.2. Axioms of the System of Integer Arithmetic WIA Modelled on Wilkosz's System WA

The primitive notions of the system of integer arithmetic WIA, modelled on Wilkosz's system WA, are the following: set I of all integers, integer zero 0 and less-than relation < in set I. The relation of weak inequality ≤ is determined by the definition (i, j, k, … run over I):

D0. $i \leq j \Leftrightarrow i < j \vee i = j$.

The axioms of system WIA which are presented by Wybraniec-Skardowska [4,5] are the following expressions:

W'1. $0 \in I$,
W'2. $i, j \in I \Rightarrow (i < j \vee i = j \vee j < i)$,

W'3. $i, j \in I \Rightarrow (i < j \Rightarrow \sim (j < i))$,
W'4. $i, j, k \in I \land (i < j \land j < k) \Rightarrow i < k$,
W'5. $\forall i \in I \; \exists j \in I \; (i < j)$ – in I there is not the greatest number,
W'6. $\forall i \in I \; \exists j \in I \; (j < i)$ – in I there is not the smallest number,
W'7. $A \subseteq I \land \exists i \in A \; \exists i \in I \; \forall j \in A \; (i < j) \Rightarrow \exists i \in A \forall j \in A \; (i \leq j)$,
W'8. $A \subseteq I \land \exists i \in A \; \exists i \in I \; \forall j \in A \; (j < i) \Rightarrow \exists i \in A \forall j \in A \; (j \leq i)$.

According to W'7, in each non-empty set of integers, which has a lower bound, there is the smallest number, while, according to W'8, in each non-empty set of integers, which has an upper bound, there exists the greatest number.

The content of axioms W'7 and W'8 is close to the principles of minimum and maximum of arithmetic *WA*. The axioms of system *WIA* state that relation < orders set I, yet do not state that it well-orders the set.

In system *WIA* one can define the notion of successor and that of predecessor of an integer as well as the notions of sets N* and *N, which are primitive notions in Iwanuś's system P^1IA. Let us note first that in system *WIA* it is possible to prove the theorem:

$$A \subseteq I \land \exists i \in A \land \exists i \in I \forall j \in A \; (i < j) \Rightarrow \exists_1 k \in A \forall j \in A \; (k \leq j). \qquad (1)$$

Condition (1) allows introducing correctly the definition of minimum in set A:
$D^W 1$. $A \subseteq I \land \exists i \in A \land \exists i \in I \forall j \in A \; (i \leq j) \Rightarrow (k = \min(A) \Leftrightarrow k \in A \land \forall j \in A \; (k \leq j))$.

It follows from Condition (1) and $D^W 1$ that there is a unique minimum, min (A), when $A \subseteq I$, $A \neq \emptyset$ and set A has a lower bound.

Let
$$G(i) = \{j \in I \mid i < j\}.$$

Hence $G(i) \neq \emptyset$ (see W'5) and $\exists_1 k \in G(i) \forall j \in G(i) \; (k \leq j)$ (see Condition (1)); then on the basis of $D^W 1$

$$\exists_1 k \in I \; (k = \min(G(i)). \qquad (2)$$

The successor of an integer i is introduced by means of the definition:
$D^W 2$. $i^* = \min(G(i))$ – i^* is the smallest integer which is greater than i.
$D^W 3$. $N^* = \{i \in I \mid 0 \leq i\}$.

The following corollary which is dual to Condition (1):

$$A \subseteq I \land \exists i \in A \land \exists i \in I \forall j \in A \; (j < i) \Rightarrow \exists_1 k \in A \forall j \in A \; (j \leq k). \qquad (3)$$

permits introducing the definition of maximum of a certain set of integers:
$D^W 4$. $A \subseteq I \land \exists i \in A \land \exists i \in I \forall j \in A \; (j \leq i) \Rightarrow (k = \max(A) \Leftrightarrow k \in A \land \forall j \in A \; (j \leq k))$.
Let $L(i) = \{j \in I \mid j < i\}$.

Hence $L(i) \neq \emptyset$ (see W'6) and $\exists_1 k \in L(i) \forall j \in L(i) \; (j \leq k)$ (see Condition (3)) then on the basis of $D^W 4$ the predecessor of integer i is defined as the greatest integer less than i, that is

$D^W 5$. $*i = \max(L(i))$,
and set *N is defined as follows
$D^W 6$. $*N = \{i \in I \mid i \leq 0\}$.

3.3. Equivalence of Systems P^1IA and WIA

Remark 3. *With the definitions of the primitive notions of system P^1IA, given in system WIA, all the axioms and definitions P^1IA become theorems of definitions in system WIA.*

The definitions of addition and multiplication, which are accepted in P^1IA are the same in system WIA, while definition D^I4 of relation < accepted in system P^1IA is a theorem in system WIA.

Thus, it follows from Remarks 3 and 2 and **MT8** that

MT10. System WIA is equivalent to those of Iwanuś P^1IA and P^2IA.

It follows from the above and **MT9** that

MT11. All the systems of integer arithmetic: WIA, P^1IA, P^2IA and SIA are mutually equivalent.

In particular,

MT12. System P^1IA modelled on Peano's system PA and system WIA modelled on Wilkosz's system WA are equivalent.

3.4. Independence of the Axioms in P^1IA and WIA

The axioms of the integer arithmetic system P^1IA can, as Iwanuś proved, be reduced by one axiom A*3 or *A3. If we found an axiom system of P^1IA on those of A*1–A*4 and *A1, *A2 and *A4, then axiom *A3 can be proved. It follows from A*1, A*2, and A5 as well as from theorems of this system: *0 $\notin$ N* and i $\in$N* $\Rightarrow$ (i $\notin$ *N $\vee$ i = 0).

MT13. The set of axioms of system P^1IA can be based on an independent set of axioms A*1–A*4 and *A1, *A2, *A4, and A5.

The independence of these axioms was proved by interpretation in integer arithmetic **IA**. The primitive terms of the tuple <N*, *N, i*, *i, 0> correspond to the elements of a tuple in the form <A, B, f(i), G(i), a^0>, respectively, which does not satisfy only one axiom of P^1IA. When we apply the denotation:

"$\mathbb{N}^+$" denotes a set of non-negative integers,

"$\mathbb{N}^-$" denotes a set of non-positive integers,

"E^+" denotes a set of even non-negative integers;

"E^-" denotes a set of even non-positive integers, then the tuple:

<$\mathbb{N}^+ \setminus \{0\}$, $\mathbb{N}^-$, i + 1, i − 1, 0> does not satisfy A*1,

<$\mathbb{N}^+$, $\mathbb{N}^- \setminus \{0\}$, i + 1, i − 1, 0> does not satisfy *A1,

<$\mathbb{N}^+ \setminus \{1\}$, $\mathbb{N}^-$, i + 1, i - 1, 0> does not satisfy A*2,

<$\mathbb{N}^+$, $\mathbb{N}^- \setminus \{-1\}$, i +1, i - 1, 0> does not satisfy *A2,

<{0, 1}, {0, 1}, $f_1(i)$, $g_1(i)$, 0>, where $f_1(i) = g_1(i) = \begin{cases} 1 \text{ for } i=0 \\ 0 \text{ for } i \neq 0 \end{cases}$ does not satisfy A*3,

<$\mathbb{N}^+$, E^-, i + 2, i − 2, 0> does not satisfy A*4,

<E^+, $\mathbb{N}^-$, i + 2, i − 2, 0> does not satisfy *A4,

<{0, 1}, $\mathbb{N}^-$, $f_2(i)$, i − 1, 0>, where $f_2(i) = \begin{cases} i+1 \text{ for } i \leq 0 \\ 1 \text{ for } i > 0 \end{cases}$ does not satisfy A5a,

<$\mathbb{N}^+$, {0, 1}, i + 1, $g_2(i)$, 0>, where $g_2(i) = \begin{cases} i-1 \text{ for } 0 \leq i \\ -1 \text{ for } i < 0 \end{cases}$ does not satisfy A5b,

when A5 is substituted by two axioms:

A5a. i $\in$ N* $\cup$ *N $\Rightarrow$ *(i*) = I; A5b. i $\in$ N* $\cup$ *N $\Rightarrow$ (*i)* = i.

It is also possible to reduce the system of the primitive notions of P^1IA system by one primitive notion—zero 0—since the following expression:

i = 0 $\Leftrightarrow$ i $\in$ N* $\wedge$ i $\in$*N

is a theorem of P^1IA.

On the other hand,

MT14. The set of axioms I1—I4 of Iwanuś's P^2IA system is an independent set.

It is so, since applying the following interpretation:

1. I $\to \mathbb{N}^+$, i* $\to |i| + 1$, 0 $\to$ 0, − I1 is not satisfied,
2. I $\to E^+ \cup \{1\}$, i* $\to$ i + 1, if i $\neq$ 0, and i* $\to$ 1, for i = 1, 0 $\to$ 0 − I2 is not satisfied,
3. I $\to \{0,1\}$, i* $\to$ 1 - |i|, 0 $\to$ 0, − I3 is not satisfied,

4. I → set of integers $\mathbb{Z}$, i* → i + 2, 0 → 0, − I4 is not satisfied.

It can also be justified that

MT15. The set of axioms of *WIA* system is an independent set.

3.5. Categoricity of the Axiomatic Systems of Integers Arithmetic IA

The classical model of P^2IA system is the triple $<\mathbb{Z}, *, 0>$, where $\mathbb{Z}$ is the set of all integers. The classical model of *WIA* system is the triple $<\mathbb{Z}, 0, <>$.

It can be proved (cf. [11]) that

MT16. Every two models of *WIA* system are isomorphic, therefore *WIA* system is categorical in power $\aleph_0$.

A model of *WIA* system is every triple $<\vartheta, 0, <>$ corresponding to that of $<I, 0, <>$ of the primitive terms of *WIA*, in which ϑ is an infinite set of cardinality $\aleph_0$, $0 \in \vartheta$, and $<$ is a binary relation satisfying axioms A'1–A'8 of *WIA* system.

The following theorem is true:

If an axiomatic system has the property that all its models are isomorphic, then each equivalent system has the same property.

Thus, from meta-theorems MT11 and MT16 follows the conclusion:

MT17. All the systems of integer arithmetic, which are presented in this work, are categorical in power of $\aleph_0$.

Thus, it is not only system *WIA* modelled on Wilkosz's system *WA* which is categorical, but also Iwanuś's system P^1IA (P^2IA) modelled on Peano's system *PA* is categorical.

A separate proof that system *SIA* is also categorical is given in the book by Sierpiński [6].

All deductive systems of integer arithmetic presented in this paper have a standard model and all their models are isomorphic (MT17), so all the theorems of these systems are true. Hence, it follows that

MT18. The systems P^1IA, P^2IA, *SLA* and *WIA* of integer arithmetic are consistent.

All these systems are mutually equivalent.

4. Final Comments

➢ Theorems of categoricity of the systems of natural numbers and the integer systems answer—in a sense—the following question: To what extent do our axioms characterize natural numbers (respectively, integers)? It follows from them that each set which has properties expressed in our axioms is the same as the set of natural numbers (resp., integers), that is it is isomorphic.

The axioms given for systems *PA* and *WA* as well as, respectively, P^1IA (P^2IA) and *WIA*, characterize very strongly natural numbers (respectively, integers).

➢ It follows from the given considerations that from the point of view of set theory, the set of axioms of integer arithmetic systems P^1IA and P^2IA, modelled on Peano's axioms of system *PA* of natural numbers arithmetic, and the set of axioms of system *WIA* modeled on Wilkosz's axioms of system *WA* of natural numbers arithmetic, have equal rights, similarly as the set of axioms of systems *PA* and *WA*. The subject of the discussion can be—as it may seem—solely one problem: Which of the set of axioms is more intuitive or more useful in the didactic process?

Wilkosz's system *WA* and system *WIA* of the similar axiomatic character seem to be of certain greater value. The former (*WA*) can be acknowledged to arise as a result of studies of the natural model—one that forms the primary study of teaching in the early years of elementary school. As it appears, though, the problem of which of the systems discussed here can play its role in a better way as the curriculum of early education may be settled exclusively through psycho-sociological research in schools.

➤ Let us also note that the built systems of integer arithmetic, modelled on the systems of arithmetic of natural numbers of Peano and Wilkosz, were treated as respective extensions of the latter, since the set of natural numbers N in Peano's system *PA*, with the function of successor * and zero 0 ∈ N, is isomorphic with a proper subset of set I of integers in the system of integer arithmetic P^1IA, that is to set N*⊂ I, function * in N* and zero 0 ∈ N*, whereas the set of natural numbers N in Wilkosz's system of natural numbers arithmetic *WA*, with zero 0 ∈ N and relation less-than < in N, is isomorphic with the proper subset N* ⊂ I in the system of integers arithmetic *WIA*, with zero 0 ∈ N* and relation < in N*.

➤ It follows from the remark above that arithmetic of integers can be defined not only through giving a set of axioms, but as an extension of arithmetic of natural numbers by the well-known method of construction, as well.

➤ So, it follows from MT6 that both integer systems P^1IA (P^2IA) and *WIA* can be treated as fragments of set theory.

Funding: This research received no external funding.

Acknowledgments: I would like to express my utmost gratitude to all referees of this article for all remarks, comments, suggestions and efforts to accomplish improvement of the content of the paper and also English-language verification of its text. Individual thanks are due to Luna Shen for help in all electronic correspondence.

Conflicts of Interest: The author declares no conflict of interest.

References

1. Peano, G. *Arithmetices Principia Nova Methodo Exposito*; Bocca: Turin, Italy, 1889.
2. Wilkosz, W. *Arytmetyka Liczb Całkowitych/System aksjomatyczny/(Arithmetic of Natural Numbers/The Axiomatic System/)*; Biblioteczka Kółka Mat. i Fiz. UJ. Nr 1: Kraków, Poland, 1932.
3. Iwanuś, B. *O Pewnych Aksjomatykach Arytmetyki Liczb Całkowitych (Some Axiomatic Systems of Integers Arithmetic)*; Teacher's Training College Press: Opole, Poland, 1965; Matematyka; Volume 4, pp. 23–61.
4. Wybraniec, U. *Pewien Układ Aksjomatów Arytmetyki Liczb Całkowitych (A Certain System of Axioms in the Integer Arithmetic)*; Teacher's Training College Press: Opole, Poland, 1965; Matematyka; Volume 4, pp. 63–83.
5. Wybraniec-Skardowska, U. O aksjomatyce Wilkosza arytmetyki liczb naturalnych (On Wilkosz's axiomatization of natural number arithmetic). In *Logika i jej Nauczanie w Dziejach Uniwersytetu Jagiellońskiego (Logic and Its Teaching in the History of the Jagiellonian University)*; Jagiellonian University Press: Kraków, Poland, 1980; pp. 41–50.
6. Sierpiński, W. *Arytmetyka Teoretyczna (Theoretical Arithmetic)*; PWN: Warszawa, Poland, 1969.
7. Dedekind, R. *Was Sind und Was Sollen Die Zalen (What Are and What Should the Numbers Be?)*; Friedr. Vieweg & Sohn, Braunschweig: Berlin, Germany, 1888.
8. Peano, G. The principles of arithmetic, presented by a new method (1889). In *Selected Works of Giuseppe Peano*; Kennedy, H.C., Ed.; trans., biographical sketch, bibliogr.; George Allen & Unwin LTD: London, UK, 1973; Chapter VII; pp. 102–134.
9. Grzegorczyk, A. *An Outline of Mathematical Logic*; PWN: Warsaw, Poland, 1981.
10. De Swart, H. *Philosophical and Mathematical Logic*; Springer: Cham, Switzerland, 2018.
11. Słupecki, J.; Hałkowska, K.; Piróg-Rzepecka, K. *Elementy Arytmetyki Teoretycznej (Elements of Theoretical Arithmetic)*; Wydawnictwa Szkolne i Pedagogiczne: Warszawa, Poland, 1980.
12. Tarski, A. Fundamental concepts of the methodology of deductive sciences. In *Logic, Semantics, Metamathematics Papers from 1929 to 1938*; Tarski, A., Woodger, J.H., Corcoran, J., Eds.; Hacket Publishing Company: Indianapolis, Indiana, 1983; Section V; pp. 60–109.
13. Słupecki, J.; Borkowski, L. *Elements of Mathematical Logic and Set Theory*; International Series of Monography in Pure and Applied Mathematics; Pergamon Press–PWN: Oxford, UK; New York, NY, USA; Toronto, ON, Canada; Warsaw, Poland, 1967.
14. Alscher, D. *Theorien der Reellen Zahlen und Interpretierbarkeit*; Logos 25; De Gruyter: Berlin, Germany, 2016.
15. Gödel, K. Über formal unentscheidbare Sätze der Principia Mathematica und verwandter Systeme, I. *Mon. Math.* **1931**, *38*, 132–213. [CrossRef]

16. Corcoran, J. Categoricity. *Hist. Philos. Logic* **1980**, *1*, 187–207. [CrossRef]
17. Corcoran, J. Teaching categoricy of arithmetic. *Bull. Symb. Logic* **1997**, *3*, 395.
18. Read, S. Completeness and Categoricity: Frege, Gödel and Model Theory. *Hist. Philos. Logic* **1997**, *18*, 79–93. [CrossRef]
19. Boolos, G.; Jeffery, R. *Computability and Logic*; Cambridge University Press: London, UK, 1982.
20. Hermes, H. *Introduction to Mathematical Logic*; Springer: London, UK, 1973.

© 2019 by the author. Licensee MDPI, Basel, Switzerland. This article is an open access article distributed under the terms and conditions of the Creative Commons Attribution (CC BY) license (http://creativecommons.org/licenses/by/4.0/).

Article

Logic of Typical and Atypical Instances of a Concept—A Mathematical Model

Jean-Pierre Desclés [1],* and Anca Christine Pascu [2],*

[1] Sens, Texte, Informatique, Histoire (STIH), Université de Paris Sorbonne, 75005 Paris, France
[2] Laboratoire des Sciences et Techniques de l'Information, de la Communication et de la Connaissance (Lab-STICC), Université de Brest, 29200 Brest, France
* Correspondence: jeanpierre.descles@gmail.com (J.-P.D.); anca.pascu@univ-brest.fr (A.C.P.)

Received: 10 July 2019; Accepted: 21 August 2019; Published: 4 September 2019

Abstract: In this paper, we give a mathematical model of the logic of determination of objects (LDO) based on preordered sets, and a mathematical model of the logic of typical and atypical instances (LTA). We prove that LTA is an extension of LDO. It can manipulate several types of "exceptions". Finally, we show that the structural part of LTA can be modeled by a quasi topology structure (QTS).

Keywords: logic of typical and atypical instances (LTA); logic of determination of objects (LDO); quasi topology structure (QTS); concept; object; typical object; atypical object; lattice; filter; ideal

1. Introduction

The difference between typical instances and atypical instances in a natural categorization process has been introduced by E. Rosh and studied by cognitive psychology [1,2] and AI. A lot of the knowledge representation systems are expressed using fuzzy concepts but a degree of membership raises some problems for natural categorizations (especially for classification problems in anthropology, ethnology, archeology, and linguistics, but also in ontologies): atypical instances of a concept cannot be apprehended adequately by different degrees from a prototype. Other formal approaches, as paraconsistent logics or non-monotonic logics, often conceptualize atypical objects as exceptions. An alternative approach was developed with the logic of typical and atypical instances (LTA) [3] and logic of determination of objects (LDO) [4]. In order to give a logical approach of typicality/atypicality associated to a concept we distinguish explicitly, in LTA, a conceptual property f ("concept" or predicate in the Frege's approach and classical logic) from a concept ˆf, associated to a conceptual property and characterized by an intension and by an essence, a part of the intension. A typical instance of a concept inherits all properties of intension; an atypical instance inherits only properties of essence, but it is a full member of the category associated to a concept and not a member with a weak degree of membership. In natural categorization, there are often exceptions which do not inherit some properties of the essence; the exceptions cannot be considered as atypical instances as they belong to the boundary of the category, that is the difference between the extension of a conceptual property and the extension of the corresponding concept (the set of all instances inheriting all properties of the essence). In LDO a typical object τf is introduced, which is canonically associated to a concept ˆf. Object τf is an abstract object such that it is the best representative object of the concept. From τf more or less objects are explicitly built that fall under the concept ˆf, in using a functional composition of different determination operators δk associated to conceptual properties k (in general not in the intension of the concept). When a property δk is the negation of a conceptual property of the intension of a concept, the generated object becomes an atypical object. All typical and atypical objects generated from τf by determination operators belong to the expansion of the concept ˆf that contains the extension of fully determinate instances of the concept. Some compositions of determination operators (incoherent compositions) can build objects that are out of the category associated to a concept.

In LTA the types of fully determined objects are extended objects to "exceptions".

This paper is organized in six sections as follows: Introduction, the logic of determination of objects (LDO), a formal description of the logic of determination of objects (LDO) as a preordered set model, the logic of typical and atypical instances (LTA), LTA as a quasi topology structure, conclusions.

The works directly related to this work are [3–7].

The novelties of this paper are:

- Theorem 1 which represents a fundamental theorem for the model of the LDO. It asserts that the LDO structure can be represented by a Galois lattice;
- The quasi topology structure (QTS) of the fully determinate objects Ext f in LTA.

2. The Logic of Determination of Objects (LDO)

2.1. Informal Description

The logic of determination of objects (LDO) is a non-classical logic of concepts and objects. It contains a theory of typicality. It is due to Jean-Pierre Desclés [8] and described as a logical model in [4]. LDO is defined within the framework of combinatory logic [9] with functional types.

LDO is inspired by the semantics of natural languages. It solves some problems that classical logic cannot describe and solve:

- It supplies a solution for the mismatch between logic categories and linguistic categories (adjectives, intransitive verbs often represented by unary predicates);
- It considers the determination as a logic operator in order to represent linguistic expression as a book, a red book, a book which is on the table;
- It reconsiders the duality of extension–intension via its theory of typicality; the entension and the intension of a concept are no longer in duality.

The LDO has a structural part and an inferential part.

Its structural part is formed by a triple of:

1. A network of concepts;
2. A set of objects;
3. A type theory.

LDO was described as an typed applicative system in the Curry's sense [9]. A concept is an operator, an object is an operand in Curry's sense [9] where a conceptual proprerty is an operator and an object is always an operand (on the applicative systems and combinatory logic as a logical formalism of operators composed and transformed by an intrinsic way, see the book of J.-P. Desclés, G. Guibert and B. Sauzay [10]). With every concept f, the following are canonically associated [4]:

- An object called typical object τf, which represents the concept f as an object. This object is completely (fully) indeterminate;
- A determination operator δf, constructing an object more determinate than the object to which it is applied;
- The intension of the concept f, Int f, conceived as the class of all concepts that the concept f «includes», that is, a semantic network of concepts structured by the relation «IS-A»;
- The essence of a concept f, Ess f; it is the class of concepts such that they are inherited by all objects falling under the concept f;
- The expanse of the concept f, Exp f, which contains all more or less determinate objects to whom the concept f can be applied;
- A part of the expanse is the extension Ext f of the concept f; it contains all fully (completely, totally) determinate objects such that the concept f applies to.

From the viewpoint of determination, in LDO, objects are of two kinds: fully (completely, totally) determinate objects and more or less determinate objects. From the viewpoint of some of their properties, LDO captures two kinds of objects: typical objects and atypical objects. The typical objects in Exp f inherit all concepts of Int f. The atypical objects in Exp f inherit only some concepts of Int f.

The inferential part of LDO contains axioms and rules of inferences. Some of the rules decide of the typicality of an object as regard with some concept [4]. In [11], we analysed the nature of these rules issued from the theory of typicality of LDO versus the paraconsistence. More precisely, we show that the rule establishing that an object that is an atypical object of a concept in the frame of the LDO is a particular case of the RA_1 rule of Da Costa [12]. We arrive at the following interpretation of the weakening of the principle of contradiction ($\neg(B \wedge \neg B)$) contained by the RA_1 rule inside the LDO: an object obtained by a LDO-rule using this form of the weakening of the principle of contradiction ($\neg(B \wedge \neg B)$) is an atypical object. From the point of view of managing negation, we can conclude that LDO is a particular case of a paraconsistent logic. For its power of description and especially for its basic notions (to emphasise the distinction between object and concept and between extention and intension), we can state that LDO is a description logic capturing at least one more cognitive feature: the typicality of objects.

2.2. Formal Description of LDO

LDO can be regarded as a formal theory of concepts and objects.

LDO = ($\mathcal{F}$, O, T) where:

$\mathcal{F}$ is the set of concepts, O is the set of objects, T is a type theory. A concept is an operator, whereas an object is always an operand. Types are associated with concepts and objects.

The types theory of LDO is a theory of functional types [9] containing:

- Primitive types are: J individual entity type, H truth value (sentence) type;
- Functional type constructor: **F**;
- Rules.

The rules of the type theory are:

- Primitive types are types;
- If α and β are types, then **F**$\alpha\beta$ is a type;
- All types are obtained by one of the above rules.

In LDO:

- All objects are operands of type J; all propositions are of type H;
- All concepts are operators of type **FJH**.

An expression X of type α is specified by: X: α.

The applicative scheme that expresses the application of a concept to an object is:

$$\frac{f: FJH \quad x: J}{f(x): H}$$

If f(x) is true, that is f(x) = T one says that "the object x falls under the concept f", if f(x) is false, that is f(x) = $\bot$ one says that "the object x does not fall under the concept f". In LDO, N_1 is the operator of negation defined as:

(N_1 f) (x) = T, if and only if (f x) = $\bot$

It has the classical logic property: (N_1 (N_1g)) = g.

In LDO, N_0 is the negation of a sentence defined as:

$N_0(f(x)) = T$, if and only if $f(x) = \perp$

LDO is an applicative language of different types of operators (on functional types, [4]) applied to operands of different types; it is composed of:

- Objects of type J and concepts of type FJH.
- Predicates defined on individual objects (concepts of type FJH) and the relators between individuals with respective types FJFJH, FJFJFJH, etc.);
- Proposition of type of H;
- Connectives between propositions are of the type FHFHH;
- Fregean quantifiers: simple quantifiers with the type FFJHH; restricted quantifiers with the type FFJHFFJHH;
- Operators of negation with the type FHH (classical negation) defined only on propositions.

2.3. Basic Operators of LDO

2.3.1. The Constructor of the "Typical Object": the Operator τ

This operator denoted by τ and called the constructor of the typical object builds an object totally indeterminate starting from a concept. Its type is FFJHJ; it canonically associates to each concept f, an indeterminate object τf, called "typical object". Its applicative scheme is:

$$\frac{\tau: \text{FFJHJ} \qquad f: \text{FJH}}{\tau f: J}$$

The object τf, is the "best representative object of the concept f; it is totally indeterminate, typical and abstractly represents the concept f in the form of an "any typical object whatever" This expression was chosen to encode the notion captured by the word "quelconque" in French. The typical object τf associated with f is unique. For example, if we take as concept f, the concept "to-be-a-man" then, the typical object associated is "a-man". For the concept f, "to-be-a-computer", τf is "a-computer":

2.3.2. The Operator of Determination: the Operator δ

The operator δ, called the constructor of determination operators, builds a determination operator, starting from a given concept.

The operator δ canonically associates a determination operator of the type FJJ to each concept f. The type of operator δ is FFJHFJJ. Its applicative scheme is:

$$\frac{\delta: \text{FFJHFJJ} \qquad f: \text{FJH}}{\delta f: \text{FJJ}}$$

A determination operator δ is an operator which being applied to an object x constructs another object y: $y = ((\delta f) x)$ (We use the prefixed notation of a function, that is (f x) for f(x)). The object y is more determinate than the object x, by means of the determination added by δ. For example, if the concept f is "to-be-red", then δf is "red"; if f is "to-be-on-the-table", then δf is "which-is-on-the-table". The determination δf, "to-be-red" applied to the object "a-book" gives the more determinate object "a-red-book".

Determinations can be composed of each other. A chain of determination Δ is a finite string of the form $\Delta = \delta g_1 \circ \delta g_2, \ldots, \circ \delta g_n$. The composition of determinations is associative and supposed to be commutative.

2.3.3. Objects in LDO

More or Less Determinate Objects

A more or less determinate object is an object recursively obtained starting from the object τf by:

- τf is a more or less determinate object;
- If δf is a chain of determinations, then $y = ((\delta g_1 \circ \delta g_2, \ldots, \circ \delta g_n) x) = (\delta g_1 (\delta g_2 (\ldots, \delta g_n(x), \ldots)))$ is a more or less determinate object;
- Each more or less determinate object is obtained by the above rules.

Fully Determinate Objects

An object x is fully (totally) determinate if and only if for each determination δg we have:
$(\delta g\ x) = x$
In LDO, objects are of two kinds from the point of view of their "determination":

- More or less determinate objects;
- Fully determinate objects.

Nevertheless, all of them are of type J.

2.3.4. Concepts and Objects

Classes of Concepts Associated with a Concept f

In LDO we postulate the existence of two classes of concepts corresponding to a given concept f:

The intension of a concept f, Int f being the class of all concepts subsumed (included) by concept f; for example, the concepts "to-have-two-legs" and "to-have-a-mind" are both in the intension of the concept "to-be-a-man"; if f is "to-be-a-bird", then in its intension there is the concept "to-fly".

The essence of a concept f, Ess f is the class of all concepts necessarily included by f.

The class Ess f is included in Int f. The concept "to-be-a-man" has the concept "to-have-two-legs" in its intension but not in its essence. As for the concept "to-have-a-mind" it is right in the essence of the concept "to-be-a-man". In other words, the essence of a concept f is the set of concepts necessarily comprised in f. If we remove a concept g from the essence of f, we "destroy" the concept f; it remains not the same. If a concept g is in the essence of a concept f, then the negation of g cannot belong to this essence. For the concept "to-be-a-bird", the concept "to-fly" is in its intension but not in its essence.

Classes of Objects Associated with a Concept f

The LDO has two classes of objects associated with a concept:

Expansion (etendue in French (the Port Royal logic talks about "etendue")). The expansion of f, denoted by Exp f is the set of all objects of (more or less determinate or fully determinate) to which f can be applied:
$Exp(f) = \{x/f(x) = T\}$

Extension. The extension of f, denoted by Ext f is the set of all objects fully determinate) to which f can be applied:
$Ext(f) = \{x/f(x) = T\}$

2.3.5. Theory of Typicality in LDO

Let us take a concept f. From the point of view of inheritance of concepts in their intension, the objects falling under it are of two kinds: typical objects of f and atypical objects of f.

Typical Object of f

An object x falling under f is a typical object of f if and only if:

For each chain of determination Δ constructs x starting from τf and for each determination along the chain we have:

- Either each determination concept is in the intension of f and its negation is not in the intension of f;
- Or if there is a determination concept such that itself and its negation are in the intension of f, then this determination concept belongs to the characteristic intension of x;
- Roughly speaking, a typical object of f is an object that inherits all the concepts of the intension of f, Int f.

Atypical Object of f

An object x falling under f is an atypical object of f if and only if:

There is a chain of determination Δ construct x starting from τf such that:

- Either there is a determination concept that it is not in the intension of f, but its negation belongs to this intension;
- Or if all determination concepts are in the intension of f, then x has an atypical "ascendant" as object.

Roughly speaking, an atypical object of f is an object that does not inherit all the concepts of the intension of f, Int f.

The typicality in LDO is based on the notion of determination.

2.3.6. The Logic of Determination of Objects (LDO) as a Deductive System

The contributions of the LDO to logic consist more in its structural part than in its deductive part. As system of deduction LDO is a natural deduction system in Gentzen sense with types associated to objects. It contains two types of inferences: typicality inferences inside the typical objects field and inferences inside the atypical objects field.

3. A Formal Description of the Logic of Determination of Objects (LDO) as a Preordered Set Model

In the LDO such as it is presented above, the set $\mathcal{F}$ can be seen as a collection of properties. From the functional point of view, the concept f, as element $f \in \mathcal{F}$ is an operator. From the structural point of view, it is a couple (Ess f, Int f), with Ess f and Int f being subsets of $\mathcal{F}$. We can organize $\mathcal{F}$ as a preordered set by the relation ($\rightarrow$) between two concepts. Between the concepts f and g one can consider that g is more primary than f. It is defined by:

$$g \rightarrow f \quad \text{if g is one of the properties of f} \tag{1}$$

Objects as elements of O are ranked by the relation $\longrightarrow$. It is defined by:

$$\text{For any objects } o_1, o_2 \in O, o_1 \longrightarrow o_2 \text{ iff the object } o_2 \text{ is more determinate than the object } o_1 \tag{2}$$

The set O is organized in this way as a preordered set.

In Figure 1, the couple $(\mathcal{F}, O)$ corresponding to a concept f and to the object τf is represented by a graph. The vertices of this graph are properties in the upper side, the objects in the lower side. The arrows correspond to the two order relations above. In the upper side the subsets Ess f (in red) and Int f (in blue) are shown. The set NInt f (in green) is the set of negations of some properties of f concerning, eventually, atypical objects corresponding to τf.

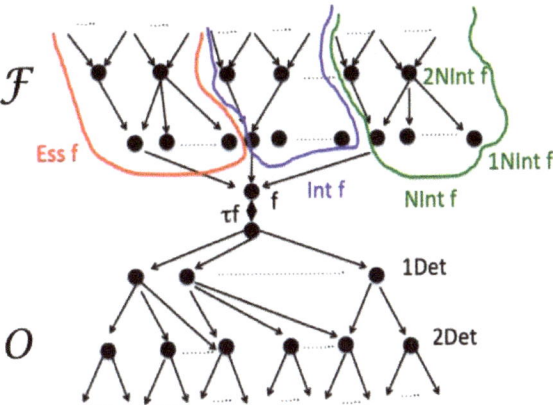

Figure 1. The preordered set model of the logic of determination of objects (LDO).

One can remark that vertices both in the concept (properties) side and in the objects side are organized on levels. In the upper side the properties on the same level are supposed to be independent. The length of a chain $g \to^n f$ can be interpreted as the number of properties in depth subsumed by the concept f.

In the lower side a level corresponds to the number of determinations between τf and the objects of the level. This is 1Det τf are the objects obtained from τf by a single determination, 2Det τf are the objects obtained from τf by two determinations, etc. These objects are more or less determinate objects. The last level of objects is the level of completely determinate objects i.e. the level of Ext f.

One can remark that the entire model of Figure 1 is a special network formed by two preordered sets. The difference between this model and the approach in [4] consists in the fact that the structural part of LDO is described now in terms of filters and ideals [13], not only in terms of sets.

Figure 2 represents the network corresponding to a concept f and its typical indeterminate object τf of LDO. This network is composed by two parts the sub-network of concepts in the upper side F and the sub-network of objects in the lower side (O). In the whole network, the vertex f corresponding to the concept f in the upper side and the vertex τf corresponding to the fully indeterminate object are identified as a single and same vertex (f, τf).

We remark that we can build a pair of a filter and an ideal (F,I) [13], the filter being in the upper side $\mathcal{F}$, the ideal in the lower side $\mathcal{O}$. Such a pair is the subset of the whole network and it represents the mathematical modeling of the relation between a concept and some of its underlying objects. Roughly speaking, a route of the network in Figure 2 is structured top-down by two sub-networks: the concepts sub-network corresponding to $\mathcal{F}$, and the objects sub-network corresponding to $\mathcal{O}$.

A route from left to right of the sub-network $\mathcal{F}$ brings out the intension of a concept f and the essence of the concept f. The intension of a concept f, Int f (in blue) is formed by the sub-network in the upper side up to the concept f included it. The essence of a concept f Ess f (in green) is a sub-network of Int f included f. The arrows correspond to the order relation from Equation (1). We can remark the sub-network NInt f (in red) which is the part corresponding to negation of concept in intension. Because of the LDO theory of typicality, a negation of a concept is described explicitly in NInt f. The vertices corresponding to negation are linked to their positive (not negated) counterparts by a green line. The arrow related with a negation is dotted in red.

Typical objects are represented in the sub-network $\mathcal{O}$ by black-green double circles. Atypical objects are represented by red-rose double circles in $\mathcal{O}$. The last level is the level of fully determinate typical objects Ext τ and of fully determinate atypical objects Ext α. Dotted arrows correspond to relation in Equation (2).

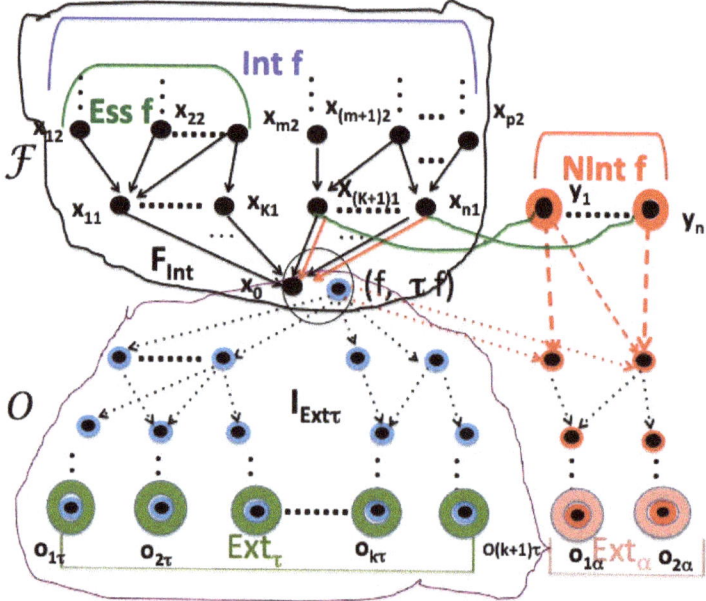

Figure 2. The network of LDO.

The Galois Connexion and the Galois Lattice of LDO

The analysis of Figure 2 carried out above leads us to define a Galois connection on the couple $(\mathcal{F}, \mathcal{O})$.

Definition 1 [14]. *A Galois connection between two preordered sets $(P \leq_P)$ and $(Q \leq_Q)$ is a couple of functions m_1 and m_2 such that:*

$$m_1: (P \leq_P) \to (Q \leq_Q),\ m_2: (Q \leq_Q) \to (P \leq_P) \text{ with}$$
$$\text{For all } p \in P,\ p \leq_P m_2(m_1(p)) \text{ and for all } q \in Q,\ q \leq_P m_1(m_2(q)). \tag{3}$$

Definition 2 [13]. *A couple $((\mathcal{P}(P), \subseteq), (\mathcal{P}(Q, \subseteq)))$ is called a Galois lattice.*

In the case of LDO formal model, one supposes both networks O and F to be finite.

Let us denote by Fil the set of all filters in $\mathcal{F}$ and by $\mathcal{P}(\text{Fil})$ the power set of Fil. In the same way, we denote by Id all the ideals in $\mathcal{O}$ and $\mathcal{P}(\text{Id})$ the power set of Id. We take $\mathcal{P}(\text{Fil})$ as P from Definition 1. The role of Q from Definition 1 is played by $\mathcal{P}(\text{Id})$.

Remark 1. *In the double network $(\mathcal{F}, \mathcal{O})$ there are several types of filters and ideals:*

- *A single filter corresponding to the intension Int f denoted by F_{Int} (in Figure 2, in black);*
- *A single ideal corresponding to all typical objects denoted by $I_{Ext\tau}$ (in Figure 2, in magenta);*
- *Several filters containing concepts from the intension Int f and, in the counterpart NInt f (see Figure 2) negations of concepts in Int f. Such a filter is denoted by $F_{Int\text{-}NInt}$;*
- *Several ideals containing some typical and some atypical objects. Such an ideal is denoted by $I_{typ\text{-}atyp}$.*

Definition 3. *Let us define* $m_1: (\mathcal{P}(\text{Fil}), \subseteq) \to (\mathcal{P}(\text{Id}), \subseteq)$ *and) by:*

$$m_1(F_{\text{Int-NInt}}) = \text{Ext}\tau \quad \text{if } F_{\text{Int-NInt}} = F_{\text{Int}} \quad (4)$$
$$m_1(F_{\text{Int-NInt}}) = I_{\text{typ-atyp}} \quad \text{otherwise}$$

and $m_2: (\mathcal{P}(\text{Id}), \subseteq) \to (\mathcal{P}(\text{Fil}), \subseteq)$ *by:*

$$m_2(I_{\text{typ-atyp}}) = F_{\text{Int}} \quad \text{if } I_{\text{typ-atyp}} = \text{Ext}\tau \quad (5)$$
$$m_2(I_{\text{typ-atyp}}) = F_{\text{Int-NInt}} \quad \text{otherwise}$$

Theorem 1. *The couple* (m_1, m_2) *is a Galois connection on the double network* $((\mathcal{P}(\text{Fil}) \subseteq), (\mathcal{P}(\text{Id}) \subseteq))$.

Corollary 1. *The double network* $((\mathcal{P}(\text{Fil}), \subseteq), (\mathcal{P}(\text{Id}), \subseteq))$ *is the Galois lattice associated to the LDO model.*

4. The Logic of Typical and Atypical Instances (LTA)

The logic of typical and atypical instances (LTA) is a logic that distinguishes the concept f from the property f. The property is an element of F, and the concept is the quadruple:

$$\hat{f} = < f, \text{Ess } f, \text{Int } f, \text{NInt } f > \quad (6)$$

A concept ^f is the quadruple formed by the property, the intension, the essence and, eventually, the set of negations of some concepts from intension. This logic extends the logic of determination of objects.

LTA allows us to make the whole problem slightly more complex by taking into account objects which being no longer atypical, are nevertheless on the external outer edges of the category, so apprehended as being related to the category but no longer belonging to it. Otherwise, ontologies of domains are structured networks of concepts and of classes of objects. Generally, in these ontologies, the problem of typical/atypical is not considered. Inside these ontologies only some objects are treated as exceptions without doing a deep "logical" analysis (especially the analysis of intensions) establishing that an object must be considered as an atypical object internal to the category or as an object on the edges of the category, and so "almost belonging" to it but not "belonging entirely".

We start with an example that is the analysis of a concept with the resulting categorization when one takes into account some properties of this concept and its typical objects. This example is about the typical and atypical objects of the class of inhabitants of a city.

Example.

The concept "living-in-X (inhabitant-of-X)" contains the property "to-be-inhabitant-of-X". This concept has an intension containing properties "to-have-rights" and "to-have-duties". These properties imply "to-be-protected-by-law" and "to-respect-the-law", respectively. All inhabitants of the city are not substitutable for one another. Some of them, the citizens, have the citizenship of the country, the others (e.g., the foreigners residing in the city) do not. All citizens must pay taxes and after age of 18 they have the right to vote. The foreigners pay taxes but, generally, they do not have the right to vote, though in some countries some foreigner residents who are not citizens also have the right to vote providing that, however, they are residents and have paid taxes for five years. One must consider also inhabitants without fixed residence (homeless) being not irregular residents ("sans papiers"). They lose (at least in some categorizations) some of their rights (such as the access to health care) and they are not considered constrained to certain duties (e.g., to pay taxes). The lattice of the concepts F corresponding to the concept f inhabitant-of-X is presented in Figure 3. The lattice of objects of the same concept f is presented in Figure 4.

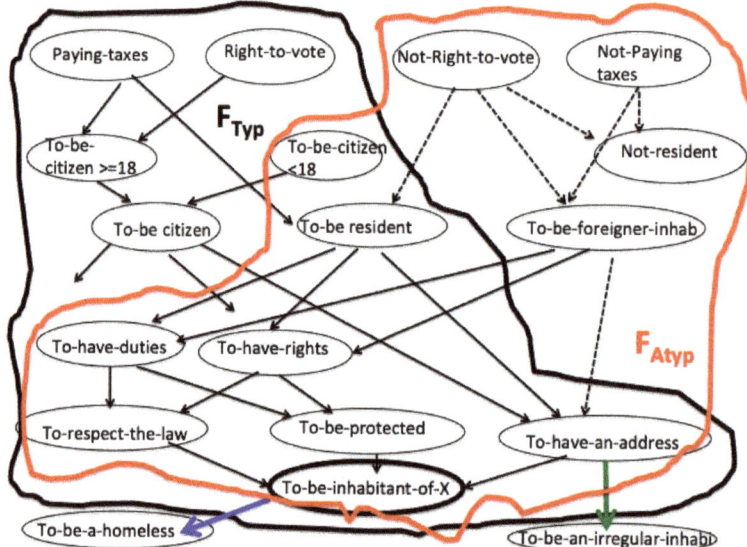

Figure 3. The concept lattice of the concept "To-be inhabitant of a city X".

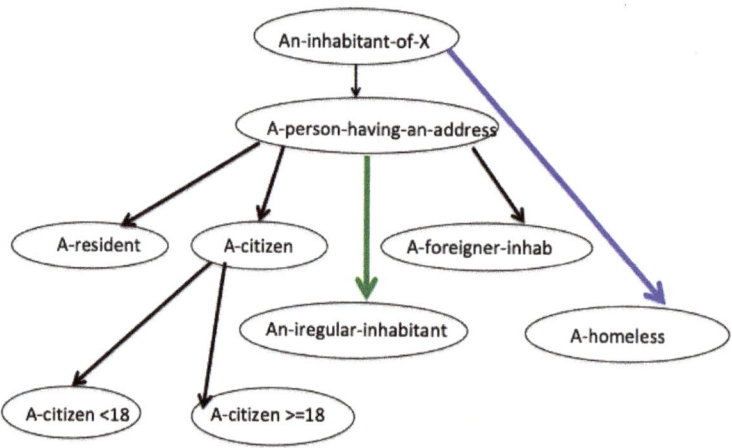

Figure 4. The objects lattice of the object "An inhabitant of a city X".

We can see from Figure 3, for example that the property to-be-a-citizen-under-18 years generates an atypical object "a-citizen-under-18-years" because it is related to the property "not-right-to-vote" represented in Figure 4.

In Figure 4, we can see two particular objects "an-irregular-inhabitant" and "a-homeless". The first one is determined only by two properties "to-have-an-address" and "to-be-an inhabitant-of-X", both belonging to the intension of "to-be-an inhabitant-of-X" (green arrow). The second one is determined only by "to-be-an inhabitant-of-X" (blue arrow). So, they both are "exceptions" but with a status different from the status of typical objects or atypical objects of the concept "to-be-an inhabitant-of-X". The object "an-irregular-inhabitant" has the property "to-be-an inhabitant of the city" and the property "to-have-an-address" both belonging to the intension of "to-be-an inhabitant-of-X". The object "a-homeless" just falls under the property "to-be-an inhabitant-of-X".

Informal Description of LTA Versus LDO

- In LTA the vertices of $\mathcal{F}$ are properties, a concept is represented by entire lattice $\mathcal{F}$. That is because it is based by the difference between property and concept.
- In LTA in the part of objects O, there are at least two categories of objects unless typical objects and atypical objects:
  - Strong exceptions (the "homeless", in Example);
  - Weak exceptions (the "irregular inhabitant" in Example).

The difference between LDO and LTA is the following:

- The LDO describes the structure of a concept f;

The LTA structure, by integrating several type of objects, considers a network having several networks $(\mathcal{F}, O)$ (Figure 2) interrelated corresponding to concepts f, g, ... , h that is $\{(f, \tau f), (g, \tau g), \ldots, (h, \tau h)\}$.

5. LTA as a Quasi Topology Structure

5.1. Quasi Topology Structure (QTS) Definition

Definition 4 [5]. *Let <X, O> be a topological space where X denotes the space and O denotes the topology. We say that a set E from this space is structured by a quasi topology or it has a quasi topology structure (QTS) if there exists two open sets O_1 and O_2 of O, and two closed sets F_1 and F_2 such that:*

$$O_2 \subset O_1 \subseteq E \subseteq F_1 \subset F_2 \tag{7}$$

with:

$$O_1 \text{ is the biggest open set contained in E, that is } O_1 = \text{Int}(E), \tag{8}$$

$$F_1 \text{ is the smallest closed set containing E, that is } F_1 = \text{Cl}(E), \tag{9}$$

$$O_2 \text{ is the biggest open set strictly contained in } O_1, \tag{10}$$

$$F_2 \text{ is the smallest closed set strictly containing } F_1. \tag{11}$$

The set O_2 is said to be the strict interior of E; the set O_1 is the large interior of E. The set F_2 is said to be the large closure of E and the set F_1 the strict closure of E.

The internal boundary, the external boundary, the strict boundary and the large boundary of E are defined by:

$$\text{Int-bound}(E) = F_1 - O_2, \tag{12}$$

$$\text{Ext-bound}(E) = F_2 - O_1, \tag{13}$$

$$\text{Large-bound}(E) = \text{Int-bound}(E) \cup \text{Ext-bound}(E), \tag{14}$$

$$\text{Bound}(E) = \text{Cl}(E) - \text{Int}(E) = F_1 - O_1. \tag{15}$$

The above definition is presented in an intuitive way in Figure 5.

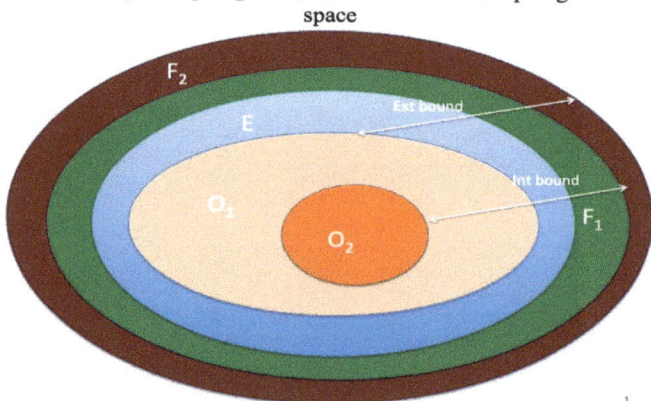

Figure 5. The quasi topology structure of a set E in a topological space.

Remark 2. *In this definition, the structure of quasi topology (QTS) is limited to topological space X being a topological space. In [15], we extended the QTS to some types of approximation spaces (rough sets spaces) by replacing the notions of openness and closure by the corresponding notions of rough sets.*

5.2. *The QTS of the LTA*

The background framework is the systems of two networks of type from Figure 2, one associated to a network F_f, corresponding to the concept ˆf = <f, Ess f, Int f, NInt f>, the other F_g, corresponding to concept ˆg = < g, Ess g, Int g, NInt g >.

The space X from Definition 4 is the entire set of fully determinate objects of O_f and O_g, namely Ext f ∪ Ext g. We define a QTS on the fully determinate objects related to the concept ˆf.

There are four relations between fully determinate objects related to ˆf. These relations are defined based on the structure of F_f.

- The objects verifying all the properties from Ess f and Int f. These objects are typical objects. They form the set O_2;
- The objects verifying all the properties from Ess f and at least one property of NInt f. These objects are atypical objects. They form the set O'_2;
- The objects verifying some properties from Ess f and some properties of Int. These objects are weak exceptions. They form the set F_1;
- The objects verifying only the property f as property. These objects are strong exceptions. They form the set F_2.

Definition 5. *Let us define the following relations on the approximation space X = Ext f ∪ Ext g:*

- typ f (o_1, o_2) *defined by:*

$$\text{for all } o_1, o_2, \text{typ } (o_1, o_2) \text{ iff } o_1, o_2 \text{ are typical objects.}$$

It is an equivalent relation. All typical objects form an equivalent class;
- atyp ^{k}f (o_1, o_2) *defined by:*

$$\text{for all } o_1, o_2, \text{typ } (o_1, o_2) \text{ iff } o_1, o_2 \text{ are atypical objects with k their degree of atypicality.}$$

There are n relations $atyp_k$, f for k = 1, ... , n, where n is the degree of atypicality. The degree of typicality of an object o is the number of properties of NInt verifying by this object. All the atypical objects having the same degree of atypicality are considered to be equivalent;

- wex_i f *(weak exception) defined by:*

For all o_1, o_2, wex_i f (o_1, o_2) iff o_1, o_2 are both objects falling under the same number of properties of Ess f.

There are m relations wex_i, f for i = 1, ... m if Ess f has m properties. All the objects falling under the same number of properties of Ess f are considered to be equivalent;

- strongex f *(strong exception) defined by:*

strongex f (o) iff o in an object verifying only the property f.

The relation strongex f *is an unary (function prototype length 1);*

Remark 3. *All the relations above are defined as relations between objects but as regarding of the LDO structure (interrelated) related to the concept ^f. For this reason, their names are followed by f.*

In LTA, a set E of objects belonging to Ext f ∪ Ext g but analyzed regarding to concept ^f contains objects that are directly related to concept ^f as the typical objects of f and the atypical objects of f, and other objects which are related to the property f but belong to the structure corresponding to another concept g.

Theorem 2. *Let us denote by* (atyp f)* = $\cup_{k=1,...n}$ (atyp$_k$ f)* *and by* (wex f)* = $\cup_{i=1,...m}$ wex$_i$ f. *A subset, set E of the approximation space* (Ext f ∪ Ext g, typ f ∪ (atyp f)*, (wex f) *∪ strongex f) *has the following QTS structure:*

$$O_2 = Ext_{typ}\ f$$

$$O_1 = (Ext_{typ}\ f \cup (Ext_{atyp}\ f)^*)$$

$$F_1 = Ext_{wex}\ f$$

$$F_2 = Ext_{strongex}\ f$$

We can easily prove that O_2 is the lower approximation of E as regarding to the relation typ f, O_1 is the lower approximation of E as regarding to the relation typ f ∪ (atyp f)*, F_1 is the upper approximation of E as regarding to the relation (wex f)* and F_2 is the upper approximation of E as regarding to the relation strongex f.

It is obvious that $O_2 \subset O_1 \subseteq E \subseteq F_1 \subset F_2$.

Remark 4. *This QTS structure in an approximation space is a hybridization built up with four relations and the lower and the upper approximation of a set regarding to them.*

We call this structure the QTS structure associated to a set of fully determinate objects in LTA.

6. Conclusions

In this paper a mathematical model of the logic of determination of objects (LDO) introduced in [4] is presented. The novelty of this model is the fact that it describes the structural level of LDO by notions of preordered sets and lattices. To represent the conceptual structure of LDO as a network allows us to extend the main theorem of formal concept analysis [13] stating that the lattice of concepts is a Galois lattice to LDO by establishing a Galois lattice associated to LDO network by Theorem 1.

A mathematical model of the logic of typical and atypical Instances (LTA) [3] is also described as an extension of LDO model. In the case of LTA we give a quasi topology structure (QTS) [6,7] to a set of objects related to a concept.

The LDO and its associated Galois lattice theorem allows a computer software for analysis and categorization inside ontologies to be built.

The QTS structure in LTA represents a type of approximation different from those existing until now. It can be also useful as a model in a computer-based tool of categorization.

Author Contributions: Conceptualization, J.P.D.; Formal analysis, A.C.P.

Conflicts of Interest: The authors declare no conflict of interest.

References

1. Rosch, E.H. Classification d'objets du monde réel: Origine et représentations dans la cognition. *Bull. Psychol. Numéro Spécial La Mémoire Sémantique* **1976**, 246–250.
2. Le Ny, J.-F. *Comment L'esprit Produit du Sens*; Odile Jacob: Paris, France, 2005.
3. Desclés, J.-P.; Pascu, A.C.; Jouis, C. The Logic of Typical and Atypical Instances (LTA). In Proceedings of the Twenty-Sixth International Florida Artificial Intelligence Research Society Conference, FLAIRS13, Miami, FL, USA, 22–24 May 2013; AAAI Press: Menlo Park, CA, USA, 2013; pp. 321–326.
4. Desclés, J.-P.; Pascu, A. Logic of Determination of Objects (LDO): How to Articulate "Extension" with "Intension" and "Objects" with "Concepts". *Log. Univers.* **2011**, *5*, 75–89. [CrossRef]
5. Desclés, J.-P.; Pascu, A.C.; Biskri, I. A topological approach for the notion of quasi topology structure. *Int. South Am. J. Log.* **2019**, submitted.
6. Desclés, J.-P.; Pascu, A.C.; Biskri, I. A Quasi-Topologic Structure of Extensions in the Logic of Typical and Atypical Objects (LTA) and Logic of Determination of Objects (LDO). In Proceedings of the 31th FLAIRS Conference, Melbourne, FL, USA, 21–23 May 2018; AAAI Press: Menlo Park, CA, USA, 2017.
7. Desclés, J.-P.; Pascu, A.C. Logique de la Détermination des Objets (LDO): Structuration topologique et quasi-topologique des extensions. In Proceedings of the Conference La logique en question/Locic in Question, Paris-Sorbonne, Paris, France, 13–14 May 2016.
8. Desclés, J.P. Categorization: A logical approach to a cognitive problem. *J. Cogn. Sci.* **2002**, *3*, 85–137.
9. Curry, H.B.; Feys, R. *Combinatory Logic*; North Holland Publishing, Co.: Amsterdam, The Netherlands, 1958.
10. Desclés, J.P.; Guibert, G.; Sauzay, B. *Logique Combinatoire et λ-Calcul: Des Logiques D'opérateurs*; Cépaduès: Toulouse, France, 2016.
11. Desclés, J.-P.; Pascu, A.C. *The Logic of Determination of Objects (LDO)—A Paraconsistenr Logic*; UNILOG: Rio de Janeiro, Brazil, 2013.
12. Da Costa, N.C.A. *Logique Classiques et non Classiques—Essai sur les Fondements de la Logique*; Traduit du Portugais et Complété par Jean-Yves Béziau; Masson: Paris, France, 1997.
13. Ganter, B.; Wille, R. *Formam Concept Analysis, Mathematical Foundations*; Springer: Berlin/Heidelberg, Germany, 1999.
14. Pawlak, Z. Rough Sets. *Int. J. Comput. Inf. Sci.* **1982**, *11*, 341–356. [CrossRef]
15. Desclés, J.-P.; Pascu, A.C.; Biskri, I. A Rough Sets Approach for defining the Notion of Quasi Topology Structure. to be submitted.

© 2019 by the authors. Licensee MDPI, Basel, Switzerland. This article is an open access article distributed under the terms and conditions of the Creative Commons Attribution (CC BY) license (http://creativecommons.org/licenses/by/4.0/).

Book Review

Review of "The Significance of the New Logic" Willard Van Orman Quine. Edited and Translated by Walter Carnielli, Frederique Janssen-Lauret, and William Pickering. Cambridge University Press, Cambridge, UK, 2018, pp. 1–200. ISBN-10: 1107179025 ISBN-13: 978-1107179028

Alfredo Roque Freire

Departament of Philosophy, University of Campinas, Campinas 13083-970, São Paulo, Brazil; alfrfreire@gmail.com

Received: 30 April 2019; Accepted: 17 May 2019; Published: 22 May 2019

Abstract: In this review, I will discuss the historical importance of "The Significance of the New Logic" by Quine. This is a translation of the original "O Sentido da Nova Lógica" in Portuguese by Carnielli, Janssen-Lauret, and Pickering. The American philosopher wrote this book in the beginning of the 1940s, before a major shift in his philosophy. Thus, I will argue that the reader must see this book as an introduction to an important period in his thinking. I will provide a brief summary of the chapters, remarking on valuable features in each of them and positions Quine abandoned in his later work.

Keywords: quine; logic; ontology

The book "Significance of the New Logic" (SNL) [1] is a translation of Quine's "O Sentido da Nova Lógica" [2]. He published the original in Portuguese as the result of a period of time spent visiting the Free School of Sociology and Politics , by that time connected to the University of São Paulo. The publication represents a stage in which the American philosopher was on the verge of a philosophical turn. Not long after this period, Quine published the important papers "Notes on Existence and Necessity", "On What There is" and "Two Dogmas of Empiricism". Most of Quine's writings were in English. Thus, it has not been difficult for scholars to have access in full to most originals. However, in the 1940s, which was a period of maturation in Quine's philosophy, he has been writing in Portuguese, and this translation fills a historical gap Quine scholars were hoping for.

Carnielli, Janssen-Lauret and Pickering explore in many details the context in which Quine wrote this book. They are successful in presenting the Brazilian philosophical background, especially as regards their relative absence in the analytical scenario. In this respect, the book intended to further introduce the Brazilians to analytic philosophy. Discussions and techniques developed by Frege, Russell, Carnap, Tarski, Gödel and others are therefore the primary topics in the volume.

We note that Quine intended SNL to be a textbook. As such, the volume fails to give an updated overview of techniques and it uses outdated language. However, SNL can now be regarded as a picture of Quine's view on logic in the early 1940s. It is wrong to regard the book only as a textbook. The way Quine develops the logical apparatus and his preparatory remarks are the result of a very distinct philosophical position. By a close examination of his writing, we realize he was arguing for an extensional, nominalistic leaning ontology and a rather reluctant logicist position.

The latter part of the book is dedicated to a discussion on themes such as ontology and its relation with philosophy of language and logic. He drafts in Portuguese the first version of his later work: "Notes on Existence and Necessity". Thus, he exits the scope of a pure textbook, including contemporary discussions on ontology and philosophy of mathematics. These topics are accompanied by the flavour of the inner conflict that suggests parts of Quine's mature philosophy.

Quine divides the book into an introduction and four parts: (1) Theory of composition, (2) Theory of quantification, (3) Identity and Existence, (4) Class, Relation and Number.

The introduction in Quine's SNL starts with a brief analysis of the new logic as opposed with Aristotelian logic. He attributes this new development to two main reasons: Cantor's set theory and Russell's paradox discovery. New developments on infinite quantities by Cantor urged mathematicians to develop reliable tools, since even good mathematical intuition could lead to error as they handle infinite sets. As Quine argues, "We must explore the ocean that Cantor discovered by navigating blindly". We thus need a precise and truth preserving tool. Russell's paradox leads to an even stronger need for further scrutiny on logical development once the proof of the paradox relies on a tacitly accepted principle. Still in this introduction, Quine expresses a logicist belief, not a position he holds in the mature phase of his work. He knew Gödel's incompleteness results and the impact it should have on the theory of classes being part of logic. However, he was confident that the virtual theory of classes avoids ontological commitments in many mathematical theories. It is interesting to find Quine defending with confidence that mathematics is reducible to logic. But even more surprising is to see his reluctance with the definition of logic. Though not conventionalist, his characterization of logic still relies on concepts as truth and the "essential occurrence" of logical terms.

In the first part, Quine exposes the theory of composition. He explores distinctions between statements and sentences that are not statements. He rejects non-declarative statements and sentences that are dependent indexical terms from logical analysis. Another notable feature of his exposition is the fact that he insists on a simple logical vocabulary with only "∼" for negation and "." for conjunction. I may attribute this, as Janssen-Lauret says in the introduction, to the influence of the *Principia* or by his parsimonious tendencies. Notwithstanding, Quine does not explore reasons for this preference.

Quine develops quantification theory in the second part of the book. He emphasises the problems of quantification in its relation to natural language. Thus, he introduces each quantificational term by first evoking misconceptions about words such as "All," "Some," and "Everything." Quine's concerns with ontology are manifest when he discusses logical pronouns, as he hints at his ontological conceptions later developed in "On What There is." A drawback of Quine's discussion is the absence of a proof-system, as logicians now do by defining the turnstile "⊢." He bases his conclusions on the truth table method and axioms introduced to quantifiers. There is an interesting section, called the Practical Aspect, in which he defends quantificational logic to insurance companies. This usage is not standard for the period. Now, logic modeling of this kind of problem is routine in computer science and engineering.

The relation between philosophical issues and logic is the main concern in the later parts of the book. The third part of the book focuses on problems about identity. It is in this part we find the original draft of the paper "Notes on Existence and necessity." I found it interesting to contrast this version with the one published in *The Journal of Philosophy*. The translators provided many clarifying notes on the main differences between the two versions. They had shown that, in some points, the undecided Quine in SNL became convinced of some positions by the time of the paper.

The last part of the book focuses on the theory of classes. Quine describes a theory of classes (now referred to as single-sorted NBG) instead of the now more standard choice for set theories. This choice may be for a weak hope that a virtual theory of classes would avoid ontological commitments. In opposition, a set theory would from the start be committed to abstract entities. He later dismisses this hope. But here we can understand the hesitant logicist influence on the American author. At this point in Quine's career, he was still adherent to logicist ideas as he held some positivist tendencies. Nonetheless, he argues for a fundamental difference between classes and aggregates, emphasising how the latter cannot account for what we may express with the former. This represents a profound downside to logicism and positivism. As Quine argues, "the theory of classes, in contrast with logic in the strict sense, implies an ontology." Particularly important, this contradicts his own statements in the introduction, where he declares himself committed to logicism and to a form of nominalism. He thus retreats from reducing mathematical theories to logic, restricting the scope to the reduction of the mathematical language. In this balance, he sets forth a tentative argument of indispensability. He later develops a virtual theory of classes, which gives hope for eliminating ontological commitment in the theory of classes. But he adds: "Arithmetic depends on the real theory of classes, with all of its ontological presuppositions."

This translation is of major importance for any Quine scholar. Apart from the good quality of the translation, the book is full of clarifying remarks. The introductory paper by Janssen-Lauret gives a valuable general picture of this specific time of Quine's thinking. Moreover, reviving the picture of now established ideas may be a good source for finding new angles to reframe old questions. Philosophers of logic, logicians, naturalist philosophers, and people interested in the history of ideas may find great insight in the ideas expressed in the book.

Funding: This research received funding from FAPESP, process number 2016/10497-8.

Conflicts of Interest: The author declares no conflict of interest.

References

1. Quine, W.V.O. *The Significance of the New Logic*; Translated by Carnielli, W., Janssen-Lauret, F., and Pickering, J.; Cambridge University Press: Cambridge, UK, 2018.
2. Quine, W.V.O. *O sentido da nova lógica*; Livraria Martins Editora Location: São Paulo, Brazil, 1944.

© 2019 by the author. Licensee MDPI, Basel, Switzerland. This article is an open access article distributed under the terms and conditions of the Creative Commons Attribution (CC BY) license (http://creativecommons.org/licenses/by/4.0/).

MDPI
St. Alban-Anlage 66
4052 Basel
Switzerland
Tel. +41 61 683 77 34
Fax +41 61 302 89 18
www.mdpi.com

Axioms Editorial Office
E-mail: axioms@mdpi.com
www.mdpi.com/journal/axioms

www.ingramcontent.com/pod-product-compliance
Lightning Source LLC
LaVergne TN
LVHW070152100526
838202LV00015B/1934